NINETEENTH-CENTURY GARDENS AND GARDENING

NINETEENTH-CENTURY GARDENS AND GARDENING

Edited by
Sarah Dewis and Brent Elliott

Volume I
Home

Routledge
Taylor & Francis Group
LONDON AND NEW YORK

First published 2024
by Routledge
4 Park Square, Milton Park, Abingdon, Oxon OX14 4RN

and by Routledge
605 Third Avenue, New York, NY 10158

Routledge is an imprint of the Taylor & Francis Group, an informa business

British Library Cataloguing-in-Publication Data
A catalogue record for this book is available from the British Library

ISBN: 978-0-367-18855-9 set
ISBN: 978-0-367-18856-6 (volume I) hbk
ISBN: 978-0-429-19880-9 (volume I) ebk

DOI: 10.4324/9780429198809

Typeset in Times New Roman
by Apex CoVantage, LLC

CONTENTS

ILLUSTRATIONS

Illustrations are anonymous except where otherwise indicated by a signed name or initials.

TYPES OF ILLUSTRATIONS

Woodblock Prints

Woodblock prints were the oldest form of printed illustration. An image would be carved in a block of wood, which would be inserted into the printing forme along with the letters for the text. The earlier prints were called woodcuts, and the images were carved along the grain of the wood. In the late 18th century, Thomas Bewick and other illustrators developed an alternative technique of carving against the grain of the wood, producing an image with a harder-wearing surface, capable of a much denser concentration of lines than the older woodcuts. This new method was called wood engraving, and it became the principal means of illustrating books, magazines, and newspapers during the 19th century. Wood engravings can easily be distinguished from woodcuts when the image is complex. But if the image is, say, a simple mathematical diagram, the difference between the techniques is difficult to determine.

Wood-engraving blocks were hard-wearing, but not hard enough to withstand the printing of thousands or even millions of copies when used in newspapers. Around 1840, a technique was developed for making a metal copy of the wood-engraving surface, which could then be tacked onto a block of wood and inserted in the printing forme as usual. By 1850, most newspaper and magazine illustrations were printed using these 'stereotypes', but they went on being called wood engravings because there was no difference on the printed page.

Engravings

Engravings were used for printing more expensive and prestigious books from the beginning of the 17th century. Lines were incised in a metal – usually copper – plate, which would be inked and wiped clean, leaving ink only in the incised lines below the surface of the plate. It required far greater pressure to produce such an image on paper, so engravings were seldom used on the same pages as letterpress; they were bound in separately, as 'plates'. Engravings could yield far more detailed images than woodblock processes. If colour plates were wanted, they had to be coloured by hand.

Etchings

Etchings tended to be used only for fine-quality books. The metal plate was covered with a waxy 'ground' onto which was pressed an image drawn on paper; the etcher then used acid-tipped needles to incise the lines for printing.

Lithographs

Around 1800, Alois Senefelder developed a process he called lithography (stone drawing). On a smooth block of stone, a drawing could be made using greasy crayon; the stone would then be inked and washed. The ink would be held by the crayon, leaving an image that could be printed without the need for any cutting or incising. Many artists liked lithography because they could make their own printable images without having to hire a technician for the purpose. Lithography became the first reliable method for colour printing: the different colours could be applied to different stones and, as long as everything was lined up correctly, the sheet of paper would be pressed to each stone in turn, producing a 'chromolithographic' image in several colours.

Before long, lithography came to be a handy tool for reproducing images from any medium, including calligraphy and letterpress. As with wood engravings, if the lines of the image are simple enough, it is not always possible to deduce from the image whether it has been reproduced by woodblock or lithograph.

Photographs and Halftone Blocks

Early photographs were printed on paper covered with a solution of egg white (albumen) and the chemicals used in photography; light was shone through the photographic glass plate onto the sheet and an image produced. The resulting images could be passed onto the pages of a book.

By the 1880s, techniques had been developed for turning the photographic image into a surface that could be used for printing along with letterpress. The halftone block was a metal block covered with gelatine onto which an image was projected through a transparent screen which would break it up into a series of dots. Varying the thickness and spacing of the dots could give a range of tones to the resulting image. By the 1890s, halftone blocks were increasingly used in newspapers and magazines for printing images from photographs.

ACKNOWLEDGMENTS

These volumes bring together a collection of 19th-century sources on gardens and gardening. We would like to thank the individuals and the institutions who have contributed to all six volumes.

We are grateful to staff at the British Library (Rare Books, Humanities, Social Sciences, Maps, Asian and African, and the Newsroom) who were helpful in supplying material and assisting with our enquiries, as were staff at the RHS Lindley Library. Staff at Harvard University Library Archive, New York Botanical Garden, and Royal Botanic Garden Edinburgh were also informative. The Manuscripts and Special Collections Library at the University of Nottingham kindly allowed us to include the Orange pamphlet *Cottage Garden Plan Considered* . . . . The Library and Archives at the Royal Botanic Gardens at Kew have permitted us to include the letters of Thomas Freeman and Thomas Meehans's paper, 'Variations in Nature'. We would also like to thank the Library of the Linnean Society and The Gardens Trust.

We are grateful to those libraries who have made their visual material free to use and to the British Library who waived permissions fees. Special thanks are due to the RHS Lindley Library who photographed material and waived fees. Thanks are also due to Klaus Gille at the Carl Hagenbeck Archive and Nogel Rothfels for making possible the use of 'Carl Hagenbeck's Zoological Paradise'.

Finally, we would like to thank our publishers for their patience.

GENERAL INTRODUCTION

This collection documents sources on gardens and gardening during the 19th century from around 1800 to 1900. The set aims to provide the reader with a variety of ways in which to read gardens – through recognition of how they were conceived and how they were experienced as they developed. The study of gardens is multidisciplinary and global. The material, however, is derived mainly from Britain, and the rest originates from Europe, the US, and Australia.[1] Topics covered include types of gardens and gardening practice, plantings, and design in a social, historical, and cultural context.

The amount of printed material available on gardens and gardening in the 19th century is considerable. Developments in technology in Britain and elsewhere led to the expansion of printed material from the 1820s onward in the form of magazines, journals, and weekly newspapers; after the 1870s, with the advent of cheap paper and ever more efficient printing methods, cheap books and pamphlets proliferated. Our selection is sourced principally from the gardening press, adverts and catalogues, and books – advice manuals, fiction, and state papers. These include forms of visual representation from wood engravings to photographs and maps and diagrams. Plans and illustrations of ideas for gardens and of the wider landscape that became more accessible through the gardening press of the later 19th century inspired and stimulated ideas for domestic landscapes which are cited throughout the collection. The materiality of the source itself might change its meaning both in terms of its intended audience and their response to it. Where possible we have given the price of the source, but we have not attempted to give the 21st century equivalent value.[2] The materiality of the *reproduction* of sources in this collection comes with the following caveats: we have not corrected printing errors. We have inserted ellipses to indicate omitted text; some cross references in original sources have been silently omitted (as have some numbering systems). Layout, type font and differences in font size in the original texts have not been replicated. Images have been repositioned.

A note on the contributors: while a number of the authors of the texts are well known, a large number were working gardeners who were not known outside horticultural circles; others, despite being published, are rarely referred to currently. Apart from authors from outside Britain, the diversity of British contributors is well represented by Welsh, Scottish, and Irish authors, although much of the data is based on surveys of England. In relation to the press, a number of contributors published anonymously. Although the practice of naming was more common in 'technical' publications, it did not become more widely established until the 1880s.[3] The naming of the 'artist' of listed illustrations was rare, with the exception of botanical plates or those images produced by a name familiar to the readership.

The multiple approaches to the history of gardens from English-speaking and European sources originating in the 18th century have been documented by Brent Elliott.[4] As an academic discipline, garden history is relatively young, emerging in the 1960s in Britain with the foundation of the Garden History Society (1965) and its journal, *Garden History* (1972), followed by a range of associated institutions.[5] Despite the relative lack of academic outlets, Elliott suggests that there has been a proliferation of scholarly publications on garden history since the 1980s, including the three Batsford histories published in the wake of the V&A Garden exhibition of 1979: John Harvey's *Medieval Gardens* (1981), David Jacques' *Georgian Gardens* (1983), and Brent Elliott's *Victorian Gardens* (1986). While *Perspectives on Garden Histories* (1999) provides alternative narratives of garden history as an academic discipline, Sarah Rutherford has argued that 'because of its junior status as a developing subject it has not yet matured its own coherent theoretical approaches but as an emerging subject it can be subject to, and so borrows from, other better-developed approaches'.[6] Histories of gardens and gardening continue to appear under the umbrella of other disciplines apart from art and architectural history, from cultural geography to social and economic history to literary studies. However, far from being a 'marginal' element of those disciplines, as John Dixon Hunt has suggested, garden history continues to be a significant area of study.[7] The aim of this collection is to reflect garden history through a prism of different approaches contained within the original sources. Moreover, the opportunity to bring these 19th century texts to a broader audience is a much-needed counterbalance to the predominance of 18th century material that has generally been the focus of garden history to the detriment of studies on the 19th and 20th centuries.[8] Since the 2000s, the number of studies on 19th-century suburbs and public landscape has increased.[9] More recently, working-class gardens have received some attention, mainly in relation to working-class homes (see below). There is currently debate about the political and social context of garden history;[10] the language of some sources may be offensive to 21st-century readers.

We have divided the sources into six volumes. Each volume considers various aspects of gardens and gardening with emphasis on different historical contexts

Volume 1 – Home – considers gardens within the context of home and the playing out of moral values attached to ideas of family, taste, and nation.

Volume 2 – Community – looks at gardens connected to local practical needs, the development of allotments and cottage gardens, and the expansion of cemeteries. Material from the makers of gardens is included in this volume.

Volume 3 – Science: Institutions – looks at how horticultural knowledge was disseminated: the gardening press was a major vehicle for this process as were the horticultural societies which flourished alongside older floral and vegetable clubs. Moreover, while the precedents of gardens for science lay in popular spectacle and expanded as part of 'leisured consumption', their creation made the study of plants accessible to a wider public.

Volume 4 – Science: Applications – considers consequences and the application of science to gardens and gardening, especially in relation to contemporary ecological concerns and the technology of gardening.

Volume 5 – Garden Design – looks at a variety of approaches to the theory and practice of garden design.

Volume 6 – Gardeners' Art. The role of gardeners in the developments of different genres of garden ranging from the flower garden to parks and woodlands are the focus of this volume. Finally, the aesthetic and social drivers for the developments of green spaces for the public and green towns and cities are looked at. All six volumes are interconnected.

Notes on headnotes: Where editorial notes to headnotes are necessary, we have kept them to a minimum, glossing the biographies of named individuals where possible. We have not annotated classical allusions, publication details of literary references, or details about suppliers of garden products and plant names, unless relevant.

Notes on sources: A number of sources are now available online, in whole or in part, through the Biodiversity Heritage Library, HathiTrust, and the Wellcome Trust.

Notes

1 We made the decision, in general, not to include material on gardens from other continents; Africa, India, and China, for example, because the range of texts and expertise available require their own set of volumes. The majority of material from non-Anglophone countries derives from translations currently available or is provided in these volumes from French and German by Brent Elliott.

2 Roderick Floud, *An Economic History of the English Garden* (London: Allen Lane, an Imprint of Penguin Books, 2019) bases the value of £1 on the average salary. The bank of England offers different methods of calculation. See also MeasuringWorth (www.measuringworth.com).

3 See Laurel Brake, *Print in Transition, 1850–1910; Studies in Media and Book History* (Basingstoke: Palgrave Macmillan, 2001), p. 15; Sarah Dewis, *The Loudons and the Gardening Press* (Farnham: Ashgate, 2014), p. 193.

4 Brent Elliott, 'The History of Garden History', *Occasional Papers*, *RHS Lindley Library*, 9 (2012), pp. 3–84, www.rhs.org.uk/about-the-rhs/pdfs/publications/lindley-library-occasional-papers/volume-9-dec-2012.pdf.

5 County Gardens Trusts in Britain; courses in garden history at the Architectural Association and other institutions of higher education.

6 Sarah Rutherford, 'The landscapes of public lunatic asylums in England, 1808–1914' (vol. 1 of 3) (PhD, De Monfort University, 2003), p. 20, www.dora.dmu.ac.uk›bitstream›handle (DX226739_1_0001.tif) [accessed 8.10.22].

7 See John Dixon Hunt, 'Approaches (new and old) to garden history' in Michal Conan (ed.), *Perspectives on Garden Histories* (Washington, DC: Dumbarton Oaks, 1999), pp. 77–90. Cited by Rutherford, p. 22. For example: Stephen Daniels, Cultural geographer whose *Fields of Vision* (1993), preceded a study, *Humphry Repton: Landscape Gardening and the Geography of Georgian England* (1999); Denis Cosgrove, *Social Formation and Symbolic Landscape* (1998); Roderick Floud, *An Economic History of the English Garden* (2019); Business history, Keith Alcorn's article 'From specimens to commodities: the London nursery trade and the introduction of exotic plants in the early nineteenth century' (2020).

8 Rutherford, pp. 23–24.

9 Hazel Conway's *People's Parks* (1991) was one of the first studies of public gardens. Barbara Simms (ed.), *London's Garden Suburbs: Community Landscape and the Urban Ideal* (London: London Historic Parks and Gardens Trust, 2001). Sophie Seifalian, 'Gardens of Metro-Land', *Garden History*, 39 (2011), 218–238. Sarah Bilston's *The Promise of the Suburbs . . .* (2019), considers suburban gardens within a literary and cultural context.

10 See, for example, Corinne Fowler, *Green and Unpleasant Land: Creative Responses to Rural England's Colonial Connections* (Leeds: Peepal Tree Press, 2020).

INTRODUCTION TO VOLUME 1, 'HOME'

Volume 1 considers gardens as an extension of the family home. The changing meanings of 'family', 'home', and 'taste' within the context of Nation and Empire are noted. The focus is generally on the gardens and gardening of the middle classes, rather than the estates of aristocrats and the landed gentry, which have historically received the most attention. There is, however, less printed material on working-class gardens from the perspective of those who gardened them.[1]

Part 1 looks at domestic gardens and garden structures, including conservatories and garden ornament. Material on fruit and vegetable gardens is included.

Part 2 examines plants cultivated under glass, the flora and fauna of home decoration, and the reconstruction of plants in different media.

Domestic gardens contained several areas depending on their size and location, including kitchen gardens, parks, pleasure gardens, woodlands, and gardens under glass. They might provide a space for self-expression, however small or impermanent (the majority of all housing was rented up to the 1900s);[2] they might also constitute a source of financial support. A snapshot of the range of functions of domestic gardens is provided with excerpts from J. C. Loudon's *The Suburban Gardener* (1838), Shirley Hibberd's *The Town Garden* (1855), and Gertrude Jekyll's *Wood and Garden* (1899), amongst others. Their position and desirability, within British culture in particular, developed out of several factors.

AGRICULTURE, INDUSTRY, AND GLOBAL TRADE

Despite the expansion of Britain's global trade and the beginnings of the industrial revolution during the 18th century, the majority of the population still lived in rural areas. By the early 19th century, the economy was still primarily agricultural. The majority of gardens at this time included a significant proportion for the growth of food, although statistics are not available to indicate how many, or what proportion of large multi-purpose gardens, were devoted to cultivation for food. Our sources reflect the variety of scale, including the advice of Elizabeth Watts on how to manage small kitchen gardens, and an account of Culford Hall, an extensive estate in Suffolk. A proportion of larger estates remained self-sufficient in terms of food production well into the mid-19th century, with some lasting into the early 20th century.[3] Roderick Floud, however, has argued that these well-funded kitchen gardens were as much a product of taste and a sign of luxury as of necessity; the maintenance of a skilled staff of gardeners, the orderly upkeep of the gardens, and the technology required resulted in extremely expensive produce.[4]

Much of Britain was transformed into a predominantly urban and industrialised society during the 19th century. The population of Swansea, for example, expanded from fewer than 7,000 in 1801 to more than 16,000 by 1841, and it exceeded 100,000 by the end of the century;[5] the population of Wales and England as a whole grew from 9 million in 1801 to 36 million by 1911. The number of houses built in the same period increased from 1.6 million in 1801 to 7.6 million.[6] The movement of people into growing towns and cities for work stimulated and was accompanied by developments in communications, the manufacture of the 'macadam' road surface (made of stones broken up to less than wheel width, aiding drainage and the smooth running of traffic) beginning in the 1820s, the expansion of rail networks from the 1830s, and the development and reform of the postal service from 1840. Building programmes (see below) and sanitary reforms were increasingly made possible by shifts in control of local areas from the Church or landowner to local councils and the central government's attempts to manage urban expansion. Increased wealth was spread across a greater percentage of the population as a result of the industrial revolution, which some commentators have argued was funded by expansion in global trade encompassing the Caribbean, the US, Africa, India, and China. The specialist triangular trade of plants into Britain which were then exported to the colonies is discussed in Volume 3. The penetration of the variety of materials available for use in gardens through technological developments and global trade is represented in the collection – from materials in the greenhouse, indoor plants, and garden ornament to the careers of traders in garden products.

DOMESTIC GARDENS, LEISURE, AND HOUSING

For some, increased wealth opened up the possibilities of 'leisure', whose meaning – 'opportunity afforded by freedom from necessary occupations' – has, since the late 14th century, remained constant.[7] Increase in leisure permitted more people to spend time on domestic and public gardens. However, until the latter decades of the 19th century, the majority of the population had little access to leisure except during religious festivals, despite the introduction of The Ten Hours Act of 1847 (which was frequently ignored by factory owners and did not apply to domestic servants).[8] The first official public holidays were not formally recognised until the Bank Holidays Act of 1871, and Sundays were generally allocated to church if a person was not working on that day. Despite these limitations, there is evidence, among mining communities at Merthyr Tydfil, for example, that gardens were cultivated.[9] Periods of severe economic downturn during the Napoleonic wars (1803–1815) and agricultural depressions of the 1830s and 1880s were other factors that had profound effects on gardens and gardening and on the makers of gardens.

Until recently, there has been a relative lack of data and research on the spatial reach of domestic gardens that accompanied 19th century housing development.[10] Zoe Crisp's study (2012) of five English towns or cities provides one of the first empirical assessments: 'private outside spaces – whether yards, gardens or shared plots – were provided for all classes from as early as the first quarter of the 19th century.'[11] With sanitary reforms from the 1840s onwards, back gardens were gradually transformed into areas worth cultivating, particularly after the Public Health Act of 1875, which extended powers to local councils in Britain to provide clean water and sewage systems. For much of the 19th century, then, in less affluent urban areas, the backyard was allocated for the privy and the front garden for display.[12] Lesley Hoskins and Rebecca Preston's recent study (2022), however, has indicated the diverse functions of backyards that included sanitation and washing space and, in common with domestic gardens of the wealthier classes, areas for play and social interaction and planting.[13]

The law of primogeniture in Britain (which established the supremacy of the oldest son) resulted in an accumulation of land which was a factor in the strength and influence of land-owners on surrounding communities. In contrast, in several European countries, laws of inheritance ensured that land was distributed, in theory, equally among all descendants. As the 'ever-diminishing plots' ceased to provide a comfortable source of income, they became a burden which those who were able to choose to leave and move to urban apartments.[14] It might be that land attached to home for leisure as well for economic gain signified more within a culture in which wealthier classes chose to maintain that lifestyle.

FAMILY

Over the course of the 19th century, as capitalism developed, concepts of property, domesticity, and leisure evolved, and patterns of land distribution changed. Domestic advice literature proliferated in the latter half of the century and is one of the forms of text that predominates in this volume, as the gardening press both absorbed it and was superseded by it with cheap books and pamphlets for non-professional male and female gardening audiences.[15] In contrast, material on allotments was, in general, orientated towards a male audience because it concerned public issues, such as the economy and how society might be run. Alongside gardening advice, domestic advice literature foregrounds ideas of home, family, religion, and identity. Home was where families lived; how the family was perceived evolved over the course of the century. From the 18th century, if not before, family was concomitant with household and included servants as well as blood relations among wealthier social groups.[16] During the 19th century, servants in general became employees, partly due to their increasing numbers and to changes in social attitudes; the presence, however, of a broad spectrum of relations who might live together was considerably larger than the 'nuclear' family of the twentieth and twenty-first centuries. The idea of family as a microcosm of nation had been circulating from the 1800s. William Cobbett, in *Cottage Economy* (1821–1822), refers to the 'economy' of all families to determine national character, a trope that is echoed late into the 19th century as evidenced by our sources.[17] From home and family, the patriotic endeavour of gardening, and of nurturing plants originating locally and, especially, from distant places, was heralded as a significant contribution to the national economy. In consequence, as Rebecca Preston has argued: 'Practically and ideologically the garden was central to private understandings of nation, Empire and of the wider world as a site of imperial potential'.[18]

RELIGION

Much of the discourse around family focuses on morality with its roots in Christian belief. Historians have suggested that 'the family had become a central feature of religious life, the home its most perfect location', and that the education of the young generally fell within the remit of the church of whatever denomination.[19] For most of the 19th century, different branches of Christianity permeated British society and politics, and legislation which restricted the access of non-Anglicans to public life at a number of levels was slowly modified. The Methodists, inspired by the Pietism which had originated in the German-speaking world in the late 17th century, moved away from the formality of Anglican Church ritual and believed in the direct relationship of the individual to God. Their faith might be manifested in helping others through education, alleviating poverty, and opposition to forms of exploitation such as the slave trade. The Wesleyan Methodist movement, which did not become a a formal Christian denomination until the later decades of the 19th century, co-existed with the Baptist Church and the independent denominations of Scotland, Wales, and Ireland. The 1851 census in England indicated that more than 10 million people out of a population of 18 million attended Christian places of worship; almost 50 percent of those who registered were non-Anglican.[20] The Evangelical revival of the early 19th century within the Anglican Church also focused on social issues, including the separation of the roles of men and women, in which the latter, as embodiments of virtue, were to be kept away from the cruelties of the marketplace.[21] In the later 19th century (preceded by the Anglo-Catholic movement in Oxford of 1840s), there was a movement back to greater formality and ritual within the Anglican Church, leading to divisions between High and Low Anglicanism. (For the effects of the former in garden design, see Volume 6.)

The presence of gardens in the Bible – the sacred, the useful, and the ornamental, starting with the garden of Eden – underpinned their significance in a Christian society at a time when access to such spaces was increasing for some. The broad trend of industrialisation and the capitalisation of labour resulted in the separation of places of work from home and family – albeit at a sporadic and geographically varied pace. It combined with the emphasis on family values and fed into attitudes towards 19th century gardens and gardening. A common trope featured in a broad range of texts and applicable across all ranks or classes (though applied with greater rigour to women and the poor) was that engaging in gardens might stimulate and perpetuate virtuous Christian behaviours such as patience, humility, and industry.

Moreover, 'Liberal Christianity' called on all people to exert themselves in their individual fields of expertise which did not preclude scientific investigation. The role of the Nonconformists in the expansion of plant species in relation to the status of nurserymen and seedsmen forms part of this volume. Their specific contributions as scientists are documented in Volume 3.

ASSOCIATIONISM AND RATIONAL RECREATION

There were secular reasons for engaging with nature and for the expansion of homes and domestic gardens. Historians whose ideas were intrinsic to the Scottish Enlightenment argued that society developed, broadly, in progressive stages from the 'rude' to the 'civilised'.[22] The predominant psychological model derived from this view of history, known as Associationism and promoted by David Hartley amongst others, suggested that as man had been born defenceless and unthinking, his condition might be improved by society.[23] The connections between impressions on the body created by external objects that link to the mind and develop into complex ideas implied that experience is what makes us human. The benefits of an improved physical environment and its emphasis on the significance of the body in determining how we think derived from Associationism and was one of the factors that contributed to concerns about housing and green spaces.[24]

One reformist ideology was 'Rational Recreation', whose proponents were concerned with how 'leisure' might be addressed. While Loudon and some others desired rational recreation for all ranks (he suggests that it might be preferable to hunting for aristocrats and the landed gentry), it was widely promoted in relation to the poor and under-occupied women. It was, perhaps, the idea of recreating all ranks in the image of how the middle classes liked to see themselves. Instead of gambling or drinking, a person might engage with nature in a way that was both practical and intellectual – hence the emphasis on the value of public gardens (see Volumes 3 and 6) and domestic gardens. In relation to the poor, the loss of the commons might be compensated for through rational recreation with the establishment of allotments and horticultural societies (see below). Some elements of horticulture originated from within artisan culture as Robin Veder, amongst others, has argued. As a key element of rational recreation, however, the pursuit of horticulture was adapted to exemplify middle-class values: 'A love of nature, it is implied, is the mark of refinement and humility. Promoters of rational recreation relied upon this characterization to justify promotion of horticulture as a remedy for idleness'.[25] The possible benefits of rational recreation are reflected in the wider-ranging debates about 'taste'.

GARDEN STRUCTURES AND ORNAMENT: TASTE AND CONSUMPTION

The plethora of garden ornaments that became accessible to non-elite consumers over the course of the 19th century was due to three main factors:

1 The expansion of empire and global trade – and greater variety and quantity of raw materials such as jute (from India), raffia (from tropical Africa, Madagascar, and Asia), and bamboo (from China) and the increased availability of goods due to more efficient means of transport (see Volume 4).

2 Developments in the manufacture of cement and other materials (preceded by lead in the 18th century) as a substitute for marble and stone and the affinity of these new materials with the moulding process which allowed for mass production (see also Volume 4).

3 The possibilities of consumption opened up by increased leisure time.

Cultural borrowings in our selection range from statuary and urns, suggesting the classical past, to a Chinese summerhouse, recalling the orient or the exotic, to wattle benches and arches, suggesting the mythologised rustic present; see Loudon's article 'Descriptive account of the Villa of Mrs Lawrence' (1834) and Hibberd's 'Garden scenery and ornaments' (1856) in particular. These garden ornaments refigured the visitor as traveller to distant lands and confirmed the proprietor as a knowledgeable person. The selection of non-organic objects for gardens embodies (albeit on a modest scale), the connections between anthropology, geography, and social formations suggested by W. J. T. Mitchell, whereby 'landscape *circulates* as a medium of exchange, a site of visual appropriation, a focus for the formation of identity'.[26]

'Taste' as a marker of what was to be valued and what was to be ignored in the 18th century appeared in the form of treatises published as papers for the learned societies or, occasionally, in a magazine or journal whose sale was limited by price to a minority. In contrast, during the 19th century, discourse on taste is wide-ranging and features in much of the material in this volume – in titles of articles, books, chapter headings, and in the body of the text. What it might mean depends on its material form, the price of the text in which it appears, the point of view of the writer, and its reception by its intended audience. That it might be an objective assessment of beauty, as some 19th century writers on gardens and gardening may have believed, was much debated at the time, as seen in Volume 5.

The argument that taste is an aesthetic value judgement made in relation to surfeit or excess rather than necessity, and suggesting a non-productive lifestyle, was made by the US economist Thorstein Veblen in *The Theory of the Leisure Class* (1899). In relation to gardens, matters of taste were foregrounded where consumption was overt. The association of plants with nature allowed for greater latitude in consumer choice than in the selection of garden ornament, which included architectural structures (staircases, balustrades, and buildings), pergolas, treillage and ironwork, fountains and grottoes, statuary, urns, vases, and garden furniture. Taste and 'appropriacy' of ornament are a common trope in our sources;

Repton, in *Observations on the Theory and Practice of Landscape Gardening* (1805), relates taste to the price in terms of labour or material of an object as well as appropriacy. Andrew Jackson Downing in *A Treatise on the Theory and Practice of Landscape Gardening* (1841) abhors the use of Palladian columns for a cottage balustrade because such mixing of cultures implies that consumers (mainly the less wealthy) do not have the education to discern different historical and aesthetic epochs. The garden gnomes featured in 'Lamport', an account in *Gardeners' Chronicle* (1897) about the estate of Sir Charles Isham, might imply, however, that if the estate was grand enough, lack of taste was acceptable. Nevertheless, the lack of contemporary material about the 'eccentric' sculptures at Whatton, Leicestershire (not included in the collection), suggests press reluctance to report on possible deviations.[27]

The evidence of our sources, then, appears to support the late 20th-century theorist Pierre Bourdieu. In *Distinction* (1979), he suggests that the relations between economic status and cultural capital are complicated; his overarching argument is, however, that the acquisition of 'taste' perpetuates class divisions because it functions to display prestige and to oppress the poor through disregard of poor culture, legitimising social difference. With an increased focus on how texts might be received in the last 30 years, Bourdieu's notion of a dominant 'high' culture has been questioned.[28] Moreover, other factors may have played a greater part in the choices that people made in their domestic setting apart from social status (and as opposed to what might be proscribed by the taste-makers), including age, gender identities, religious affiliations, and family connections.[29]

The democratisation of taste might be beneficial. Free entry to some museums and art galleries in the 21st century derives from the principle that the acquisition of taste should be accessible to all and that the appreciation of beauty gives pleasure and improves quality of life. Isobel Armstrong has argued for the energising effect of taste as a symptom of curiosity and discovery, a process exemplified in the work of J. C. Loudon.[30] In *Rustic adornments for Homes of Taste* (1st ed., 1856), Shirley Hibberd followed Loudon's trajectory to advise the less wealthy. Although taste is part of Hibberd's vocabulary, ultimately, he suggested it was up to his readers to choose what pleased them, which was also generally true of Loudon. What might have constituted taste for Hibberd, however, was derided by a number of his younger contemporaries, exemplified by the work of William Robinson (1838–1935) and a decade or so later by Gertrude Jekyll (1843–1932), as is documented elsewhere in the collection. A plurality of attitudes to taste co-existed: *Rustic Adornments* was being republished into the 1880s, ten years after Robinson's *Wild Garden* was published, each catering to different audiences. It might be that there are two phases in the evolution of taste during the 19th century: initially associated with science, excitement, and curiosity and then, paradoxically, as science and technology increasingly informed domesticity and garden culture, associated with the austere and the simple. The excerpt from Josiah Condor's *The Flowers of Japan and The Art of Floral Arrangement* (1891) codifies austere flower arrangements.

PLANTS UNDER GLASS, GREENHOUSES, AND COMMODITIES OF EMPIRE

The greenhouse was intended for the protection, cultivation, and forcing of plants, including sources of food, and the conservatory for the more permanent display of plants. Greenhouses were generally separated from the house while the conservatory might be attached to it, although the latter might serve both functions depending on the means of the owner. Other kinds of 'green-house' (Loudon's umbrella term for glasshouses), hot houses or 'stoves', evolved over the course of the century from simple brick buildings heated with open fires and equipped with wooden shutters in the colder months to architectural statements coveted by the elite to show off rare or exotic plants. Changes in the manufacture of glass, metal, and wood and the repeal of the glass tax (1845) contributed significantly to the transformation of the construction of glasshouses for the home (see also Volume 4). Paxton's 'Hothouse for the Million' project of 1860, in which he developed prefabricated construction discovered through the building of the Crystal Palace, gives the impression of the democratisation of glasshouse ownership. Floud, however, has argued that the price of a 30-by-16-foot greenhouse in 1860, advertised at £30 (which, he suggests, is the current equivalent of £1,600), opened up the market for the middle classes but not for the poor.[31] The 1860s and 1870s also saw the growing popularity of cool-house treatment for orchids, reducing the need for heating.[32] The general trend, indicated in adverts and articles across the spectrum of the gardening press, suggests the widespread construction and sales of greenhouses and conservatories towards the end of the 19th century in Britain.

The term 'exotics' describes the recently imported true exotic plants from tropical regions; the *Brugmansia suaveolens*, for example, a native of South America, was a true exotic that was cultivated in the hothouse, overwhelming visitors with its plethora of blooms and intoxicating with its evening scent due to its membership in the Solanaceae (nightshade) family.[33] The presence of 'exotics' in the home, either in glasshouses or in Wardian cases (and in gardens generally), offered several possibilities articulated in our texts. The bringing of the Other into the home made it seem exciting and connected those circulating in it or restricted to it to distant worlds. In the excerpt from *The Greenhouse Companion* (1824), Loudon championed domestic greenhouses in this respect as an emancipatory space for women. It might be that consumer demand, stimulated by the cultivation of Other landscapes, was as significant a factor in the justification and perpetuation of empire as the activities of plant collectors and imperial explorers.[34] John Dillwyn Llewelyn's 'Some account of an orchideous house . . . at Penllegare', *Journal of the Horticultural Society* (1846), references the German explorer Robert Schomburgk, who was employed by the British to map British Guiana (see Volume 3). The evocation of a jungle display in Edouard André's 'Conservatories in the natural style', *Gardeners' Chronicle* (1872), suggests the intrinsic connections of exotic landscape to British culture. Finally, the 'psycho-sexual' element of the conservatory was a trope of late 19th-century fiction, as Michael Waters has suggested in *The Garden in Victorian Literature* (1988), and that trope is evoked in the source from the novel by Rhoda Broughton, *Not wisely but Too Well* (1867).

As a result of the housing boom (see above), large numbers of people were massed together rapidly in limited spaces. From mid-century, glass windows became a general feature of urban buildings, providing light for smaller terraced houses, with taller sash windows suggesting a hint of grandeur to terraces intended for middle-class occupants. The availability of glass was due to a combination of reduction of taxes on glass, building requirements, and developments in glass manufacture.[35] Entries from Nathaniel Bagshaw Ward (*On the Growth of plants in closely glazed Cases,* 1842) and John Lindley ('A Belgian Window Garden', *Horticulturalist*, 1849) suggest that the landscape of window boxes or plants under glass might provide a screen from prying eyes. Moreover, the miniaturisation of landscape embodied in window boxes, the Wardian case, and aquariums offered possibilities of play and acts of the imagination. Ward's own case, in which he planted ferns, was designed 'as a window to Tintern Abbey' and featured as the frontispiece for the second edition of *On the Growth of Plants . . .* (1852; not included). In his Preface to *Rustic Adornments*, Hibberd is explicit about their value as 'toys'.[36] These miniature worlds were made possible partly by technological developments, but their production was stimulated by the desires of consumers whose personal efforts in their creation were substantial.

LIVING INDOOR PLANTS AND CUT FLOWERS

The difficulties of keeping plants alive indoors were extensive during the 19th century (though their prospects might seem better than outdoors in urban areas). They needed protection from the smoke and ash of coal fires and, later in the century, the noxious fumes of gaslight; maintaining the right temperature for them to flourish and keeping them sufficiently watered were other challenges. The energy, industriousness, and patience required to nurture indoor plants support the exhortations of the proponents of rational recreation and Christian morality on the beneficial effects of horticulture for leisured women, exemplified by entries from E. A. Maling, *Flowers and Foliage for In-door Plant Cases* (1861), and Annie Hassard, amongst others, and, explicitly in relation to the poor, in the excerpt by Ward, *On the Growth of plants* (1842). These sources indicate that the vogue for evergreens such as ferns as indoor plants flourished because they were able to withstand the noxious fumes of 19th-century interiors. Sand, moss, and clay were the advised environment for cultivating indoor plants (and those under glass). The avid collection of mosses and ferns resulted in the disappearance of certain species and is documented in Volume 4.

The bringing of living plants indoors for their sensory and visual effects on a temporary 'entertainments' basis was a tradition among elite households as indicated by Loudon's entry on 'management of town green-houses, and of plants in chambers at routs' (1824).[37] Their use as more permanent furnishings developed through the 19th century as the range of plants increased and became more widely available. As indoor plants became accessible to humbler households and to invalids without access to a garden, greenhouse, or conservatory, they remained an indicator of leisure and social status. The variety of decorative arrangements in which the indoor plants might be displayed – from a portable trellis to multiple-potted stands – was another factor in their desirability and suggested in articles by Louis Van Houtte, 'Movable trellis' ('Treillis mobile'), *Flore des Serres* (1848), George McEwen, 'A plea for pyramidal pelargoniums', *Gardener's Magazine of Botany* (1850), and an excerpt 'Pot plants in rooms' from Annie Hassard's *Floral Decorations for the Dwelling House* (1875). Increased demand for indoor plants stimulated the expansion of the network of commercial outlets, nurseries, and florists who could supply them and offer assistance in looking after them.[38] In this way the burden on the amateur of nurturing plants was considerably reduced.

In Britain and the US, advice on plants for interior decoration before 1862 focused primarily on living plants – generally in relation to the use of houseplants rather than temporary display. The new competitive category of 'Floral Arts' made by the newly 'Royal' Horticultural Society of 1861 signified change.[39] Competitions and Florists' Societies were long established; the existence of some, however, was blighted by industrialisation, as Margaret Willes has documented.[40] The cultural authority of the RHS led to the foundation of national decorative florists' societies specialising in particular genera of plants, partly replacing or absorbing the former, generating much-needed income for the societies involved (see Volume 3).

Another stimulus to the introduction of the floral arts into the home came from changes in styles of entertaining by wealthier households exemplified by John Lindley's leader on table decoration, *Gardeners' Chronicle* (1861), and an excerpt by Thomas C. March from *Flower and Fruit Decoration* (1862). Excerpts from John Perkins, *Floral Designs for the Table* (1877), and Edward Luckhurst's article 'The arrangement of cut flowers', *Journal of Horticulture* (1881), indicate the role of the head gardener. Advice given in an excerpt by Maling from *Flowers for Ornament, and Decoration and How to Arrange them* (1862) and Hassard's article 'Artificial light of dining-rooms', *The Garden* (1874), suggest that women played their part in smaller households.

Jules Lachaume's entry 'Decoration for balls and soirées' ('Decoration de bals et soirées'), *Les Fleurs Naturelles* (1847), anticipated the process that took place in Britain almost four decades later as lavish floral displays extended from the table to whole rooms for special events in elite households. These were reported in the general press as well as the horticultural press. Moreover, with the intensification of the culture of domestic floral decoration, the conflation of the gardening press with lifestyle publications continued apace – exemplified in the expansion of this topic in both the *Floral World* and in *The Garden* ('The garden in the house' weekly column) and the foundation of *Country Life* (1897–present). From the 1860s, the *Gardeners' Chronicle* began to include items for amateurs including on-table decoration, as did the *Journal of Horticulture*. The topic entered the discourse of the wider press including *The Queen*, the *Illustrated London News*, the *Daily Graphic*, and *The Lady*.[41]

THE LANGUAGE OF FLOWERS, WAX MODELLING, AND LEAF PRINTS

Domestic advice manuals on floral decoration advised on how to cultivate, maintain, and present growing plants and cut flowers in the home, often omitting the different concepts of permanent houseplants and plants for temporary display, as we have seen. They covered a wide range of subjects including bouquets, garlands, wreaths (for church as well as Christmas), flowers for clothing (in the buttonhole and on hats, for men and women), perfumes, and designs for dried flowers. Flower arrangements, floral motifs generally, and copies of the forms of flowers held other meanings, which unless made explicit, as in some of our texts, would not be evident to 21st-century readers. The proliferation of books on the language of flowers between the 1840s and the 1880s, however, may not have clarified matters for contemporaries as definitions of what a particular flower might mean or its associations varied. Such books were, in contrast to domestic advice manuals, predominantly gift books rather than practical guides.[42]

The main influence on the language of flowers tradition in English during the 19th century was Charlotte de la Tour, whose *Language des Fleurs* was first published in 1819. It did not appear in translation until 1834, but it had already been partially absorbed into Henry Phillips' *Floral Emblems* in 1825; an excerpt from the latter is included in the collection.[43] Some sources acknowledged global sources for the language, including Turkish, Indian, and Japanese, among others. The broad trend, however, within what became a trans-Atlantic publishing phenomenon, was to place the origins within the framework of a European and English tradition, exemplified by the excerpt from Pratt and Miller's *Language of Flowers* (1870s). They were symptomatic of a wider movement towards the reimagining of national identity through history and reflected in 'historical revival' fashions in garden design (see Volume 5). The definitions provided were often contradictory, and the increasingly complex rules for positions of flowers in a bouquet, for example, were confusing, as our sources indicate. In his paper of 2013, Brent Elliott concludes that (given the indifference of the gardening press in general) the full impact of a floral language was not significant on the practicalities of floral decoration, although it entered the culture of the middle classes in the form parlour games evolving into 'a practical theme for the publishers of greeting cards and picture postcards, genres of publishing that operated at a cultural level below the horizon of educated audiences'.[44]

The language of flowers is historically connected to representations of floral decoration. During the 19th century, floral motifs continued to appear in crafts practised in middle-class homes – in general by women – for example, embroidery, tapestry-work, crochet, and wool-work. These activities fell under the umbrella of 'fancywork'; 'ornamental, as opposed to plain, work' (*OED*) and were understood to have beneficial effects on under-occupied women.[45] There was, however, debate about the value of fancywork; some sections of the press suggested that such 'perpetual *stitchery*' was unhealthy and bad for the eyes, and it was unfavourably compared with the benefits of fresh air and the wholesome effects of gardening.[46] The excerpt by Emma Peachey from her *Royal Guide to Wax Flower Modelling* (1851) defends her craft in relation to textile work. She was one of a number of women who practised (and made a business out of) the craft with as long a history as fancywork but which reached its apogee in 19th-century Britain and the US. Leaf-skeletonisation was another craft of home decoration and was closely related to the photographic process developed by Henry Fox Talbot. The entry by Edward Parrish, a chemist, from his pamphlet *The Phantom Bouquet* (1862) documents the first description of the process in the US, and F. W. Burbidge included it in his *Domestic Floriculture* (1875).

Notes

1 See Margaret Willes, *The Gardens of the British Working Class* (New Haven and London: Yale University Press, 2014). See also, Lesley Hoskins and Rebecca Preston, 'Chickens, Ducks, Rabbits and Me Dad's Geraniums: The Use and Meanings of Yards, Gardens and Other Outside Spaces in Urban Working-Class Homes, 1890–1930' in Joseph Harley, Vicky Holmes and Laika Nevalainen (eds.), *The Working Class at Home, 1790–1940* (London: Palgrave MacMillan, 2022), pp. 80–92.

2 See Stephan Muthesius, *The English Terraced House* (New Haven and London: Yale University Press, 1982), p. 17.

3 See Mary Keen, *Paradise and Plenty: A Rothschild Family Garden* (London: Pimpernel Press, 2015).

4 See Floud, p. 257.

5 See Tim Lambert, *A History of Swansea*, https://localhistories.org/a-history-of-swansea/ [accessed 08.10.22].

6 See Muthesius, p. 17.

7 www.etymonline.com/word/leisure/ [accessed 08.10.22].

8 See Helena Chance, *The Factory in a Garden: A History of Corporate Landscapes from the Industrial to the Digital Age* (Manchester: Manchester University Press, 2017), p. 94.

9 A row of these gardens has been relocated from Merthyr Tydfil to St. Fagan's Natural History Museum, Cardiff. See Willes, pp. 131–132.

10 See Muthesius, pp. 74–78 for a succinct account of garden development. See, Lesley Hoskins and Rebecca Preston, 'The House: Inside and Outside Villas and Terraces in England' in *A Cultural History of the Home in the Age of Empire*, vol. 5 (London: Bloomsbury Academic, 2022).

11 Zoe Crisp, 'Housing and private outside spaces in nineteenth century England', Cambridge Working Papers in Economic and Social History (CWPESH 3), 2011, www.econsoc.hist.cam.ac.uk/docs/CWPESH%20number%203%20March%202012.pdf [accessed 30.09.22].

12 Crisp (2011).

13 See Hoskins and Preston (2022).

14 Anne Wilkinson, *The Victorian Gardener: The Growth of Gardening and the Floral World* (Stroud: Sutton Publishing, 2006), p. 5.

15 See Wilkinson, pp. 49–50.
16 For an introduction to the multiple concepts of family in the period see Kelly Hager and Talia Schaffer, 'Special issue: Extending families', *Victorian Review*, vol. 39 (Fall 2013), www.jstor/stable/24497057 [accessed 07.10.22].
17 William Cobbett, *Cottage Economy* (London: C Clement, 1822), p. 2.
18 Rebecca Preston, 'Home landscapes: Amateur gardening and popular horticulture in the making of personal, national and imperial identities, 1815–1914' (PhD, University of London, Royal Holloway, 1999), p. 252. DX2094455 [accessed 06.02.2021]. See also, '"The scenery of the torrid zone": Imagined travels and the culture of exotics in nineteenth-century British gardens' in Felix Driver and David Gilbert (eds.), *Imperial Cities: Landscape, Display and Identity* (Manchester: Manchester University Press, 1999), pp. 194–209.
19 See Leonore Davidoff, Megan Doolittle, Janet Fink and Katherine Holden, *Family Story: Blood, Contract and Intimacy: 1830–1960* (London and New York: Longman, 1999), p. 109.
20 See www.english-heritage.org.uk/learn/story-of-england/victorian/religion/ [accessed 08.10.22].
Research into the census suggests instead that a more likely figure for church attendance was 20% (and almost half were Nonconformists), see Jeffrey Cox, 'Were victorian nonconformists the worst imperialists of all?', *Victorian Studies*, 46, No. 2. Papers from the Inaugural Conference of the North American Victorian Studies Association (Winter, 2004), 243–255 (pp. 244–245), www.jstor.org/stable/3830292 [accessed 08.10.2022].
21 See Catherine Hall, *White, Male and Middle-Class: Explorations in Feminism and History* (Cambridge: Polity Press, and Oxford: Blackwell Publishers, 1992), pp. 75–93.
22 See Christopher Berry, *Social Theory of the Scottish Enlightenment* (Edinburgh: Edinburgh University Press, 1997); the idea of progressive stages of human development is promulgated in the work of William Robertson, Adam Smith, John Millar and Adam Ferguson.
23 David Hartley, *Observations on Man* (London, 1749).
24 See Dewis, pp. 170–171.
25 Robin Veder, 'How gardening pays: Leisure, labor and luxury in nineteenth-century transatlantic culture' (PhD, College of William & Mary, Williamsburg, 2000), p. 77, https://dx.doi.org/doi:10.21220/s2-cvgv-zy92 [accessed 08.10.22]; See also, Anne Secord, 'Artisan Botany' in *Cultures of Natural History* (Cambridge: Cambridge University Press, 1996), pp. 378–393 and Willes, pp. 96–192.
26 William John Thomas Mitchell, *Landscape and Power*, 2nd ed. (Chicago: University of Chicago Press, 2004), p. 2. Cited by Preston, 'Home Landscapes . . .', p. 17.
27 The collection included a number of Chinese Bronze figures of 'bizarre and unexpected appearance' alongside more conventional statuary, George Plumptre, *Garden Ornament: Five Hundred Years of History and Practice* (London: Thames and Hudson, 1989), p. 115.
28 See Dewis, p. 172.
29 See Judith Neiswander, *The Cosmopolitan Interior: Liberalism and the British Home, 1870–1914* (New Haven and London: Yale University Press, 2008) and Deborah Cohen, *Household Gods: A History of the British and Their Possessions* (New Haven and London: Yale University Press, 2006). Note that Bourdieu's argument is developed by David Wall, 'Andrew Jackson downing and the tyranny of taste', *American Nineteenth Century History*, 8, No. 2 (June 2007), 187–203.
30 See Armstrong, *Victorian Glassworlds, Glass Culture and the Imagination* (Oxford: Oxford University Press, 2008), pp. 167–202.
31 See Floud, pp. 210–211.

32 See Elliott, 'The Royal Horticultural Society and its orchids', *RHS Lindley Library Occasional Papers*, 2 (2010), 3–53, www.rhs.org.uk/about-the-rhs/pdfs/publications/lindley-library-occasional-papers/volume-2-march-2010.pdf.
33 See Preston, *Home Landscapes*, pp. 260–261.
34 Ibid, see p. 250.
35 See Muthesius, pp. 49–51.
36 See Laura Fosberg, *Worlds Beyond: Miniatures and Victorian Fiction* (New Haven and London: Yale University Press, 2021), pp. 1–23, for an introduction to a range of theories in relation to miniaturisation.
37 See Mary Rose Blacker, *Flora Domestica: A History of Flower Arranging, 1500–1930* (London: National Trust Enterprises, 2000).
38 See Keith Alcorn, 'From specimens to commodities: The London nursery trade and the introduction of exotic plants in the early nineteenth century', *Historical Research*, 93 (November 2020), 715–733, https://doi.org/10.1093/hisres/htaa025 [accessed 11.12.22].
39 See Elliott, *The Royal Horticultural Society: A History, 1804–2004* (Chichester: Phillimore, 2004), p. 307.
40 See Willes, pp. 109–111.
41 See Blacker, p. 155.
42 See Beverly Seaton, *The Language of Flowers: A History* (Charlottesville and London: University Press of Virginia, 1995), p. 2.
43 Elliott, 'The Victorian Language of Flowers', *RHS Lindley Library Occasional Papers*, 10, (2013), 3–94 (pp. 26–27), www.rhs.org.uk/about-the-rhs/pdfs/publications/lindley-library-occasional-papers/volume-ten.pdf.
44 Elliott, p. 48.
45 See Veder, 'How Gardening Pays: . .', pp. 211–213; also, C. Bornhorst-Winslow, 'The important role played by household crafts in the lives of nineteenth century women in Britain and America' (Master's thesis, Wright State University, Dayton, Ohio, 2012). https://corescholar.libraries.wright.edu/cgi/viewcontent.cgi?article=1008&context=humanities [accessed 11.10.22].
46 Jackson Downing, 'Reviews', *Horticulturalist*, 3 (October 1848), p. 190.

Part 1

GREEN HOMES

a)

DOMESTIC GARDENS

Editorial Headnote

1. J. C. Loudon, 'Introduction', *The Suburban Gardener and Villa Companion* (1838), [1–11] excerpts, pp. 1–6, 8–10
2. Andrew Jackson Downing, 'Preface', *A Treatise on the Theory and Practise of Landscape Gardening*, 1st edn (1841)
3. William Howitt, 'Favourite Pursuits of English Cottagers and Workmen', *The Rural Life of England* (2 vols) 2 (1838), pp. 305–311. [first published in 'Stepping-stones in our progress towards a great Christian republic', *Taits Edinburgh Magazine* (November 1835), 731–742]
4. Shirley Hibberd, 'Introduction', *The Town Garden: A Manual for the Management of City and Suburban Gardens* (1855)
5. Charles Dickens, ch. 6, *Great Expectations* (3 vols) 2 (1861), [88–104] excerpt, pp. 95–102
6. Shirley Hibberd, 'Villa Gardening', *Amateur Gardening* (23 June 1888), p. 85
7. Gertrude Jekyll, ch. 1 Introductory' and ch 14, 'Large and Small Gardens', *Wood and Garden, Notes and Thoughts, Practical and Critical, of a Working Amateur* (1899), pp. 1–6, [171–188] excerpt, pp. 171–179
8. Eugène Noel, 'Small Gardens' ('Petits Jardins'), *Le Jardin* (1887), pp. 50–51

This selection includes accounts of domestic gardens and ideas about their value. They share common tropes in which gardens are imagined as spaces of self-fulfilment for men and women: as havens for moral behaviour either in connection to God's work and/or in closeness to the natural world; as refuge for family together in shared moments of leisure; or as relief from the busyness of daily life. There was, however, a general shift in discourse around gardening (and gardens) from patriotic duty and form of self-expression to a more individualised occupation in the final decades of the century. Domestic gardens as a source of food are touched on in some sources here; for more on this topic, see Ib. For sources concerning the design and planting of particular areas of domestic gardens including parks, pleasure gardens, and woodlands, see Volumes 3 and 6.

John Claudius Loudon (1783–1843), journalist, editor, publisher, and landscape gardener, was born in Scotland, the son of a farmer. He trained in nurseries there and attended Edinburgh University. His first article appeared in 1803; journalism and books followed in tandem with a career in practical management and design of agricultural and ornamental landscape. In 1822 he was commissioned by Longman to produce *An Encyclopaedia of Gardening* – the first of a series. Consequently, he started the *Gardener's Magazine* (one of several journals that he founded). *The Suburban Gardener and Villa Companion* (1838) was an attempt to recoup the losses he incurred through the publication of the *Arboretum et Fruticetum Britannicum* (1838); see Volume 3. In the *Suburban Gardener* he advises on how gardens, with homes attached, might be created and managed within a hierarchy of the first to fourth rate and gives examples, including his own property at Bayswater. He also refers to his earlier publication, *An Encyclopaedia of Cottage, Farm and Villa Residences* (1833). The minimum annual income of £200 to £300 required to maintain such establishments was high in comparison with the average wage which appears to have been around 9 shillings a week (for a farm labourer in the 1830s and 1840s, for example). Loudon emphasises the democratisation of space and good citizenship, as does Andrew Jackson Downing (who also includes the design of his own house and garden) in *A Treatise on the Theory and Practice of Landscape Gardening* first published in 1841. Jackson Downing (1815–1852) was a professional nurseryman and had just started his architectural and landscaping career. *A Treatise*, his first book, contained coloured landscape images and was, despite its high price, a commercial success. The geographical and political contexts in the US and Britain were different. However, both Jackson Downing and Loudon were responding to the chaos and brutal conditions arising in newly formed or expanding towns and cities and argued for the reformist powers of managed landscape.

Both William Howitt and Shirley Hibberd suggest that gardens and gardening might benefit working people. The excerpt from *The Rural Life of England* (1838) on artisans' gardens in Nottingham by William Howitt (1792–1879) was first published in 1835 in *Tait's Edinburgh Magazine* (1832–1861), the radical rival to the Tory *Blackwoods* (1817–1980). Howitt argues that the relentless demands of the recently industrialised factories in Manchester were subjecting their workers to a form of slavery worse than that of 'slaves of the lamp', the genii that appear in *Arabian Nights*. The first English language edition of *Arabian Nights* appeared circa 1706–1721, three years before plantation slavery was abolished in British colonies.) Howitt, who held the office of alderman for Nottingham (1833–1837), provides a snapshot of what he called 'allotments', although with the cultivation of fruit and flowers and high rents these do not conform to allotments in the strictest sense. The article served as good publicity for the city and to persuade readers of *Tait's* to support the allotment movement. Note that the reprint (1835) concludes with a call to Christianity and to patriotism.

Two decades later, Shirley Hibberd comments on the state of gardens in the city as well as the suburb and points out the disparity of space available to the poor as well as the pollution of city air in *The Town Garden* (1855). This very small publication (10 × 15 cm) was one of many by Hibberd that were intended for an amateur readership and earned him hostility from some sections of the professionally orientated gardening press (see Volume 3). Hibberd (1825–1890) was born into a Nonconformist household. An autodidact, he published a volume of poetry (1851) before moving into a career as a journalist and writer on environmental issues, gardens, and gardening. He also gave public lectures on science and natural history.

Mr. Wemmick's garden in Charles Dickens's novel *Great Expectations* (1861) is located in the poorer southern outskirts of London. He shares the garden with his deaf father, and the cannon that he fires on his return from work is one of his ways of communicating. The representation of Mr. Wemmick's garden is, in part, a comment by Dickens (who, by the 1860s, was a celebrity novelist and social critic) on the ideology of separate spheres.

The illustration in 'Villa gardening' features the Butters family garden in East London engraved for the printer and publisher of *Amateur Gardening* (W. H. and L. Collingridge) from a painting of 1874 (artist unknown). It is an image that readers might aspire to in the creation of their own gardens and would require the employment of others. The point for amateurs, however, is to manage their gardens independently as far as possible and to observe and to learn from their own experience. Gertrude Jekyll (1843–1932) also advocates a practical and informed approach to garden-making and allows for the breaking of rules and of serendipity framed within Christian service. She concludes her introduction with a quote from Corinthians. *Wood and Garden Notes and Thoughts, Practical and Critical, of a Working Amateur* (1899) is a lavish volume that includes photographs and Jekyll's first published collection of writing based on articles from the *Guardian*. Jekyll was born into an upper-class family which, to a degree, freed her from conventional restraints in both her designs for gardens and the way she lived her life.

Eugène Noel (1816–1899), a French journalist and nature writer based in Rouen (the city of Flaubert's birth, see Ib), celebrates small gardens in his piece 'Les petits jardins' for *Le Jardin* (1887). He concludes with an argument for self-sufficiency in terms of food production around the world, unaware perhaps, of the interruption of small-scale horticulture practiced on the African continent caused by colonisation. The reference to the 'conjunction of America and India' suggests anxiety concerning export of food from non-French territories.

Further Reading

Roy A. Church, *Economic and Social Change in a Midlands Town: Victorian Nottingham, 1815–1900* (1966)

1

INTRODUCTION TO *THE SUBURBAN GARDENER AND VILLA COMPANION*

J. C. Loudon

Source: J. C. Loudon, 'Introduction', *The Suburban Gardener and Villa Companion* (1838), [1–11 [excerpts], pp. 1–6, 8–10

INTRODUCTION

THE enjoyments to be derived from a suburban residence depend principally on a knowledge of the resources which a garden, however small, is capable of affording. The benefits experienced by breathing air unconfined by close streets of houses, and uncontaminated by the smoke of chimneys; the cheerful aspect of vegetation; the singing of birds in their season; and the enlivening effect of finding ourselves unpent-up by buildings, and in comparatively unlimited space; are felt by most people: but it requires some little knowledge of the progress or decline of vegetation throughout the year, and of rural nature generally, to be able to derive much enjoyment from the recurrence of the seasons, and their influence on plants; and much more knowledge of vegetation, botany, natural history, and the art of gardening, to derive the greatest amount of advantages which a country house and garden are calculated to afford.

For these reasons, it is our intention to endeavour to give our readers a more intimate knowledge of the subjects treated of, than has hitherto been attempted in works of this kind. Instead of a mere didactic calendar, or a dictionary of plants to be cultivated, directing what works are to be done in each month, or what operations are to be performed in the culture of particular plants or trees, we shall endeavour to convey such a notion of the structure and nature of a plant, as to show the necessity and advantages of cultivating it; and such an idea of the nature of soils, as to show the improvements which may be made in them, by mechanical operations, and the application of manures. We shall treat of the various operations of gardening, whether performed on plants, or on the soil; and for the performance of every operation, from the most mechanical to the most scientific, we shall assign a reason. Let not the reader be alarmed, however, lest we should go into minute scientific details, only capable of being understood by the botanist or chemist, or by the professional gardener: so far from contemplating this, our intention is to confine ourselves to the simplest and most important features; to

DOI: 10.4324/9780429198809-3

treat every subject in such a manner as to be understood by those who have little knowledge of either gardening or country affairs; and, more especially, to adapt it to the instruction and amusement of ladies. For the sake of those who wish to pursue any particular topic further than we have done, we shall always refer to what are considered the best works on each subject of which we shall treat.

Before proceeding to the garden itself, we shall endeavour to excite some interest in its favour, and to show the solid advantages which may be derived from a suburban residence; though on both these topics we shall bestow very few words.

There is a great deal of enjoyment to be derived from performing the different operations of gardening, independently altogether of the health resulting from this kind of exercise. To labour for the sake of arriving at a result, and to be successful in attaining it, are, as cause and effect, attended by a certain degree of satisfaction to the mind, however simple or rude the labour may be, and however unimportant the result obtained. To be convinced of this, we have only to imagine ourselves employed in any labour from which no result ensues, but that of fatiguing the body, or wearying the mind: the turning of a wheel, for example, that is connected with no machinery, or, if connected, effects no useful purpose; the carrying of a weight from one point to another and back again; or the taking of a walk without any object in view, but the negative one of preserving health. Thus, it is not only a condition of our nature, that, in order to secure health and cheerfulness, we must labour; but we must also labour in such a way as to produce something useful or agreeable. Now, of the different kinds of useful things produced by labour, those things, surely, which are living beings, and which grow and undergo changes before our eyes, must be more productive of enjoyment than such as are mere brute matter; the kind of labour, and other circumstances, being the same. Hence, a man who plants a hedge, or sows a grass-plot in his garden, lays a more certain foundation for enjoyment, than he who builds a wall or lays down a gravel walk; and, hence, the enjoyment of a citizen whose recreation, at his suburban residence, consists in working in his garden must be higher in the scale, than that of him who amuses himself, in the plot round his house, with shooting at a mark or playing at bowls.

To dig, to hoe, and to rake, are not operations requiring much skill; and the amateur gardener will, perhaps, chiefly value them for their use in preparing for crops, or in encouraging the growth of crops already coming forward: but the operations of pruning and training trees, when well performed, are not only interesting to the operator at the time, but the plants so pruned or trained afford him pleasure every time he sees them afterwards throughout the season, till the period returns when they must be again pruned and trained. The operation of striking plants from cuttings is performed in a variety of ways, according to the nature of the plants; and may truly be called one of intense interest, both in its performance, and in the expectation of its results. By the great majority of amateur gardeners, cuttings are made and planted at random; and their failure or success is, in consequence, a matter of chance: but a very little scientific light thrown on the subject leads to rules for operating, which will turn chance into certainty in almost every case that can occur to ordinary practitioners; and, consequently, will greatly enhance the pleasure of performing the operation, from the consciousness that the labour bestowed will not be thrown away. We need not here refer to the operations of grafting, layering, or sowing seeds; nor need we mention innumerable other operations which require to be performed in the course of the year, even in the very smallest garden; but we must be allowed to notice the watering of plants, which all persons can enjoy from the earliest infancy upwards. What pleasure have not children in applying their little green watering-pans to plants in pots, or pouring water in at the roots of favourite flowers in borders? And what can be more rational than the satisfaction which the grown up amateur, or master of the house, enjoys, when he returns from the city to his garden in the summer evenings, and applies the syringe to his wall trees, with refreshing enjoyment to himself and the plants, and to the delight of his children, who may be watching his operations? What can be more refreshing than, in a warm summer's evening, to hear, while sitting in a cool parlour, with the windows open, or in a summer-house, the showering of water by the syringe upon the leaves of the vines or fig trees trained under the adjoining veranda, or upon the orange trees and camellias, or other exotic shrubs, planted in the conservatory connected with it? What more delightful than to see the master or the mistress of a small garden or pleasure-ground, with all the boys and girls, the maids, and, in short, all the strength of the house, carrying pots and pails of water to different parts of the garden; and to see the refreshment produced to the soil and plants by the application of the watering-pan and the syringe?

Even the search after insects is a great enjoyment in a garden; and, in fact, opens up an entirely new field of exertion and interest to those who have not before made minute observations in this department of nature. Fifty years ago, the subject of destroying insects was scarcely considered as belonging to gardening; and their eggs, which now every young gardener recognises, in winter glued in rings to the branches of his fruit trees, or in spring deposited on the back of his gooseberry leaves, passed unheeded through their different stages of developement; and the ravages the larvæ committed on crops were considered as inevitable blights, produced by the atmosphere. In the present day, so much of the beauty and the value of the products of all gardens is known to depend on subduing insects, that a knowledge of the subject is considered essential to every gardener: but it is more especially necessary that the possessor of a suburban garden should know how to keep insects in subjection; both because he is generally his own gardener, and because insects are more abundant in such gardens than in those situated farther in the country, sometimes from the comparatively weak and crowded state of the plants, and, in other instances, from the absence of those natural enemies of insects, the small birds.

One of the greatest of all the sources of enjoyment resulting from the possession of a garden is, the endless variety which it produces, either by the perpetual progress of vegetation which is going forward in it to maturity, dormancy, or decay, or by the almost innumerable kinds of plants which may be raised in even the smallest garden. Even the same trees, grown in the same garden, are undergoing perpetual changes throughout the year; and trees change, also, in every succeeding year, relatively to that which is past; because they become larger and larger as they advance in age, and acquire more of their characteristic and mature forms. The number of plants, and especially of trees, which can be cultivated in a suburban garden at one time is necessarily circumscribed; but, if a suburban amateur chose to limit the period during which he cultivated each tree or plant to the time of its flowering with him for the first time, he might, in the course of a few years, more or less in number according to the size of his garden, have had growing in it all the plants in cultivation in the open air in Britain, with the exception of a few of the larger of the forest trees; and even these he might also have flowered, by making use of plants raised from cuttings or layers, or of miniature trees, made by ringing and rooting the branches of old trees in the Chinese manner. Independently, however, of the variety and change resulting from the plants cultivated, every month throughout the year has its particular operations and its products: nay, it would not be too much to say, that during six months of the year a change takes place, and is perceptible, in the plants of a garden, every day; and every day has, in consequence, its operations and its products. Even in winter, there is still something to do in every garden, however small may be its extent: the walks require to be kept in order, and some plants must be protected by litter or matting; and, if there should be no trees to prune, no ground to dig, no manure to collect or to barrow out, no dung to turn and prepare for hotbeds, there is, at all events, the preparation of names or numbers for plants; the cutting and painting of rods to tie them to; the sorting of seeds; the making of baskets; and the search after information on the subjects of plants and their culture, in books.

But imagine that to the suburban garden there is added a small green-house, or a flued pit! What a source of amusement and interest does not either of these garden structures hold out to the amateur gardener, during the winter and spring! Exactly in proportion as, in autumn, the out-door operations become fewer, the in-door operations of the green-house or pit become more numerous; and, in midwinter, the citizen amateur, if he is detained in his shop or his counting-house till after sunset, will be under the necessity of shifting, cleaning, and watering his plants, and otherwise operating with them (as some of our friends are obliged to do), by candle-light. A green-house, from the quantity of glass that it requires, is, for some suburban residences, too costly to erect: but much of the produce of the green-house may be procured, at half the expense, by the use of a pit, which requires no other glass than the sashes which form its roof. The amusement and the products which such a pit, in the hands of an ingenious amateur, is calculated to afford, are almost without end. Small salading may be produced in it throughout the whole winter. Chicory roots (though this may be accomplished in a common cellar) may be made to throw out their blanched leaves, which form the most delightful of all winter salads, at least to our taste; tart rhubarb or sea-kale may be forced in pots; as may parsley, mint, and other herbs. Bulbs may be forced; and a bloom of China roses may be kept up throughout the winter. But, perhaps, the most important use to which such a pit can be applied, in a small suburban garden, is to preserve throughout the winter, and to bring forward in spring, pelargoniums, fuchsias, salvias, calceolarias, verbenas, and other fine exotic flowers, and also half-hardy and tender annuals, for turning out into the flower-garden, or into the miscellaneous border, in the beginning of summer. We are, however, going too much into detail; we shall, therefore, only further allude to the enjoyment which can be had in every suburban house, and even town house, without the aid of a green-house or a pit, by keeping plants through the winter in a garret (always the warmest, and, when furnished with windows, much the lightest, part of a house), immediately under a skylight, or other window; or by keeping them near a window in the ordinary rooms.

These are a few of the absolute enjoyments to be derived from a suburban house and garden; and we shall next notice another, which flows from the same source, but which may be called relative or incidental. The opportunity which a garden affords to its possessor of acquiring a scientific and practical knowledge of plants is a source of great interest, not only in his own garden, but wherever else plants may come in his way; whether in a wild state, in gardens, exposed for sale in markets, or delineated and described in books. Another source of incidental enjoyment is that which will arise from the acquirement of some knowledge of gardening, and of rural architecture as an art of design and taste. As it forms a part of our plan to assist our readers in the choice of a suburban house and garden, or in the building, laying out, and planting of these, when he prefers, or is obliged, to undertake the forming of them himself; so we trust we shall, in some degree, render him a man of taste, and, consequently, a critic, both in landscape-gardening and rural architecture. How great a source of enjoyment this is, and how great an interest it enables its possessor to take in suburban scenery, in landscape, and in architecture, generally; or, in short, wherever he sees a house or a tree; those only can know who have gone through the necessary preparation. We anticipate, in particular, that it will afford a great source of enjoyment to ladies; and that it will tend to the improvement, as far as relates to matters of taste, not only of the gardens under their care, but of the architecture of their houses, and of the style and taste of the furniture and finishing of their interiors

. . .

We might enlarge, here, on the great advantages which would result from bringing up children with a taste for garden pursuits and natural history; and the vast influence which this is calculated to have on their future happiness, and on the welfare of society, by enabling them, instead of passing their leisure hours in a manner degrading to human nature, to interest themselves in recreations both agreeable and useful: but the field is too wide to be entered on within our limits, and we must therefore leave the subject to be worked out by the imagination of our readers.

Much of the enjoyment of a suburban residence depends on knowing what to expect from it; what, in short, is consistent, and what is inconsistent, with its limits and its local situation. We have shown, in the *Encyclopædia of Cottage, Farm, and Villa Architecture* (p. 8.), that all, in the way of house accommodation, that is essential to the enjoyment of life, may be obtained in a cottage of three or four rooms, as well as in a palace; and we shall prove, in this work, that a suburban residence, with a very small portion of land attached, will contain all that is essential to happiness, in the garden, park, and demesne of the most extensive country residence. Let us briefly make the comparison. The objects of the possessors of both are the same: health, which is the result of temperance and exercise; enjoyment, which is the possession of something which we can call our own, and on which we can set our heart and affections; and the respect of society, which is the result of their favourable opinion of our sentiments and moral conduct. No man in this world, however high may be his rank, great his wealth, powerful his genius, or extensive his acquirements, can ever attain more than health, enjoyment, and respect. The lord of an extensive demesne seeks after health by hunting, shooting, or other field sports, or by superintending the general management and cultivation of his estate; the lady seeks recreation in her pleasure-ground, or in airings in her carriage: and both find their enjoyment in their children, and in their house and garden, and other surrounding objects. Now, the master of a suburban villa finds health in the change it affords from his occupation as a citizen; or, if he has retired from business, in the personal cultivation of his garden. He also finds enjoyment, not only in his family, friends, and books, but in his garden, and in the other rural objects which he can call his own, and which he can alter at pleasure, at a trifling expense, and often with his own hands. It is this which gives the charm of creation, and makes a thing essentially one's own. Every one must have felt the infinitely greater pleasure which is enjoyed from the contemplation of what we have planned and executed ourselves, to what can be experienced by seeing the finest works belonging to, and planned by, another. Our own work is endeared to us by the difficulties we have met with and conquered at every step: every step has, indeed, its history, and recalls a train of interesting recollections connected with it. The master of a suburban residence, however small may be his demesne, may thus procure health and enjoyment at the same time, with more certainty than the possessor of a larger property; because his grounds lie more in his hands, and he can superintend every change himself. His wife's exercise consists in the personal management of her household affairs; and her enjoyments are the same as those of her husband. The respect, or the good opinion, of his friends and neighbours, or of the inhabitants of the parish in which he resides, are, to the citizen retired to the suburbs, of as great value as the respect of the inhabitants of a district, or of the whole nation, is to the wealthy landholder or the senator. The difference of the happiness of the parties will therefore depend almost entirely on the difference in the degrees of their ambition; for in every other respect they are equal. All the necessaries of life may be obtained in as great perfection by the occupier of a suburban residence in the neighbourhood of London, who possesses 200*l.* or 300*l.* a year, as by the greatest nobleman in England, and at a mere fraction of the expense. Nay, the markets of London will supply the citizen, at all times, with the larger and more important articles of vegetable and animal food, of a better quality than can be raised, with certainty, in the garden or farm of any private individual in the country, where it may reasonably be calculated on, that every year, from unforeseen casualties, some crop will be found to fail, and others to be of inferior quality. Whatever advantages are enjoyed by the possessor of a residence in the suburbs of London, may be enjoyed to nearly the same extent in the suburbs of any large town in Europe and America, where the communication between such town and the sea is by good roads, and where the interchange of goods is free from heavy and vexatious government duties. Hence it follows (though it would occupy too much time to explain the process in detail), that, ultimately, the possessors of suburban residences in every country, and in every part of that country, will have nearly equal advantages; and that such residences may be considered as the ultimatum, in point of comfort and enjoyment, of the great mass of society; not only at present, but even after society has advanced to a much higher degree of civilisation, and to a comparative equalisation of knowledge, wealth, and taste.

2

PREFACE TO *A TREATISE ON THE THEORY AND PRACTICE OF LANDSCAPE GARDENING*

Andrew Jackson Downing

Source: Andrew Jackson Downing, 'Preface', *A Treatise on the Theory and Practise of Landscape Gardening*, 1st edn (1841)

PREFACE

A TASTE for rural improvements of every description is advancing silently, but with great rapidity in this country. While yet, in the far west the pioneer constructs his rude hut of logs for a dwelling, and sweeps away with his axe the lofty forest trees that encumber the ground, – in the older portions of the Union bordering the Atlantic, we are surrounded by all the luxuries and refinements that belong to an old and long cultivated country. Within the last ten years especially, the evidences of the growing wealth and prosperity of our citizens have become apparent in the great increase of elegant cottage and villa residences on the banks of our noble rivers, throughout our rich valleys, and wherever nature seems to invite us by her rich and varied charms.

In all the expenditure of means in these improvements, amounting in the aggregate to an immense sum, professional talent is seldom employed, in Architecture or Landscape Gardening, but almost every man fancies himself an amateur, and endeavours to plan and arrange his own residence. With but little practical knowledge, and few correct principles for his guidance, it is not surprising that we witness much incongruity and great waste of time and money. Even those who are familiar with foreign works on the subject in question, labour under many obstacles in practice, which grow out of the difference in our soil and climate, or our social and political position.

These views have so often presented themselves to me of late, and have been so frequently urged by persons desiring advice, that I have ventured to prepare the present volume, in the hope of supplying, in some degree, the desideratum so much required at present. While we have treatises in abundance on the various departments of the arts and sciences, there has not appeared even a single essay on the elegant art of Landscape Gardening. Hundreds of individuals who wish to

 DOI: 10.4324/9780429198809-4

ornament their grounds and embellish their places, are at a loss how to proceed, from the want of some *leading principles*, with the knowledge of which they would find it comparatively easy to produce delightful and satisfactory results.

In the following pages I have attempted to trace out such principles, and to suggest practicable methods of embellishing our Rural Residences, on a scale commensurate to the views and means of our proprietors. While I have availed myself of the works of European authors, and especially those of Britain, where Landscape Gardening was first raised to the rank of a fine art, I have also endeavoured to adapt my suggestions especially to this country and to the peculiar wants of its inhabitants.

As a people descended from the English stock, we inherit much of the ardent love of rural life and its pursuits which belongs to that nation; but our peculiar position, in a new world that required a population full of enterprise and energy to subdue and improve its vast territory, has, until lately, left but little time to cultivate a taste for Rural Embellishment. But in the older states, as wealth has accumulated, the country become populous, and society more fixed in its character, a return to, and fondness for, those simple and fascinating enjoyments to be found in country life and rural pursuits, is witnessed on every side. And to this innate feeling, out of which grows a strong attachment to natal soil, we must look for a counterpoise to the great tendency towards constant change, and the restless spirit of emigration, which form part of our national character; and which, though to a certain extent highly necessary to our national prosperity, are, on the other hand, opposed to social and domestic happiness. "In the midst[1] of the continual movement which agitates a democratic community," says the most philosophical writer who has yet discussed our institutions, "the tie which unites one generation to another, is relaxed or broken; every man readily loses the trace of the ideas of his forefathers, or takes no care about them."

The love of country is inseparably connected with the *love of home*. Whatever, therefore, leads man to assemble the comforts and elegancies of life around his habitation, tends to increase local attachments, and render domestic life more delightful; thus not only augmenting his own enjoyment, but strengthening his patriotism and making him a better citizen. And there is no employment or recreation which affords the mind greater or more permanent satisfaction, than that of cultivating the earth and adorning our own property. "God Almighty first planted a garden; and, indeed, it is the purest of human pleasures," says Lord Bacon. And as the first man was shut out from the *garden*, in the cultivation of which no alloy was mixed with his happiness, the desire to return to it seems to be implanted by nature, more or less strongly, in every heart.

In Landscape Gardening the country gentleman of leisure finds a resource of the most agreeable nature. While there is no more rational pleasure than that derived from its practice by him, who

> "Plucks[2] life's roses in his quiet fields,"

the enjoyment drawn from it, (unlike many other amusements,) is unembittered by the after recollection of pain or injury inflicted on others, or the loss of moral rectitude. In rendering his home more beautiful, he not only contributes to the happiness of his own family, but improves the taste, and adds loveliness to the country at large. There is, perhaps, something exclusive in the taste for some of the fine arts. A collection of pictures, for example, is comparatively shut up from the world, in the private gallery. But the sylvan and floral collections, – the groves and gardens, which surround the country residence of the man of taste, – are confined by no barriers narrower than the blue heaven above and around them. The taste and the treasures, gradually, but certainly, creep beyond the nominal boundaries of the estate, and reappear in the pot of flowers in the window, or the luxuriant, blossoming vines which clamber over the porch of the humblest cottage by the way side.

. . .

A. J. D.
Botanic Garden and Nurseries, }
Newburgh, N. Y. April, 1841.

Editorial Endnotes

1 "In the midst of continual movement"– Alexis de Toqueville, *Democracy in America* (1835–1840), Ch. 1, 'The Philosophical Method of the Americans'. The recent publication analysed the early decades of the US political and social systems.

2 "Plucks life's roses in his quiet fields" – Ebenezer Elliott, 'Love'. Elliott (1781–1849) was an English poet and factory owner who campaigned to repeal the Corn Laws. This was an early poem dedicated to a surgeon in Sheffield who was, like himself, self-taught. Elliott's verses are quoted elsewhere in the collection.

3

'FAVOURITE PURSUITS OF ENGLISH COTTAGERS AND WORKMEN'

William Howitt

Source: William Howitt, 'Favourite Pursuits of English Cottagers and Workmen', *The Rural Life of England* (2 vols) 2 (1838), pp. 305–311

Lastly, the general pleasures of a garden form a grand item in the enjoyments of the poor man. To shew what these pleasures are, to what an extent they are enjoyed in some districts, even by town mechanics, and how much further they may be extended, I shall quote a portion of a paper published by me in November 1835, in Tait's Magazine.

There are, in the outskirts of Nottingham, upwards of 5000 gardens, the bulk of which are occupied by the working class. A good many there are belonging to the substantial tradesmen and wealthier inhabitants; but the great mass are those of the mechanics. These lie on various sides of the town, in expanses of many acres in a place, and many of them as much as a mile and a half distant from the centre of the town. In the winter they have rather a desolate aspect, with their naked trees and hedges, and all their little summer-houses exposed, damp-looking and forlorn; but in spring and summer, they look exceedingly well, – in spring all starred with blossoms, all thick with leaves; and their summer-houses peeping pleasantly from among them. The advantage of these gardens to the working-class of a great manufacturing town, is beyond calculation; and I believe no town in the kingdom has so many of them in proportion to its population. It were to be desired that the example of the Nottingham artizans was imitated by those of other great towns; or rather, that the taste for them was encouraged, and in fact, created, by the example of the middle classes, and by patriotic persons laying out fields for this purpose, and letting them at a reasonable rate. A wide difference in the capability of indulging in this healthful species of recreation, must of course, depend on the species of manufacture carried on. Where steam-engines abound, and are at the foundation of all the labours of a place, as in Manchester, for instance, there you will find few gardens in the possession of the mechanics. The steam-engine is a never-resting, unweariable, unpersuadable giant and despot; and will go on thumping and setting thousands of wheels and spindles in motion; and men must stand, as it were, the slaves of its unsleeping energies. O! what was the fate of the ancient genii to the

DOI: 10.4324/9780429198809-5

fate of our modern mechanics! What was the fate of "the slaves of the lamp," or the slaves of talismanic ring, to that of the slaves of the steam-engine! *They* could vanish and lie at rest till came the irresistible call; they could sport over ocean and desert, through the air and the clouds; they could speed into the depths of space and wander amid the inconceivable mysteries and miracles of unknown worlds, till the omnipotent spell recalled them to execute some temporary wish of their tyrant, and then return to a wide liberty. But the slave of the steam-engine must be at the beck of *his* tyrant night or day, with only such intervals as barely suffice to restore his wearied strength and faculties: – therefore you shall not see gardens flourish and summer-houses rise in the vicinity of this hurrying and tremendous power. But where it is not, or but partially predominates, there may the mechanic enjoy the real pleasures of a garden. And how many are those pleasures!

Early in spring – as soon, in fact, as the days begin to lengthen, and the shrewd air to dry up the wintry moisture – you see them getting into their gardens, clearing away the dead stalks of last year's growth, and digging up the soil; but especially on fine days in February and March are they busy. Trees are pruned, beds are dug, walks cleaned, and all the refuse and decayed vegetation piled up in heaps; and the smoke of the fires in which it is burnt, rolling up from many a garden, and sending its pungent odour to meet you afar off. It is pleasant to see, as the season advances, how busy their occupants become; bustling there with their basses in their hands and their tools on their shoulders; wheeling in manure; and clearing out their summer-houses; and what an air of daily increasing neatness they assume, till they are one wide expanse of blossomed fruit-trees and flowering fragrance. Every garden has its summer-house; and these are of all scales and grades, from the erection of a few tub staves, with an attempt to train a pumpkin or a wild-hop over it, to substantial brick houses with glass windows, good cellars for a deposit of choice wines, a kitchen, and all necessary apparatus, and a good pump to supply them with water. Many are very picturesque rustic huts, built with great taste, and hidden by tall hedges in a perfect little paradise of lawn and shrubbery – most delightful spots to go and read in of a summer day, or to take a dinner or tea in with a pleasant party of friends. Some of these places which belong to the substantial tradespeople have cost their occupiers from one to five hundred pounds, and the pleasure they take in them may be thence imagined; but many of the mechanics have very excellent summer-houses, and there they delight to go, and smoke a solitary pipe, as they look over the smiling face of their garden, or take a quiet stroll amongst their flowers: or to take a pipe with a friend; or to spend a Sunday afternoon, or a summer evening, with their families. The amount of enjoyment which these gardens afford to a great number of families, is not easily to be calculated – and then the health and the improved taste! You meet them coming home, having been busy for hours in the freshness of the summer morning in them, and now are carrying home a bass brimful of vegetables for the house. In the evening thither-ward you see groups and families going; the key which admits to the common paths that lead between them is produced; a door is opened and closed; and you feel that they are vanished into a pure and sacred retirement, such as the mechanic of a large town could not possess without these suburban gardens. And then to think of the alehouse, the drinking, noisy, politics-bawling alehouse, where a great many of these very men would most probably be, if they had not this attraction, – to think of this, and then to see the variety of sources of a beautiful and healthful interest which they create for themselves here: – what a contrast! – what a most gratifying contrast! There are the worthy couple, sitting in the open summer-house of one garden, quietly enjoying themselves, and watching their children romping on the grass-plot, or playing about the walks; in another, a social group of friends round the tea-table, or enjoying the reward of all their spring labours, picking strawberries fresh from the bed, or raspberries, gooseberries, and currants from the bush. In one you find a grower of fine apples, pears, or plums, or of large gooseberries; in another, a florist, with his shew of tulips, ranunculuses, hyacinths, carnations, or other choice flowers that claim all his leisure moments, and are a source of a thousand cares and interests. And of these cares and interests, the neat awning of white canvas, raised on its light frame of wood; the glasses, and screens of board and matting, to defend those precious objects from every rude attack of sun, wind, or rain – all these are sufficient testimonies; and tell of hours early and late, in the dawn of morning and the dusk of evening, when the happy man has been entranced in his zealous labours, and absorbed in a thousand delicious fancies, and speculations of perfection. Of late, the splendid dahlia and the pansy have become objects of attention; and I believe of the latter flower, till recently despised and overlooked, except in the old English cottage garden, there are now more than a hundred varieties, of such brilliance and richness of hue, and many of them of such superb expanse of corolla, as merit all the value set upon them.

This is the allotment system of the manufacturing town; to the full as desirable as that for the country, and which may be facilitated, fraught as it is with abundant physical and moral good, by philanthropic individuals to a great extent. At Nottingham, as I have observed, the taste seems to have grown up originally of itself, and then, exciting the attention of speculators, has been extended to its present growth by them. The mechanics there have not their gardens at a cheap rate. They all say that they could purchase their vegetables in the market for the amount of their rental and incidental expenses; but then, they get the health and the enjoyment, and their fruit and vegetables they get so fresh.

There are, according to a personal examination made by myself, now, upwards of 5000 of these gardens, containing, as single gardens, 400 square yards each, – the general scale of a garden; though a good many are held as double, and even treble gardens. These let at from a half-penny to three half-pence per yard; but averaged at three farthings, make a rental of 1*l*. 5s. per garden, or a total of 6250*l*. Five thousand gardens of 400 yards each of clear garden ground, independent of fences and roads, give 413 acres and about a rood. Now, if we add one-fifth for fences and roads, the total quantity of land occupied is 496 acres, or we may say, in round numbers, 500 acres. Here then, 500 acres, which, at fifty shillings an acre – a good rent for ordinary purposes, would yield a rent of 1250*l*.; yield, by being converted into gardens, a rent of 6250*l*., or a clear profit of 5000*l*.

Thus, it is evident, that any persons willing to promote the taste for gardening in the neighbourhood of towns, might double, in many instances, the ordinary rent of the land, and yet let it in gardens at half the price of these Nottingham ones. Even where land in the vicinity of a large town is very highly rented, a half-penny a yard, and ten gardens to the acre, fences and roads included, would produce 8*l*. 6s. 8d. per acre; no contemptible sum; to say nothing of the real kindness of the accommodation, and the health, pleasure, and pure taste communicated to their fellow men. Whilst, against the increased risk of loss, and the increased trouble of the collection of rent, are to be set the value of the garden stock, fruit-trees, shrubs, and flower-roots, and the summer-houses, which enhance the value to the next tenant.

Here I close this chapter, and this department of my work, – the habits and amusements of the people. It is a subject to which I attach no common importance. The people make the majority of our race; and if they are all equally the objects of that divine care which created them, they must be equally the objects of our truest sympathies. This has not hitherto been sufficiently considered: but every day that consideration must be forced more and more upon us; and we shall be made to feel that no philosophy is good which does not include the poor in its theory; no religion is sound, which does not recognize their kinship; no legislation is wise which does not operate for their physical and intellectual benefit; and no country can be said to be truly prosperous, where the multitude is not respectable, enlightened, moral, and happy.

Let us all endeavour to hasten this period, as a living proof that Christianity is really preached to the poor; and that our knowledge has produced the most felicitous of its genuine fruits, in peopling this great nation with a race such as no nation has yet possessed; such as may eat,

> Well earned, the bread of service, yet may have
> A mounting spirit; – one that entertains
> Scorn of base action, deed dishonorable,
> Or aught unseemly.
>
> *Charles Lamb.*[1]

Editorial Endnote

1 "Well earned, the bread of service, yet may have" – 'The Grandame' by Charles Lamb (1775–1834).

4

INTRODUCTION TO *THE TOWN GARDEN*

Shirley Hibberd

Source: Shirley Hibberd, ch. 1, 'Introduction', *The Town Garden: A Manual for the Management of City and Suburban Gardens* (1855)

INTRODUCTION

It is generally thought that a city garden is an impossibility! that vegetation cannot be reconciled to the close air, the darkness, and the smoke of towns; and that all attempts to mingle the rural with the urban must, like Brummel's[1] forty neckcloths, turn out failures. There can be no greater mistake than this; and though city gardening has its disadvantages and difficulties, the man determined to succeed, may produce not only green stuff but flowers, in a north aspect, and under the very shadow of a gasometer. The problem is, how to grow flowers in a soil of cinders, and an atmosphere of smoke? – flowers which, of all things, revel in sunshine, and demand it as their daily food.

Town gardening, as at present practised, is indeed a sorry affair. Neither the scientific gardener, nor the enthusiastic lover of natural beauty, carries his garden with him to the town when necessity compels him to a sojourn in Smoke-land. Agricultural ingenuity and amateur patience seem neither of them to dream that city life might be vastly improved if a few flowers were sprinkled about it, so as to restore to the heart the social warmth which money-getting had crushed out of it. Yet the love of flowers seldom dies out in man; and though the eye get hard and the mouth puckered by long sacrifice of life to commerce, yet the one is capable of a joyous twinkle, and the other of a sudden smile, when a violet-girl crosses the path in February, or when the nose catches a momentary odour of mignionette from the soot-box on the window-sill in May.

In the musty courts and alleys, wall-flowers, stocks, and musk-plants are purchased every spring, and set to flourish in broken teapots, saucepans, flower-pots – damned for ever by green or brown paint – or rotten boxes filled with stuff called mould, but which looks like the dust of a perished mummy. These go black in the face in four days from the date of planting, and die three days after that from sheer suffocation, gasping up to the last moment for light and air. Geraniums pass a torpid life on window-sills and in dark parlours, where none but the housekeeper

DOI: 10.4324/9780429198809-6

can aver they are geraniums – such naked, smoke-dried sticks do they appear. In summer they become herbaceous, and put forth a few shoots of any tint but green, and sometimes a blossom or two of wretched quality, much to the joy of the dame who lavishes her skill upon them; but, after a brave attempt to grow and flourish, they once more take the soot-sickness, become limp and leafless, and pass the winter again in oblivious hybernation.

Thousands of beautiful plants are every spring and summer brought from the nurseries round London, and sold in the city to undergo the slow death of suffocation – dying literally of asphyxia, from an absorption of soot in the place of air – their demise being accelerated by copious supplies of water at improper times, or the withholding altogether of the refreshing element. The wonder is, not that such plants perish miserably, but that they last so long, when plunged, without hope of relief, into such a "Black Hole of Calcutta."[2]

As we approach the suburbs, gardening matters look up a bit. As shop fronts disappear, forests of green iron spikes spring up, enclosing small plots of garden-ground in the fronts of suburban residences, where bank clerks, thrifty traders, agents, actuaries, poor authors, and *hoc genus omne*, seek evening repose with their families, and ridicule the innate love of man for things rural by giving countenance to the labours of the gardener, who calls once a fortnight to dig and rake the sour mould round the paternal laurel trees, and shave the small lawn into a condition more smooth than green.

Here overgrown box borderings give shelter to sooty patches of London-pride, consumptive heartsease, and nasturtiums that rival pumpkins in size and rankness. In spring, a few clumps of primroses and cowslips appear, planted by some hopeful Alcinous or phlegmatic Semiramis, which would refresh the eye of the man of business as he hurries out at half-past nine, did they not droop as soon as planted, and vanish altogether in a week, no one knows why or whither. When these are gone, wall-flowers, early stocks, sweet-williams, and double-daisies, appear in lively variety; and as to arrangement, in every degree of bad taste. These are usually purchased of some itinerant florist – who carries the joy of Midas in a wheelbarrow – at various prices, ranging from one to ten a penny. Forthwith the garden breaks into bloom again, and displays as much colour as the ten thousand variegated oil lamps at Vauxhall. But, alas! the lamps go out, and leave the wicks sticking in the ground, in a state of very pure charcoal.

Towards June a few geraniums are to be seen, arrayed in papistical scarlet, as if they were blushing for the gardener; and with these are, perhaps, intermingled a few verbenas, and a blue iris, some marigolds, catstails, crysanthemums, and hollyhocks, all more or less blighted by smoke or ill treatment; dingy hollies, rising out of pyramids of stones; stiff laurels, that look as if they were cast in bronze, and have not been known to grow an inch since they were planted; dilapidated lilacs, forming Arcadian retreats for cats and slugs; lime trees, that are every year cut into the form of square boxes on pedestals, and sufficiently accurate in outline to have pleased Euclid; with occasionally a spasmodic Virginian creeper, which perishes every autumn, purple with asphyxia, plastered on the wall like a half-finished fresco; a palisade of ivy claiming no kindred with "the ivy green;" or a jasmine, sticking to a wall or portico, its long thread-like fingers drooping to the ground, as if its blood had long ago dried up and left it too weak even for an act of suicide: – these are the elements of what are called front gardens, each of which would be a Pæstum, were it not that the roses, hitherto unmentioned, never blossom. Nevertheless, go from the heart of the coal-black Babylon in whatever direction we may, we meet with the same spectacle of sooty gardens run to seed, the plants being converted into smoke filters, and the soil into a warren of worms.

Yet this Abdera[3] of suffocated greens may be converted into a Campania of luxuriance, provided the right means are adopted, and this, too, without excessive toil or exorbitant expense. It is, indeed, one of the peculiarities of gardening that a man may spend upon it just as much as he can or will, and – short of absolute waste of money – always with a satisfactory result; yet excessive expenditure and great loss of time are neither of them necessary sacrifices, when a few parterres are to be filled with flowers, and maintained in such order as shall not only please the eye but instruct the mind and refine the heart. Men love flowers, or they would never tolerate those nurseries of soot to which we have just adverted; and because these Stagyrian retreats are abandoned to the cats and the sparrows, only proves that their owners have not sufficient knowledge of their management, not that flowers are obnoxious, or gardening in town essentially a bugbear.

The point we wish to insist upon here is, that the wretched aspect which town gardens usually present is by no means necessary, but in every respect inexcusable and disgraceful. In proof of this, observe with what success some few of the gardens in the centre of London are cultivated, particularly those near the Inns of Law, which exhibit a freshness as Arcadian as the surrounding enclosures do a darkness quite chaotic. It is on the verge of the town, where the suburban villages – such as Edgeware on the west, Bromley and Plaistow on the east, Clapton on the north, and Dulwich on the south – commence, that we find the best examples of suburban gardening. But the improved appearance here is owing rather to the better means, and more rural inclinations of the inhabitants, than to any very great advantages in point of situation; although, of course, every increase of distance from the influence of the city dust and smoke is an advantage not to be forgotten. The profusion of floral beauty, the exquisite freshness of vegetation, and the superior taste of arrangement, are as striking in these localities as the sooty blight and deformity are in the terraces nearer town. Yet in the midst of the urban dreariness there are some strange exceptions, which prove the capabilities of a London atmosphere in a pleasing and satisfactory manner. In the midst of a close row of houses, adorned in front with marigold and chrysanthemum wildernesses, and at back with elder trees and wet clothes busily imbibing blacks, we suddenly come upon a garden, so exquisitely arranged and cultivated, that we are positively startled at its appearance in a region which seems eaten up with dinginess and dun colour. There are two such examples in the New Road at Pentonville. One is on the right hand from town, a few houses past the Angel Inn, the other is on the left, a few yards on the city side of Pentonville Chapel. Both these gardens are models of grace and beauty, and the freshness of the foliage and brilliancy of the flowers are marvels to one who associates with a town garden the idea of necessary barrenness. We ourselves have succeeded in a district where the chimnies are more numerous than the trees, and the sky not of the most Italian blue, in rearing and maintaining a profusion of gay flowers, the healthiness of which match with their beauty, and stand in strange contrast with the perennial dulness of the gardens of our neighbours. The result of our experience is to convince us that, while there may be some few localities where the atmosphere is so impregnated with the fumes from chemical works as to render floriculture a very hopeless enterprise, yet, as a rule, every citizen may have his garden, in doors or out of doors, at a small expenditure of time and money, provided he knows the best way of going to work, and is not bitten with any madness to grow such plants as are known not to bear the smoke complacently.

To remove such difficulties as are peculiar to town gardening is the object of the present work; and if the reader has sufficient enthusiasm for the task, we can assure him that, by following our instructions, he may improve the elegance of his city residence, and increase his resources of innocent and pleasurable recreation, by surrounding his home with vegetable life and beauty, of a perfection but little inferior to that attained under circumstances the most favourable.

As might be expected, the expense and trouble attending the management of a city garden are greater than those which attend the management of one in the country. To succeed to such an extent as to derive something more than a child's pleasure from the pursuit, needs care and patience, and a settled determination to set small difficulties at defiance. But then how great the remuneration – to be able to subdue nature thus, and compel to a healthy, and (as we believe in Wordsworth) we will say a happy life, such fragile and delicate existences as flowers, under circumstances the very opposite of those to which, in their free state, they are accustomed. In the dusky town, of all places in the world, the presence of flowers must be needed to call us back, now and then, from the feverish haste of business, and remind us of our childhood and early home, of our first loves and first rambles, of our mothers' smiles and kisses, and of the golden California we then found in a meadow of buttercups. Or if our whole life, from infancy onward, has been spent between these begrimed walls, there is the greater need to be continually reminded that there is a world of perennial loveliness beyond the influence of our commercial hurry, to which, if we will sometimes turn for solace and refreshment, we shall escape having our souls crushed out of us by the sharp edge of a shilling. And even if they serve us no such poetical uses, it is still worth a small sacrifice to take breakfast with the cheerful prospect of blooming parterres, redolent with odour and glowing with rainbow colours, pleasantly smiling before the parlour window; and there is little fear that, when this has once been enjoyed, the flowers will soon become a source of much higher gratification; the more pure because it appeals to our love of beauty, the more enduring from its incessant change on change – each season bringing its own troop of floral attendants, which, like the hours, revolve in a circle, and keep pace with the varying emotions of the heart.

Editorial Endnotes

1 'Brummel's 40 neck-cloths'. Beau Brummel (1778–1840) was famous for his invention of the neck tie. The process of preparing and starching the fabric, however, was complicated and led to his valet carrying a pile of 'our failures' every morning (see Louis Melville, *Beau Brummell, His Life and Letters*, p. 50).

2 Black hole of Calcutta – A dungeon in Calcutta in which British prisoners of war were held in cramped conditions by the Nawab of Bengal in 1757; the majority died.

3 John Lindley suggested in a critical review of *The Town Garden* in *Gardeners' Chronicle* (14 April 1855), p. 40, that 'gardeners' would not make classical literary references; 'abderas' (Abdera was a Greek colony founded to escape the Persian control, 'Stagyrian retreats' (retreat of the philosopher Aristotle). These are two of a number of classical references Hibberd makes in the excerpt.

5

GREAT EXPECTATIONS

(Excerpt)

Charles Dickens

Source: Charles Dickens, ch. 6, *Great Expectations* (3 vols) 2 (1861), [88–104] excerpt, pp. 95–102

At first with such discourse, and afterwards with conversation of a more general nature, did Mr. Wemmick and I beguile the time and the road, until he gave me to understand that we had arrived in the district of Walworth.

It appeared to be a collection of back lanes, ditches, and little gardens, and to present the aspect of a rather dull retirement. Wemmick's house was a little wooden cottage in the midst of plots of garden, and the top of it was cut out and painted like a battery mounted with guns.

"My own doing," said Wemmick. "Looks pretty; don't it?"

I highly commended it. I think it was the smallest house I ever saw; with the queerest gothic windows (by far the greater part of them sham), and a gothic door, almost too small to get in at.

"That's a real flagstaff, you see," said Wemmick, "and on Sundays I run up a real flag. Then look here. After I have crossed this bridge, I hoist it up – so – and cut off the communication."

The bridge was a plank, and it crossed a chasm about four feet wide and two deep. But it was very pleasant to see the pride with which he hoisted it up and made it fast; smiling as he did so, with a relish and not merely mechanically.

"At nine o'clock every night, Greenwich time," said Wemmick, "the gun fires. There he is, you see! And when you hear him go, I think you'll say he's a Stinger."

The piece of ordnance referred to, was mounted in a separate fortress, constructed of lattice-work. It was protected from the weather by an ingenious little tarpaulin contrivance in the nature of an umbrella.

"Then, at the back," said Wemmick, "out of sight, so as not to impede the idea of fortifications – for it's a principle with me, if you have an idea, carry it out and keep it up – I don't know whether that's your opinion – "

I said, decidedly.

"At the back, there's a pig, and there are fowls and rabbits; then, I knock together my own little frame, you see, and grow cucumbers; and you'll judge at supper what sort of a salad I can raise. So, sir," said Wemmick, smiling again, but

 DOI: 10.4324/9780429198809-7

seriously too, as he shook his head, "if you can suppose the little place besieged, it would hold out a devil of a time in point of provisions."

Then, he conducted me to a bower about a dozen yards off, but which was approached by such ingenious twists of path that it took quite a long time to get at; and in this retreat our glasses were already set forth. Our punch was cooling in an ornamental lake, on whose margin the bower was raised. This piece of water (with an island in the middle which might have been the salad for supper) was of a circular form, and he had constructed a fountain in it, which, when you set a little mill going and took a cork out of a pipe, played to that powerful extent that it made the back of your hand quite wet.

"I am my own engineer, and my own carpenter, and my own plumber, and my own gardener, and my own Jack of all Trades," said Wemmick, in acknowledging my compliments. "Well; it's a good thing, you know. It brushes the Newgate cobwebs away, and pleases the Aged. You wouldn't mind being at once introduced to the Aged, would you? It wouldn't put you out?"

I expressed the readiness I felt, and we went into the Castle. There, we found, sitting by a fire, a very old man in a flannel coat: clean, cheerful, comfortable, and well cared for, but intensely deaf.

"Well aged parent," said Wemmick, shaking hands with him in a cordial and jocose way, "how am you?"

"All right, John; all right!" replied the old man.

"Here's Mr. Pip, aged parent," said Wemmick, "and I wish you could hear his name. Nod away at him, Mr. Pip; that's what he likes. Nod away at him, if you please, like winking!"

"This is a fine place of my son's, sir," cried the old man, while I nodded as hard as I possibly could. "This is a pretty pleasure-ground, sir. This spot and these beautiful works upon it ought to be kept together by the Nation, after my son's time, for the people's enjoyment."

"You're as proud of it as Punch; ain't you, Aged?" said Wemmick, contemplating the old man, with his hard face really softened; "*there's* a nod for you;" giving him a tremendous one; "*there's* another for you;" giving him a still more tremendous one; "you like that, don't you? If you're not tired, Mr. Pip – though I know it's tiring to strangers – will you tip him one more? You can't think how it pleases him."

I tipped him several more, and he was in great spirits. We left him bestirring himself to feed the fowls, and we sat down to our punch in the arbour; where Wemmick told me as he smoked a pipe that it had taken him a good many years to bring the property up to its present pitch of perfection.

"Is it your own, Mr. Wemmick?"

"Oh yes," said Wemmick, "I have got hold of it, a bit at a time. It's a freehold, by George!"

"Is it, indeed? I hope Mr. Jaggers admires it?"

"Never seen it," said Wemmick. "Never heard of it. Never seen the Aged. Never heard of him. No; the office is one thing, and private life is another. When I go into the office, I leave the Castle behind me, and when I come into the Castle, I leave the office behind me. If it's not in any way disagreeable to you, you'll oblige me by doing the same. I don't wish it professionally spoken about."

Of course I felt my good faith involved in the observance of his request. The punch being very nice, we sat there drinking it and talking, until it was almost nine o'clock. "Getting near gun-fire," said Wemmick then, as he laid down his pipe; "it's the Aged's treat."

Proceeding into the Castle again, we found the Aged heating the poker, with expectant eyes, as a preliminary to the performance of this great nightly ceremony. Wemmick stood with his watch in his hand, until the moment was come for him to take the red-hot poker from the Aged, and repair to the battery. He took it, and went out, and presently the Stinger went off with a Bang that shook the crazy little box of a cottage as if it must fall to pieces, and made every glass and teacup in it ring. Upon this, the Aged – who I believe would have been blown out of his arm-chair but for holding on by the elbows – cried out exultingly, "He's fired! I heerd him!" and I nodded at the old gentleman until it is no figure of speech to declare that I absolutely could not see him.

6

'VILLA GARDENING'

Shirley Hibberd

Source: Shirley Hibberd, 'Villa Gardening', *Amateur Gardening* (23 June 1888), p. 85

Villa Gardening. – No phase of gardening has probably made itself more popular or of greater value to the horticultural trade than this. It comes in intermediate between the higher and more advanced gardening of the affluent and the humbler efforts of the cottager, and is, perhaps, carried out with proportionately greater zeal than in either of the other cases. A large garden, for instance, requires a good income to keep it well furnished and in excellent order, and we know from experience that there are very few cases in which the latter is equal to the former. The natural consequence of this is much of the garden has to be indifferently kept, unless, as is sometimes the case, the owner disposes of his surplus produce and devotes the money so obtained to its maintenance. Then, with regard to the cottage garden, we do not see the cultivation of flowers or fruit carried out very largely from motives of economy, but in the matter of vegetables the cottager can undoubtedly hold his own. A villa garden, as a rule, contains just enough space to be within the means of the owner or occupier to keep in order, and it happens more often than not that he does so even to a lavish extent. In doing so he is only acting in perfect harmony with the natural order of things, and everyone who has the means should certainly strive to have not only his house but his garden beautiful also. As well make a lady's dress without its lace or other trimmings, or a bonnet without its exterior adjuncts and call it finished, as to consider a home complete in the absence of a neatly laid out garden. Happily there are few who fail to recognise the importance of a beautiful garden, and hence, go where we will in the regions of villadom, we are sure to find that great attention is paid to this subject. As we have previously intimated, villa gardening has done more towards the support of the horticultural trade than any other class, as since it has assumed such large proportions, there has been an increasing demand for plants and small shrubs. Of course some decorate their gardens much more lavishly and beautiful than others. Those who have a taste or hobby for this sort of thing go in for it enthusiastically, and often afford a considerable amount of glass in which to grow the greater part of the plants required for garden and house decoration,

DOI: 10.4324/9780429198809-8

EXAMPLE OF VILLA GARDEN IN FORMAL STYLE

and, consequently, to do this thoroughly, must keep one or more gardeners. Of this class it is not within our province to dwell upon, since we are in duty bound to confine our observations to those who come within the distinction of an amateur, and supervise the working of their own gardens themselves. If much assistance has to be had, the garden becomes an expensive luxury, and a fair-sized one laid out in the usual way is sure to require this. It is necessary therefore to, in some measure, prevent this occurring by having the garden laid out in such a way as to dispense with the necessity of growing many expensive or tender plants. This is not a difficult matter at all now we have a large selection of lovely hardy plants at our command. A garden can be stocked with these and hardy shrubs so that only a very few bedding plants are required. This will be a great saving in expenditure as compared with having to fill large areas with expensive plants. Take for example the villa garden, which forms the subject of our plate in this week's issue. This represents an example of the latter laid out and planted in the formal style, with a number of beds filled with carpet and other valuable foliage and flowering plants. On one side is a straight walk with shrubs at the back, and on the other, shrubs and trees and a border with a curved outline, the boundary of which is formed of clipped ivy. Then, in the centre there is a fountain and water, and one or two standard trees. Now the whole of this arrangement may, when kept in trim order, look exceedingly grand, but can anyone say that it is beautiful? Everything is formal, and bears the stamp of artificiality about it very prominently. It may, nevertheless, suit many who wish to have grand displays of spring and summer bedding plants, and can afford to employ one or more jobbing gardeners for several days in the week to keep the whole in order. We may at no distant date furnish an example of an opposite character, in which the more natural style predominates – that is, where fewer formal beds and a greater quantity of permanent hardy plants are used. The great aim now-a-days should be to have the garden in as beautiful a condition as possible at the least expenditure, and this can only be done by going in more largely for permanent bedding plants. Lists of plants have been frequently given in these pages, and articles and notes on the most noteworthy of them are contributed more or less every week. Useful plans have also been given for laying out a villa garden in a style suitable for growing hardy plants, in our last volume, and more will follow at the proper season for dealing with such matters. Any information, too, not embodied in our pages in regard to this can always be supplied though the Doubts and Difficulties column by simply taking the trouble to write and ask for it. Those of our readers who contemplate making alterations in their garden should preserve this plate until the autumn, when we again take up the subject of Laying Out Small Gardens, as we shall then refer to it from a more practical point of view than we have now done.

7

WOOD AND GARDEN

(Excerpts)

Gertrude Jekyll

Source: Gertrude Jekyll, ch. 1 'Introductory' and ch. 14. 'Large and Small Gardens', *Wood and Garden, Notes and Thoughts, Practical and Critical, of a Working Amateur* (1899), pp. 1–6, [171–188] excerpt 171–179

INTRODUCTORY

THERE are already many and excellent books about gardening; but the love of a garden, already so deeply implanted in the English heart, is so rapidly growing, that no excuse is needed for putting forth another.

I lay no claim either to literary ability, or to botanical knowledge, or even to knowing the best practical methods of cultivation; but I have lived among outdoor flowers for many years, and have not spared myself in the way of actual labour, and have come to be on closely intimate and friendly terms with a great many growing things, and have acquired certain instincts which, though not clearly defined, are of the nature of useful knowledge.

But the lesson I have thoroughly learnt, and wish to pass on to others, is to know the enduring happiness that the love of a garden gives. I rejoice when I see any one, and especially children, inquiring about flowers, and wanting gardens of their own, and carefully working in them. For the love of gardening is a seed that once sown never dies, but always grows and grows to an enduring and ever-increasing source of happiness.

If in the following chapters I have laid special stress upon gardening for beautiful effect, it is because it is the way of gardening that I love best, and understand most of, and that seems to me capable of giving the greatest amount of pleasure. I am strongly for treating garden and wooded ground in a pictorial way, mainly with large effects, and in the second place with lesser beautiful incidents, and for so arranging plants and trees and grassy spaces that they look happy and at home, and make no parade of conscious effort. I try for beauty and harmony everywhere, and especially for harmony of colour. A garden so treated gives the delightful feeling of repose, and refreshment, and purest enjoyment of beauty, that seems to my understanding to be the best fulfilment of its purpose; while to its diligent worker

 DOI: 10.4324/9780429198809-9

its happiness is like the offering of a constant hymn of praise. For I hold that the best purpose of a garden is to give delight and to give refreshment of mind, to soothe, to refine, and to lift up the heart in a spirit of praise and thankfulness. It is certain that those who practise gardening in the best ways find it to be so.

But the scope of practical gardening covers a range of horticultural practice wide enough to give play to every variety of human taste. Some find their greatest pleasure in collecting as large a number as possible of all sorts of plants from all sources, others in collecting them themselves in their foreign homes, others in making rock-gardens, or ferneries, or peat-gardens, or bog-gardens, or gardens for conifers or for flowering shrubs, or special gardens of plants and trees with variegated or coloured leaves, or in the cultivation of some particular race or family of plants. Others may best like wide lawns with large trees, or wild gardening, or a quite formal garden, with trim hedge and walk, and terrace, and brilliant parterre, or a combination of several ways of gardening. And all are right and reasonable and enjoyable to their owners, and in some way or degree helpful to others.

The way that seems to me most desirable is again different, and I have made an attempt to describe it in some of its aspects. But I have learned much, and am always learning, from other people's gardens, and the lesson I have learned most thoroughly is, never to say "I know" – there is so infinitely much to learn, and the conditions of different gardens vary so greatly, even when soil and situation appear to be alike and they are in the same district. Nature is such a subtle chemist that one never knows what she is about, or what surprises she may have in store for us.

Often one sees in the gardening papers discussions about the treatment of some particular plant. One man writes to say it can only be done one way, another to say it can only be done quite some other way, and the discussion waxes hot and almost angry, and the puzzled reader, perhaps as yet young in gardening, cannot tell what to make of it. And yet the two writers are both able gardeners, and both absolutely trustworthy, only they should have said, "In my experience *in this place* such a plant can only be done in such a way." Even plants of the same family will not do equally well in the same garden. Every practical gardener knows this in the case of strawberries and potatoes; he has to find out which kinds will do in his garden; the experience of his friend in the next county is probably of no use whatever.

I have learnt much from the little cottage gardens that help to make our English waysides the prettiest in the temperate world. One can hardly go into the smallest cottage garden without learning or observing something new. It may be some two plants growing beautifully together by some happy chance, or a pretty mixed tangle of creepers, or something that one always thought must have a south wall doing better on an east one. But eye and brain must be alert to receive the impression and studious to store it, to add to the hoard of experience. And it is important to train oneself to have a good flower-eye; to be able to see at a glance what flowers are good and which are unworthy, and why, and to keep an open mind about it; not to be swayed by the petty tyrannies of the "florist" or show judge; for, though some part of his judgment may be sound, he is himself a slave to rules, and must go by points which are defined arbitrarily and rigidly, and have reference mainly to the show-table, leaving out of account, as if unworthy of consideration, such matters as gardens and garden beauty, and human delight, and sunshine, and varying lights of morning and evening and noonday. But many, both nurserymen and private people, devote themselves to growing and improving the best classes of hardy flowers, and we can hardly offer them too much grateful praise, or do them too much honour. For what would our gardens be without the Roses, Pæonies, and Gladiolus of France, and the Tulips and Hyacinths of Holland, to say nothing of the hosts of good things raised by our home growers, and of the enterprise of the great firms whose agents are always searching the world for garden treasures?

Let no one be discouraged by the thought of how much there is to learn. Looking back upon nearly thirty years of gardening (the earlier part of it in groping ignorance with scant means of help), I can remember no part of it that was not full of pleasure and encouragement. For the first steps are steps into a delightful Unknown, the first successes are victories all the happier for being scarcely expected, and with the growing knowledge comes the widening outlook, and the comforting sense of an ever-increasing gain of critical appreciation. Each new step becomes a little surer, and each new grasp a little firmer, till, little by little, comes the power of intelligent combination, the nearest thing we can know to the mighty force of creation.

And a garden is a grand teacher. It teaches patience and careful watchfulness; it teaches industry and thrift; above all, it teaches entire trust. "Paul planteth and Apollos watereth, but God giveth the increase." The good gardener knows with absolute certainty that if he does his part, if he gives the labour, the love, and every aid that his knowledge of his craft, experience of the conditions of his place, and exercise of his personal wit can work together to suggest, that so surely as he does this diligently and faithfully, so surely will God give the increase. Then with the honestly-earned success comes the consciousness of encouragement to renewed effort, and, as it were, an echo of the gracious words, "Well done, good and faithful servant."

LARGE AND SMALL GARDENS

THE size of a garden has very little to do with its merit. It is merely an accident relating to the circumstances of the owner. It is the size of his heart and brain and goodwill that will make his garden either delightful or dull, as the case may be, and either leave it at the usual monotonous dead-level, or raise it, in whatever degree he may, towards that of a work of fine art. If a man knows much, it is more difficult for him to deal with a small space than a larger, for he will have to make the more sacrifice; but if he is wise he will at once make up his mind about what he will let go, and how he may best treat the restricted space. Some years ago I visited a small garden attached to a villa on the outskirts of a watering-place on the south coast. In ordinary hands it would have been a perfectly commonplace thing, with the usual weary mixture, and exhibiting the usual distressing symptoms that come in the train of the ministrations of the jobbing-gardener. In size it may have been a third of an acre, and it was one of the most interesting and enjoyable gardens I have ever seen, its master and mistress giving it daily care and devotion, and enjoying to the full its glad response of grateful growth. The master had built with his own hands, on one side where more privacy was wanted, high rugged walls, with spaces for many rock-loving plants, and had made the wall die away so cleverly into the rock-garden, that the whole thing looked like a garden founded on some ancient ruined structure. And it was all done with so much taste that there was nothing jarring or strained-looking, still less anything cockneyfied, but all easy and pleasant and pretty, while the happy look of the plants at once proclaimed his sympathy with them, and his comprehensive knowledge of their wants. In the same garden was a walled enclosure where Tree Pæonies and some of the hardier of the oriental Rhododendrons were thriving, and there were pretty spaces of lawn, and flower border, and shrub clump, alike beautiful and enjoyable, all within a small space, and yet not crowded – the garden of one who was a keen flower lover, as well as a world-known botanist.

I am always thankful to have seen this garden, because it showed me, in a way that had never been so clearly brought home to me, how much may be done in a small space.

Another and much smaller garden that I remember with pleasure was in a sort of yard among houses, in a country town. The house it belonged to, a rather high one, was on its east side, and halfway along on the south; the rest was bounded by a wall about ten feet high. Opposite the house the owner had built of rough blocks of sandstone what served as a workshop, about twelve feet long, along the wall, and six feet wide within. A low archway of the same rough stone was the entrance, and immediately above it a lean-to roof sloped up to the top of the wall, which just here had been carried a little higher. The roof was of large flat sandstones, only slightly lapping over each other, with spaces and chinks where grew luxuriant masses of Polypody Fern. It was contrived with a cement bed, so that it was quite weather-tight, and the room was lighted by a skylight at one end that did not show from the garden. A small surface of lead-flat, on a level with the top of the wall, in one of the opposite angles, carried an old oil-jar, from which fell masses of gorgeous Tropæolum, and the actual surface of the flat was a garden of Stonecrops. The rounded coping of the walls, and the joints in many places (for the wall was an old one), were gay with yellow Corydalis and Snapdragons and more Stonecrops. The little garden had a few pleasant flowering bushes, Ribes and Laurustinus, a Bay and an Almond tree. In the coolest and shadiest corner were a fern-grotto and a tiny tank. The rest of the garden, only a few yards across, was laid out with a square bed in the middle, and a little path round, then a three-feet-wide border next the wall, all edged with rather tall-grown Box. The middle bed had garden Roses and Carnations, and Mignonette and Stocks. All round were well-chosen plants and shrubs, looking well and happy, though in a confined and rather airless space. Every square foot had been made the most of with the utmost ingenuity, but the ingenuity was always directed by good taste, so that nothing looked crowded or out of place.

And I think of two other gardens of restricted space, both long strips of ground walled at the sides, whose owners I am thankful to count among my friends – one in the favoured climate of the Isle of Wight, a little garden where I suppose there are more rare and beautiful plants brought together within a small space than perhaps in any other garden of the same size in England; the other in a cathedral town, now a memory only, for the master of what was one of the most beautiful gardens I have ever seen now lives elsewhere. The garden was long in shape, and divided about midway by a wall. The division next the house was a quiet lawn, with a mulberry tree and a few mounded borders near the sides that were unobstrusive, and in no way spoilt the quiet feeling of the lawn space. Then a doorway in the dividing wall led to a straight path with a double flower border. I suppose there was a vegetable garden behind the borders, but of that I have no recollection, only a vivid remembrance of that brilliantly beautiful mass of flowers. The picture was good enough as one went along, especially as at the end one came first within sound and then within sight of a rushing river, one of those swift, clear, shallow streams with stony bottom that the trout love; but it was ten times more beautiful on turning to go back, for there was the mass of flowers, and towering high above it the noble mass of the giant structure – one of the greatest and yet most graceful buildings that has ever been raised by man to the glory of God.

It is true that it is not every one that has the advantage of a garden bounded by a river and a noble church, but even these advantages might have been lost by vulgar or unsuitable treatment of the garden. But the mind of the master was so entirely in sympathy with the place, that no one that had the privilege of seeing it could feel that it was otherwise than right and beautiful.

Both these were the gardens of clergymen; indeed, some of our greatest gardeners are, and have been, within the ranks of the Church. For have we not a brilliantly-gifted dignitary whose loving praise of the Queen of flowers has become a classic? and have we not among churchmen the greatest grower of seedling Daffodils the world has yet seen, and other names of clergymen honourably associated with Roses and Auriculas and Tulips and other good flowers, and all greatly to their bettering? The conditions of the life of a parish priest would tend to make him a good gardener, for, while other men roam about, he stays mostly at home, and to live with one's garden is one of the best ways to ensure its welfare. And then, among the many anxieties and vexations and disappointments that must needs grieve the heart of the pastor of his people, his garden, with its wholesome labour and all its lessons of patience and trust and hopefulness, and its comforting power of solace, must be one of the best of medicines for the healing of his often sorrowing soul.

I do not envy the owners of very large gardens. The garden should fit its master or his tastes just as his clothes do; it should be neither too large nor too small, but just comfortable. If the garden is larger than he can individually govern and plan and look after, then he is no longer its master but its slave, just as surely as the much-too-rich man is the slave and not the master of his superfluous wealth. And when I hear of the great place with a kitchen garden of twenty acres within the walls, my heart sinks as I think of the uncomfortable disproportion between the man and those immediately around him, and his vast output of edible vegetation, and I fall to wondering how much of it goes as it should go, or whether the greater part of it does not go dribbling away, leaking into unholy back-channels; and of how the looking after it must needs be subdivided; and of how many side-interests are likely to steal in, and altogether how great a burden of anxiety or matter of temptation it must give rise to. A grand truth is in the old farmer's saying, "The master's eye makes the pig fat;" but how can any one master's eye fat that vast pig of twenty acres, with all its minute and costly cultivation, its two or three crops a year off all ground given to soft vegetables, its stoves, greenhouses, orchid and orchard houses, its vineries, pineries, figgeries, and all manner of glass structures?

But happily these monstrous gardens are but few – I only know of or have seen two, but I hope never to see another.

Nothing is more satisfactory than to see the well-designed and well-organised garden of the large country house, whose master loves his garden, and has good taste and a reasonable amount of leisure.

I think that the first thing in such a place is to have large unbroken lawn spaces – all the better if they are continuous, passing round the south and west sides of the house. I am supposing a house of the best class, but not necessarily of the largest size. Immediately adjoining the house, except for the few feet needed for a border for climbing plants, is a broad walk, dry and smooth, and perfectly level from end to end. This, in the case of many houses, and nearly always with good effect, is raised two or three feet above the garden ground, and if the architecture of the house demands it has a retaining wall surrounded by a balustrade of masonry and wrought stone. Broad and shallow stone steps lead down to the turf both at the end of the walk and in the middle of the front of the house, the wider and shallower the better, and at the foot of the wall may be a narrow border for a few climbing plants that will here and there rise above the coping of the parapet. I do not think it desirable where there are stone balusters or other distinct architectural features to let them be smothered with climbing plants, but that there should be, say, a *Pyrus japonica* or an Escallonia, and perhaps a white Jasmine, and on a larger space perhaps a cut-leaved or a Claret Vine. Some of the best effects of the kind I have seen were where the bush, being well established, rose straight out of the grass, the border being unnecessary except just at the beginning.

The large lawn space I am supposing stretches away a good distance from the house, and is bounded on the south and west by fine trees; away beyond that is all wild wood. On summer afternoons the greater part of the lawn expanse is in cool shade, while winter sunsets show through the tree stems. Towards the south-east the wood would pass into shrub plantations, and farther still into garden and wild orchard (of which I shall have something to say presently). At this end of the lawn would be the brilliant parterre of bedded plants, seen both from the shaded lawn and from the terrace, which at this end forms part of its design. Beyond the parterre would be a distinct division from the farther garden, either of Yew or Box hedge, with bays for seats, or in the case of a change of level, of another terrace wall. The next space beyond would be the main garden for hardy plants, at its southern end leading into the wild orchard. This would be the place for the free garden or the reserve garden, or for any of the many delightful ways in which hardy flowers can be used; and if it happened by good fortune to have a stream or any means of having running water, the possibilities of beautiful gardening would be endless.

8

'SMALL GARDENS'

Eugène Noel

Source: Eugène Noel, 'Small Gardens' ('Petits Jardins'), *Le Jardin* (1887), pp. 50–51

Has anyone collected statistics about the small gardens created in France in the past 80 years? I don't know. But I can say what I have seen taking place over the past 60 years in the neighbourhood where I live, in Seine-Intérieure.

The farms, important enough 50 years ago, for the most part had nothing but a narrow kitchen garden most often badly cultivated, badly kept. Today, it is not only the wealthy growers who have a proper well-kept garden, rich in choice vegetables, planted with vigorous and healthy fruit trees, but the majority of buildings and walls are covered with espaliers of fruit. And you won't find such pretty little gardens, orchards, espaliered trees only among the wealthy, you will find them equally among the poorer folk. There, for the rural population, is a source of foodstuffs which won't appear in the statistics because, being eaten on the spot, they do not appear in any market and are neither weighed, measured, or counted. But in fruits – apples, pears, prunes, plums, cherries, gooseberries, raspberries, strawberries – and vegetables – peas, beans, cabbage, carrots, swedes, onions, salad crops, potatoes, salsify, artichokes, sorrel, etc. etc. – their contribution is nonetheless considerable. The rural populations find here a notable part of their subsistence. The poorest labourers are attracted by this pretty and skillful cultivation to the study, the observation of the laws of nature; from this they derive a taste for beauty, for most of them combine pleasure and utility in their activities. What beautiful plants I have often seen around the humblest cottage! And those who cultivate these little gardens find there not only a ready path to study, to a taste for the beautiful, they find morality.

A wealthy industrialist never took on a new labourer with asking him the question: – Do you have a garden? If the labourer said no, he was seldom accepted.

The possessors of gardens, more numerous every year, form among them a sort of free association for the exchange of grafts, buds, floral and vegetable seeds. Seldom, in Normandy, will one find a village without gardens and gardeners of this stamp. There are many localities where every little house is surrounded by full-length espaliers and by laughing and fertile parterres.

DOI: 10.4324/9780429198809-10

This will remain for our country one of the characteristic traits of the 19th century; the devotion of a large portion of our ground to vegetable, fruit, and floral cultivation will be the great social task accomplished in our days by horticulture, without noise, without trouble, without cries, without fakery, without deception, without anyone being ruined, almost without anyone noticing. How many hectares have passed from field cultivation, or even from being uncultivated, to development as gardens, nurseries and orchards? I don't have the figures in front of me, but I can see the effects in Normandy, and hints for the future one of the best industries of France. Already we are sending excellent fruits to England and the rest of Europe. We know what is being exported every year from Duclair, the little port halfway between Le Havre and Rouen.

But in France, and especially in Rouen, classic country of excellent jams, what development there will be in confectionery the day that sugar reaches its most competitive price. It will then be a production to rival that of gooseberries and strawberries: jellies, pastes and syrups will multiply, and we will know what a garden dedicated to this production can bring its owner.

Jam-making, the making of fruit preserves, is an industry whose importance no one could have suspected; but once we have cheap sugar (which is to jam what salt is to the preservation of meat and fish), you will see what a simple cherry tree or gooseberry bush can provide.

Already some shrewd people are investing in this improvement, the development of this culture, and have made a tidy income from it. If there aren't millions to be made, there are, which is much better, a life to win, and morality, intelligence, and knowledge to develop. All this is the road to happiness, and it runs through our small gardens. But if the reader is willing, we will talk about this again

. . .

This question of small gardens becomes more important with thinking on it. The small garden not only contributes considerably to the food of the rural population and the working classes around our big cities; it has contributed over the last fifty years to the development of the floral industry. Our horticulturists, our nurserymen, ever more numerous, have found a good outlet for their ornamental and culinary plants, for their young fruit trees and decorative shrubs. A walk in our towns and even in our villages will suffice to show that today the floral industry has its markets, that plants are being sold, everywhere. Flowers have become everyone's luxury. Rich and poor decorate their houses in front and at rear. There is hardly a fete or a ceremony without flowers. They cover the coffins at funerals. In the big cities you will find few windows or balconies without flowers. Look up in the urban promenades and you will see hanging gardens almost everywhere. And how often they are pretty and well looked after!

And house plants, where will you not find them? and when the use of the *mousse fertilisante* of our friend and collaborator M Alfred Dumesnil of Vascoeuil, and when we hear everywhere that people are using the *papyrine Cuisinier-Dumesnil* for living and cut flowers, then we will finally realise the dream of François de Sales: *Perpetual flowers and fruits*.[1]

Flowers, fruit, culinary plants, bulbs and edible roots, we must acknowledge, have assumed an importance in our society which they have never before had.

Compare the fruits and vegetables of a mere fifty years ago with those you find today in our arcades and markets. And what new varieties! Small gardens, small-scale horticulture of every sort are one of the principal causes of this unexpected development. Florists and nurserymen have found an excellent clientele. Rich and great gardens are perhaps less profitable to them.

Most of our horticultural societies have taken insufficient note of this phenomenon. They seem to appeal too much to the great and good. No doubt they deserve attention, encouragement and applause, but let's keep our eye on the small gardens above all. Which of us has not discovered charming things in our travels and promenades! Go into these pretty little gardens, chat for a moment with the owners of these little parterres, and you will find a wealth of knowledge and advice. A weaver I know raises carnation seedlings so well publicised that a summer rarely goes by without some Belgian horticulturist paying his well for them. I know a modest pensioner who, through skillful and patient propagation, has built up a collection of the most remarkable auriculas, but which she does not offer for sale: her joy is in keeping them for herself. These auriculas of a marvellous beauty are the enchantment of a tranquil and retired old age . . . But why do I speak of old age, there's no such thing when there's a happy passion for gardening.

No art can make one happier; I believe it has formed a number of good people, of simple manners and modest habits, of good and healthy minds.

It is never among the enthusiastic owners of small gardens, that you will find the perpetrators of the crime of the rue Montaigne. The singers of gardens, in prose and verse, are almost always beside the point, celebrating as they do royal and princely gardens. The owners of those gardens are sometimes complete strangers to their cultivation. The great garden may do nothing spiritual for its lord and master, who often lives at a distance, and when he does visit it knows nothing of the wonders of his estate.[2]

The earth neither loves, favours, fortifies, consoles, nor instructs any but the one who knows how to look after his little portion.

Small gardening has made over thousands of years the prosperity, morality and wisdom of the vastest, most populated, most pacific of empires. China is the land where the earth cultivated, nourished, pampered, caressed (that's the right word) by millions of smallholders, has shown best what is can gives in terms of physical and moral riches.

Gardening, culinary, floral, and fruit growing, has been practised in this country for centuries and centuries. But who in the middle ages, in our Europe dying of hunger, thought to ask how so many millions of people were fed in the immense Asiatic empire?

Ah! They were right, these Chinese, to surround themselves with a wall such as had never been seen, to protect themselves from the stupidity of the rest of the world!

They had the inestimable secret of life for all, at least of life for all sorts of working people. They feared that they would be taught something else, they feared above all invasion by the starving populations of the rest of the globe. Their secret was small-scale horticulture, small gardens.

Oh! If we imagine Europe and the fertile regions of Africa deliver the same sort of cultivation: what abundance! How little we would have to worry about the conjunction of American and India! Small-scale horticulture, small gardens, are the beginning of this happy reform. We will recognise this later.

Editorial Endnotes

1 Alfred Dumesnil (1821–1894) was an art historian who had an estate at Chateau de Vascoeuil. Papyrine Cuisinière-Dumesnil refers to a tough paper for wrapping plants, so called after the 'paper-like' moss that bore Dumesnil's name.

2 When St Francis de Sales wrote of 'perpetual flowers and fruits', he was referring to the gifts of the spirit, not to the cultivation of plants. Noel's reference to him is a joke. In the late nineteenth century, an anthology called "The Mystical Flora of St. Francis de Sales" was created by assembling gardening-inspired passages.

The 'crime in the rue Montaigne' refers to a recent murder case of 1887 in which a man named Pranzini killed his mistress and her servants.

b)

KITCHEN AND FRUIT GARDENS

Editorial Headnotes

9. Walter Nicol, 'The Culinary Garden, Introduction', 'On Situations Fit for Kitchen Gardens', 'The Forcing Garden, Introduction', *The Gardener's Kalendar; or, monthly directory of operations in every branch of horticulture* (1810) 2nd edn. (1812), pp. 3–9, 265–268
10. J. C. Loudon, pt. 3 ch. 1, 'Laying Out and Planting the Kitchen and Fruit-garden', pt. 3 ch. 2, 'Cropping and General Management of a Kitchen-garden', *The Suburban Horticulturist* (1842), [416–433] excerpt, pp. 416–421, [434–441] excerpt, 434–435
11. Elizabeth Watts, 'Preface and Dedication', ch.1 'Economy of Making Gardens Pay', ch. 10 'Roots Less General in Use', *Vegetables and How to Grow Them* (1866), pp. 1–5, 47–51
12. Elizabeth Watts, 'Preface', ch. 1 'The Orchard', *Orchard and Fruit Garden* (1867), pp. 1–5
13. Gustave Flaubert, ch. 2 (Experiments in Agriculture), *Bouvard and Pécuchet* (1881), trans. by D. F. Hannigan (1896), [29–72] [excerpts] pp. 29, 43–48, 56–60
14. J. Douglas, 'Culford Hall, Suffolk', *Journal of Horticulture* (25 February 1875), pp. 164–166
15. Gertrude Jekyll, ch. 14, 'Large and Small Gardens', *Wood and Garden Notes and Thoughts, Practical and Critical, of a Working Amateur* (1899), [171–188] excerpt, pp. 179–183
16. George Gordon, 'Introduction', Anon. 'Kent', Anon.'Staffordshire', *The Gardeners' Magazine; The Wasted Orchards of England* (1896), pp. v–viii, 1–5, 84–88

The majority of gardens at the beginning of the 19th century included a significant portion for the growth of food. The flourishing of vegetable- and fruit-growing competitions from the 18th century, particularly amongst artisan communities, indicates the ubiquity of the practice (see Volume 3). Up until the second half of the 19th century, hedgerow fruit, gooseberries, redcurrants and blackcurrants, cherries, plums, and apples were cultivated by rural workers; pineapples, grapes, and peaches – varieties that required warmer temperatures under glass – were tended by the wealthier classes.

The rate of expansion of food production in commercial gardens varied considerably from region to region in Britain throughout the 19th century. The repeal of the glass tax in 1845 which stimulated the development of glasshouses in domestic gardens also affected commercial market gardens, resulting in an increase in the quantity and variety of fruit and vegetables. Transport developments were another factor as the movement of food across the country speeded up and the variety of fruit and vegetables sold at local markets and in shops increased. The presence of a select number of large estates who retained their well-funded kitchen and fruit gardens beyond the 19th century and the development of allotments were the exceptions to the overall trend towards the industrialisation of food production. In the process the range of cultivars expanded but the variety of species declined. The reduction of fruit and vegetable gardening in domestic gardens is reflected in the gardening press whereby coverage disappears from the pages of the more upmarket titles orientated towards lifestyle.

These sources exemplify some aspects of the overall trends. Walter Nicol (1769–1811), son of a gardener and born outside Edinburgh, produced several publications on practical gardening. He worked on a number of Scottish estates, specialising in the design of hothouses. *The Gardener's Kalendar* (1812) was intended for practising gardeners and landed gentry with staff. It offered horticultural advice (and design suggestions) for the pleasure garden, hothouse, greenhouse, and conservatory as well as for fruit and vegetables. The position of the 'Culinary Garden' chapter at the front of the book suggests the importance of the topic. J. C. Loudon's *Suburban Horticulturalist* (1842), a companion to the *Suburban Gardener* (with a projected third volume on suburban floriculture), contained practical horticultural advice. It was addressed to a generally male and less wealthy audience that included 'the mechanic, the labourer, the colonist or the emigrant' ('Preface', not included). The plans featured in the excerpt represent plots of around an acre which was considerably more than the most common size for allotment plots of 10 rods or 1/16 of an acre. This more technical volume did not achieve the same sales as *Suburban Gardener* partly because by 1842 some of the middle and upper-middle classes were buying a proportion of fruit and vegetables from shops or the market, and the less wealthy among the intended audience could not afford it.

Over two decades later, the sales of *Vegetables and How to Grow Them* (1866) by Elizabeth Watts (n/d) must have been high, as the 1-shilling octavo volume was reissued singly and in combination with her other books on gardening. As editor of the *Poultry Chronicle* (later absorbed into the *Cottage Gardener*) and of the 'Poultry Department' in the *Field* (1853 to present), Watts was active in the dissemination of horticultural knowledge. She does not signal the gender of her audience, and it is clear from the price and the particular emphasis on 'Economy' that she is writing for a readership of relatively modest income. The Jerusalem artichoke and 'old-fashioned' root crops including salsify, scorzonera, and chicory, which had been recommended by Loudon as part of a general range of vegetables, fall under the heading of 'Roots in less general use' in Watt's survey. Moreover, the skirret (*Sium sisarum*), mentioned by Watts, disappeared from the seed catalogues within the next two decades indicating the reduction of cultivated plant species in Britain. The popularity of her publications – *The Orchard and the fruit garden* followed in 1867 – indicates that, as large-scale kitchen gardens became an elite prerogative over the course of the century, a market remained for the cultivation of food in smaller gardens.

Bouvard and Pécuchet (1881), the posthumously published novel of Gustave Flaubert (1821–1880), looks back to the 1840s. In the excerpts (1896 English edition), the eponymous anti-heroes have just settled in rural France. The account of their disastrous attempts at gardening satirises the experts on garden management who offer contrasting advice and those who become confused when they attempt to follow it. *The Good Gardener* is a translation of *Le Bon Jardinier* – one of the most well-established horticultural publications in France and is one of several references made to existing publications on the subject.

The thousands of pounds needed to maintain a fruit and vegetable garden on an elite estate, including the upkeep of extensive hothouses and the wages of numerous gardeners to manage them, are enumerated in James Douglas's account 'Culford Hall, Suffolk', *Journal of Horticulture* (1875). Douglas (1837–1911) specialised in the breeding of florists' flowers including auriculas at his nursery at Great Bookham. The head gardener referred to in the article is Peter Grieve (1811–1895), known for his experiments in the hybridisation of pelargoniums (geraniums) to produce coloured leaves. The illustration features the conservatory and flower-garden (signed with the monogram BCL) rather than the kitchen and fruit gardens. It suggests the different roles of gardeners on the estate: the head gardener or foreman (long jacket and cravat); an aproned woman seated and sorting plants on her lap; a man and a boy – gardener's apprentices generally started work from the age of 12 – operating a lawnmower. Note that the 'orchard house' was a specific type of greenhouse construction that was easily assembled for the cultivation of fruit, promoted by the nurseryman Thomas Rivers (1798–1877). The excerpt from Gertrude Jekyll's *Wood and Garden* (1899) touches on the maintenance of estate gardens whereby the kitchen garden was part of the working garden of male staff. Her vision of a 'wilded' orchard differs somewhat from the more orderly layout of earlier experts.

The *Gardeners' Magazine* was a cheaper horticultural weekly (1860–1913) revamped by Shirley Hibberd from 1865. It soon became a rival to the *Gardeners' Chronicle,* and George Gordon (1841–1914) was its editor when it produced *The Wasted Orchards of England* (1896). In light of the prolonged consequences of the agricultural depression of the 1880s, Gordon proposes that revived orchards might provide additional income to farmers and other gardeners. The variety of apples available compared with those of the 21st century is notable. The survival of orchards was patchy in parts of Staffordshire because of industrial pollution. Moreover, the profitability of damson orchards, which had been cultivated in the region for their use in the natural dyeing process, was reduced (and their cultivation abandoned) by the use of 'aniline' dyes extracted from coal tar; 'mauveine', for example, was among the first of the mass-produced chemical dyes discovered in the mid-19th century.

Further Reading

Bunyard & Thomas, *The Fruit Garden* (1904)
Bill Laws, *Spade, Skirret and Parsnip: The Curious History of Vegetables* (2004)

9

THE GARDENER'S KALENDAR

(Excerpts)

Walter Nicol

Source: Walter Nicol, 'The Culinary Garden, Introduction', 'On Situations Fit for Kitchen Gardens', 'The Forcing Garden, Introduction', *The Gardener's Kalendar; or, monthly directory of operations in every branch of horticulture* (1810) 2nd edn. (1812), pp. 3–9, 265–68

THE CULINARY GARDEN

Introduction

THE cultivation of culinary vegetables is certainly the most important branch of gardening. It occupies the attention of a large proportion of the community, of the fruits of whose labours all daily partake. To the palace, and to the humble shepherd's cot, the kitchen-garden is a necessary appendage. Every city is surrounded by culinary gardens, whose productions are matter of very considerable interest to the inhabitants.

In beholding a well cultivated garden, every one is capable of feeling a certain degree of pleasure. Those indifferent feel satisfied, without perhaps perceiving the reason why. The lover of horticulture is delighted, be the work his own, or that of another. The advantages to the proprietor are great: if it be a public garden, to the community they are considerable. The labours of the industrious man yield peace; of the scientifically industrious man, wealth. But the garden of the sluggard is a reproach to him, and to the public a certain loss. It behoves us, therefore, to be at all due pains in the cultivation of the earth, and to cherish its products; well knowing, that the more we do for the soil, the more grateful will it be in return, and yield us fruits in abundance.

A knowledge of the quality of soils; how to improve them; the value, effect, and proper application of manures; is the basis of every horticultural improvement: and an acquaintance with the nature of the different plants to be cultivated; their duration; manner of growth, and the soils they most affect, is necessary to him who would excel in gardening.

Hence the necessity of close application and study, and of keen observation in the young gardener. Nothing should escape him; he should mark every occurrence. His situation is different from that of the mechanic, the operations of

DOI: 10.4324/9780429198809-12

whose business revolve daily or weekly, who has frequent opportunities to correct mistakes. The operations in gardening revolve more seldom; many of them but once a-year. They are all liable to accidents; to be affected by the changes in the weather, and are subject to the depredations of insect enemies.

The efforts of those most skilful and attentive in the cultivation of the delicate kinds of vegetables, are frequently baffled; more particularly in cases where the soil is not congenial for the purpose. It requires, therefore, no ordinary degree of patience and industry, in bad seasons, to bear up and make head against the casualties of a climate so changeable, in which we often experience spring in mid-winter, and the effects of winter at mid-summer*.

In order to remedy, then, as much as lies in our power, these disadvantages, let us endeavour to improve the climate, by the high cultivation of the soil. To drain it of superabundant water, should be the first object of the improver; by which noxious vapours may be expelled, and the air may be rendered more pure. To drain it of latent water, is necessary to the production of wholesome vegetables; for in a soil surcharged with moisture, neither fruits nor kitchen esculents will thrive, but languish.

Next to draining, effectual digging and trenching, aërating and incorporating the soil, according to its quality, and the purposes for which it may be intended, demand our particular attention. By periodical and judicious trenching, subtrenching, and digging; by the discreet application of fit manures; and by a proper rotation of crops, the soil of the garden may be preserved fresh and sound for a long series of years, so as to be capable of producing wholesome and plentiful crops of vegetables.

In the following Sections, I shall endeavour to set forth these matters in a clear point of view; and in the Kalendar, I shall endeavour to exhibit, in a plain and simple manner, methods of cultivating the various culinary vegetables at present in ordinary use, that may instruct the novice, and, perhaps, may assist the adept in gardening.

ON SITUATIONS FIT FOR KITCHEN-GARDENS.

In a great place, the kitchen-garden, considered merely as such, should be so situated as to be convenient, and, at the same time, be concealed from the house. It should be sheltered by plantations, but should by no means be shaded, or be crowded by them. If walled round, it should be open and free on all sides, or at least to the south, east, and west, that the walls may be clothed with fruit trees on both sides. Round these walls should be a slip of ground and an outer fence, capable of keeping off cattle, and of excluding hares. This slip of ground should be at least twenty feet broad, in order to afford a sufficient border for the trees, and a walk; but it may be as much more in breadth as may be necessary to give ground, with that inclosed by walls, for the supply of the family; and it may be enlarged on all sides, or on any particular side, for that purpose.

The exposure should be towards the south, and the aspect at some point between south-east and south-west; the ground sloping to these points in an easy manner. If quite flat, it seldom can be laid sufficiently dry; and if very steep, it is worked under many disadvantages. It may have a fall, however, of a foot in twenty, without being very inconvenient; but a fall of a foot in thirty is most desirable, by which the ground is sufficiently elevated, yet not too much so. If there be no natural stream that can be conducted through it, water should be conveyed from the nearest river, lake, or pond; soft water being most desirable for the use of the garden.

Under other circumstances than the above-mentioned, and even in places of considerable extent, the Kitchen-Garden is often combined with the Shrubbry or Pleasure-Garden, and also placed nearer to the house. There can be no impropriety in this, provided it be kept in good order, and that the walls be screened by shrubbry from the immediate view of the public rooms. Indeed, it has been found, that there is both comfort and economy in having the various gardens of a place combined, and placed at no great distance from the house. In stepping from the parlour to the shrubbry or flower-garden, thence to the orchard, and, lastly, to the culinary garden, there is a gradation both natural and pleasant. With such an arrangement, in cases where the aspect of the ground is answerable, and the surface, perhaps, is considerably varied, few faults will be found.

Sometimes we find the kitchen-garden placed immediately in front of the house, which I consider the most awkward situation of any; especially if placed near, and so that it cannot be properly screened by some sort of plantation. Generally speaking, the kitchen-garden should be placed in the rear, or on the flank of the house, by which the lawn may not be broken, and rendered unshapely where it is required to be most complete. The necessary traffic with this garden, if placed in front, is always offensive.

Descending to the consideration of more humble gardens, circumstances are often so arbitrary with respect to their situations, as that they cannot be placed either so as to please, or give satisfaction by their products. These are cases where the kitchen-garden is necessarily thrust into a corner, and perhaps is shaded by buildings, or by tall trees, from the sun and air; where they are placed on steep hangs, in a northern aspect; and in all cases where the subsoil is a till or a cankering gravel, and the site cold and bleak.

Such situations as these are to be avoided, and should be considered among the worst possible. Next are open, unsheltered plains. But even there, if the soil be tolerably good, and the subsoil be not particularly bad, shelter may be reared, so as that in a few years the garden may produce a return for the expence laid out in its improvement.

With respect to the situation of market-gardens, those in the vicinity of great towns must have more regard to soil and to local conveniencies, than those at a distance need to have, who can choose, perhaps, a free situation, and a better aspect; such as we find many of the village-gardens, lying on sloping banks, by the sides of streams, that water and render them both pleasant and productive. He who is to pay a high rent should consider well before putting down and improving a garden, whether the situation be eligible, the distance from town convenient, and such as to enable him to carry his goods to market on a footing with his neighbours.

THE FORCING GARDEN.

INTRODUCTION.

THE forcing of fruits, flowers, and some kinds of culinary vegetables, constitutes one of the principal branches of modern gardening. Since the first introduction of hot-houses into this country, this branch has made regular progress, and has uniformly extended to all parts of the island, insomuch as, that a garden is not now reckoned complete without being provided with several compartments for forcing fruits, with a green-house or conservatory, with flued walls, and with frames and lights.

I here allude to the gardens of the great and wealthy, who find, in this department, a source of real pleasure and rational amusement. They are gratified by the production of the finest exotic fruits, in a very considerable degree of perfection; of many native fruits, and rare esculents, at an early and untimely season; and of the finest exotic plants and flowers in the world, many of which could otherwise be known only by their histories. The botanist and florist find here a fund of constant enjoyment. The horticulturist never tires: his pursuits are endless. The agriculturist finds here relaxation, amusement, and often instruction. And all find a garden, blooming fair, and flourishing at midwinter!

But even many market-gardeners, from the demand they have for early fruits and culinary vegetables, and nurserymen in general, for the rearing of exotic plants and flowers, are under the necessity of having hot-houses of various descriptions, green-houses, and hot-beds, in order to enable them to carry on business, and answer the demands of their customers. So much for the wealth of a great nation, in which every luxury is cherished, and every science is fostered!

This has been deemed the most expensive branch of gardening; and many, for want of success in it, have no doubt had cause to be dissatisfied. The following remarks, the hints contained in the succeeding sections, and the general tenor of the directions throughout the kalendar, are intended to set forth methods of forcing, simple and easy, and as little expensive as the different cases will admit of. They are the result of considerable experience, and of much observation, for a period of twenty-five years.

The success attendant on forcing, that is, producing fruits and flowers at an untimely season, and in an artificial climate, depends very much in how near we approach to nature in the process. Of course, it is necessary to introduce the imposed climate as it were by stealth; bringing on, as in nature, first spring, and then summer; making artificial dews and showers, to refresh the foliage and nourish the root; admitting air freely, unless of an impure kind; and ventilating the hot-house frequently, to purify it of noxious damps, which are to all fruits pernicious, especially as they approach to maturity.

The plants being forced out of their natural inclinations in a great measure, do not perspire so copiously, nor imbibe nutritive matter from the atmosphere, so abundantly as if they did grow in the open air. Hence it is necessary to compose for them a soil more rich and deep than might otherwise be required, which must be kept in good heart, by the addition of fresh compost, and by the application of fit manures, from time to time, according to necessity.

In the culture of plants in an artificial climate, regard ought to be had to that of their nativity, which should, as nearly as possible, be imitated in the hot-house; introducing the natural changes of the seasons with equal care. In our variable climate, however, this is done with considerable difficulty, and only by a strict attention to the state of the weather; which may be conceived by considering the difference of latitude, and the variation of the altitude of the sun throughout the year; which variation is greater the nearer we approach to the pole.

Hence, in the forcing of fruits in succession, the difference of trouble attendant on early and on late forcing. Also, the difference of injury done the plants in the process, and in order that they may sustain as little injury as possible, the necessity of a more moderate treatment in the one than in the other. Late forcing, though mild, is, to a certain degree, constraining nature; early forcing is committing an outrage against her.

Sudden changes from heat to cold, or the contrary, ought carefully to be avoided, especially while the bloom or fruit is in a tender state. And this is more necessary to be observed in the forcing of stone than other fruits, these being impatient of violent checks, particularly in the time of setting and stoning.

To the production of fruits in high perfection and flavour, large portions of fresh air are necessary and essential, more particularly as they approach towards maturity. And although moderate, frequent, and plentiful waterings, according to the nature of the plants in question, be advisable, when in a state of free growth, and while the fruit are swelling; yet water should almost entirely be withheld from the time it begins to colour, that the beauty of the fruit may not be lessened nor its flavour be deteriorated.

Source Endnote

* In 1802, June 3, the snow fell, and lay till mid-day on the Lomonds in Fife. The same year new-fallen snow lay till two P. M. on the south sides of the hills to the west of Inverness, on the 20th July.

10

THE SUBURBAN HORTICULTURIST

(Excerpts)

J. C. Loudon

Source: J. C. Loudon, pt. 3. ch. 1 'Laying Out and Planting the Kitchen and Fruit-garden', pt. 3 ch. 2 'Cropping and General Management of a Kitchen-garden', *The Suburban Horticulturist* (1842), [416–433] excerpt, pp. 416–421, [434–441], excerpt, 434–435

LAYING OUT AND PLANTING THE KITCHEN AND FRUIT-GARDEN.

Laying out the Kitchen-Garden.

The *situation* relative to the other parts of a residence, should be as near the house as is consistent with other details. In general the kitchen, stable-offices and kitchen-garden should be on one side of the mansion or dwelling, and so placed as to admit of intercommunication without bringing the operations or operators into the view of the family or their visitors. As the stable-offices are generally near the kitchen-offices, so the kitchen-garden may be near the stables; and in such a situation it will generally be found that the kitchen-garden is less seen from the windows of the mansion, than if it were placed at a much greater distance. A very little reflection will convince any one that this must necessarily be the case. Relatively to *surface*, one which is level, open, and airy, is the best; because it is least liable to be affected by high winds. The next best surface is one gently sloping to the south, or south-east; and the worst is one sloping to the north-east. The surface of a hill is to be avoided on account of its exposure to high winds; and equally so one in a valley on account of the cold air which descends from the adjoining heights and settles there. The *extent* is regulated by the wants of the family, and may vary from a quarter of an acre to several acres; every thing depending on the quantity and quality of the produce required. The best *soil* is a loam, rather sandy than clayey, on a subsoil moderately retentive. The *form* of the garden should be rectangular, as better adapted than any other for the operations to be carried on within. The *area* is enclosed by walls, in general forming a parallelogram with its longest side in the direction of east and west, in consequence of which the greater length of walling has a surface exposed to the south. When the situation is such as to require

 DOI: 10.4324/9780429198809-13

artificial shelter, plantations are formed exterior to the garden for this purpose, but they should never, if practicable, be nearer the walls than 100 feet or 150 feet; for though science has not yet satisfactorily assigned the reason, yet it is certain that nothing is more injurious to culinary vegetables and fruits, than the exclusion of a free current of air in every direction. The sole object of shelter ought to be to break the force of high winds. *Water* should never be wanting in a garden, and as we have already observed it should always be exposed in a basin for some time before being used. The garden *walls* should if possible be of brick; or if they are formed of stone, or of mud or compressed earth, which in some parts of the country make excellent walls, retaining much heat and lasting a long time, they ought to be covered with a wooden trellis on which to train the trees. It has been recommended by Hitt and others to build the walls on piers, for the sake of allowing the roots of the trees to extend themselves on both sides of the wall.[1] As however the branches of the trees are constrained so ought to be the roots, in order that the one may be proportionate to the other. Besides, as there are generally trees on both sides of every garden wall, it does not appear that, under ordinary circumstances at least, anything would be gained by this mode of building walls, excepting the saving of a small proportion of materials. Where walls are not built of brick, stone or earth, they may be formed of boards, which when properly seasoned and afterwards saturated with boiling tar, will endure many years, and produce as much heat in the summer season as brick or stone. They are indeed colder in winter and spring, but that circumstance is often an advantage by retarding the blossoming of the trees, and lessening the risk of their being injured by spring frosts. If a cavity were formed by the boarding, and filled with pounded clinkers, or charcoal, or coke, much heat would be absorbed from the sun heat, and thus form a source for giving out heat at night. Where the walls are formed of brick they may always be built hollow, to save material; and as very little additional expense will be required to form the hollows into flues or channels for hot-water pipes, such an arrangement should not be neglected in the colder parts of the island. The *walks* in the interior of the garden are laid out in a direction parallel to the walls, and *espalier rails* are commonly formed parallel to the walks. Exterior to the walls, a narrow portion of ground is inclosed which is technically called *the slip*, the object of which is to admit of getting the full benefit of the wall on the outside as well as within.

In *trenching* and *levelling* the surface of the kitchen-garden, care must be taken to form a complete system of underground *drainage;* not only by having drains formed of tiles to carry off subterraneous water, but by having the surface of the subsoil parallel to the exposed surface, both being inclined towards the situation of the drains; so that the water in sinking down from the surface may not rest in hollows. The best situation for these drains will generally be under the walks. The depth of the soil of a garden should seldom be less than two feet, this depth being penetrated by the roots of even the smallest kinds of culinary vegetables when growing vigorously. The depth of the soil, however, ought to bear some relation to its quality, and to the climate. A loamy or clayey soil in a humid climate need not be trenched to the same depth as if it were in a warm and dry climate; because the use of the soil to plants being to retain moisture, a small body not liable to lose by evaporation, may be as effective as a larger one so constituted as to lose a great deal. The *borders for fruit-trees* form an important part of the kitchen-garden, and should always be prepared with a due regard to the soil, the climate, and the kinds of trees to be planted. The bottom should generally be prepared so as to prevent the roots from penetrating into the subsoil: though as this naturally limits the supply of water to the roots in dry seasons, and consequently gives occasion for artificial waterings, a better mode than making the borders very shallow, is never to dig them, and to spread the manure always on the surface. By this means the roots will not be forced downwards, as they necessarily must be when the surface is loosened and exposed to the drying influence of the sun and winds, or the exhaustion of crops of vegetables. The subsoil of the borders, however, ought in every case to be drained. *In planting fruit-trees in the kitchen garden*, we would on no account whatever introduce standards, or any description of fruit-tree, in those parts of the open garden which are to be cropped with herbaceous vegetables; because such trees injure the surrounding crops by their shade, and never produce much fruit, or fruit of good quality, in consequence of their roots being forced down into the subsoil by the necessary stirring of the soil among the herbaceous crops. We have enlarged on this subject elsewhere, and we therefore only add that we recommend no fruit-trees to be planted in the kitchen-garden excepting against the walls, against espalier rails, in rows along the walks, or in compartments by themselves. It may be objected to what we recommend, that it is contrary to the practice of market-gardeners, who in general grow fruit-trees among their culinary crops; but to this we reply, that the fruit of such trees, and the flavour of the crops which grow under them, must necessarily be far inferior to that of fruit grown on trees which draw their nourishment from the surface of the soil, and of vegetables which enjoy the full benefit of the sun and air. Market-gardeners know this, though their customers may not. A *forcing department*, a *frame ground* and a *reserve ground*, are accompaniments to every complete kitchen-garden, and even the smallest has at least a reserve and frame ground. The two latter accompaniments are generally placed exterior to the walls of the garden, in that part of the slip which is nearest the stables, and the forcing department is sometimes placed there also; though more generally it consists of glass structures placed against the north wall of the garden. The best outer fence for a garden is a sunk wall, the ditch in which it is built serving as a main drain, into which all the drains in the interior may discharge themselves. The wall of this fence may be carried up three feet or four feet above the surface of the ground, to render it more formidable as a fence, without at the same time producing too much shelter and shade in the slip. In many places it is customary to surround the slip with a shrubbery bounded by a hedge, which has a very good effect for a few years while the trees are young, but when they grow large they produce an injurious degree of shelter and shade. The main entrance to a kitchen-garden should always be so placed as to look towards the main feature within, this feature necessarily being the south side of the north wall, not only because that wall supports the hot-houses when there are any within the garden, but because on it are grown the finest fruits. As an example of a kitchen-garden arranged agreeably to the foregoing observations, but combining also a flower-garden, as being frequently required in a suburban villa, we refer to fig. 330. It contains one acre within the walls, and half an acre in the slips; and the following references will explain the details.

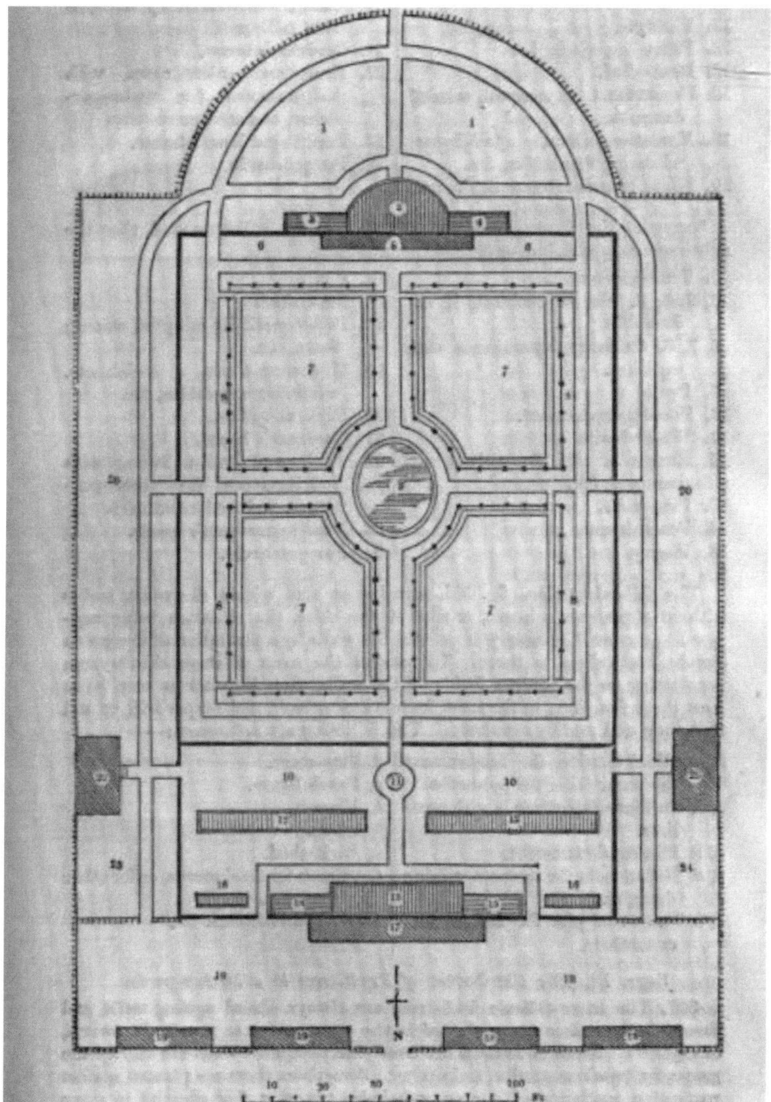

Fig. 330 Plan of a kitchen garden containing one acre within the walls, and half an acre in the surrounding slips.

1. Flower-garden.
2. Conservatory.
3. Green-house.
4. Forcing-house for flowers.
5. Back-shed.
6. Area for setting out green-house plants in summer.
7. Culinary departments with espaliers.
8. Espalier-borders.
9. Pond, surrounded with a stone margin.
10. Forcing department.
11. Water-basin.
12. Ranges of pits for melons, cucumbers, &c.
13. Pine-stove.
14. Peach-house.
15. Vinery.
16. Pits.
17. Back-shed.
18. Department for compost, mixing dung, &c.
19. Mushroom-sheds, tool-house, wintering vegetables, &c.
20. Slips, bounded by a sunk wall fence, surmounted by an open iron railing.
21. Gardener's house.
22. Fruit and onion room, with lodging-room for under-gardener, and seed-room over.
23. Yard to gardener's house.
24. For pot-herbs.

Supposing the flower-gardens and hothouses are to be omitted, then the references may stand as under: –

1. Fruit-garden.

2, 3, 4, 5. To be omitted, if not desirable.

6, 7, 8. Culinary departments with espaliers.

9. Pond.
10. Forcing department.
11. Water-basin.
12. Ranges of pits, for melons, cucumbers, &c.
13. Pine-stove.
14. Peach-house.
15. Vinery.
16. Pits.
17. Back-shed.
18. Department for compost, mixing dung, &c.
19. Mushroom-sheds, tool-house, wintering vegetables, &c.
20. Slips, as before.
21. Gardener's house.
22. Fruit and onion room, with lodging-room for under-gardener, and seed-room over.
23. Yard to gardener's house.
24. For pot-herbs.

The following plan, fig. 331, contains an acre within the walls, and is without a gardener's house, or slips at the sides, the situation being supposed to render it necessary to conceal the walls by a plantation of evergreen shrubs made close to them. To prevent the roots of these shrubs from penetrating to the borders inside of the walls, their foundations must be at least three feet deep in the most

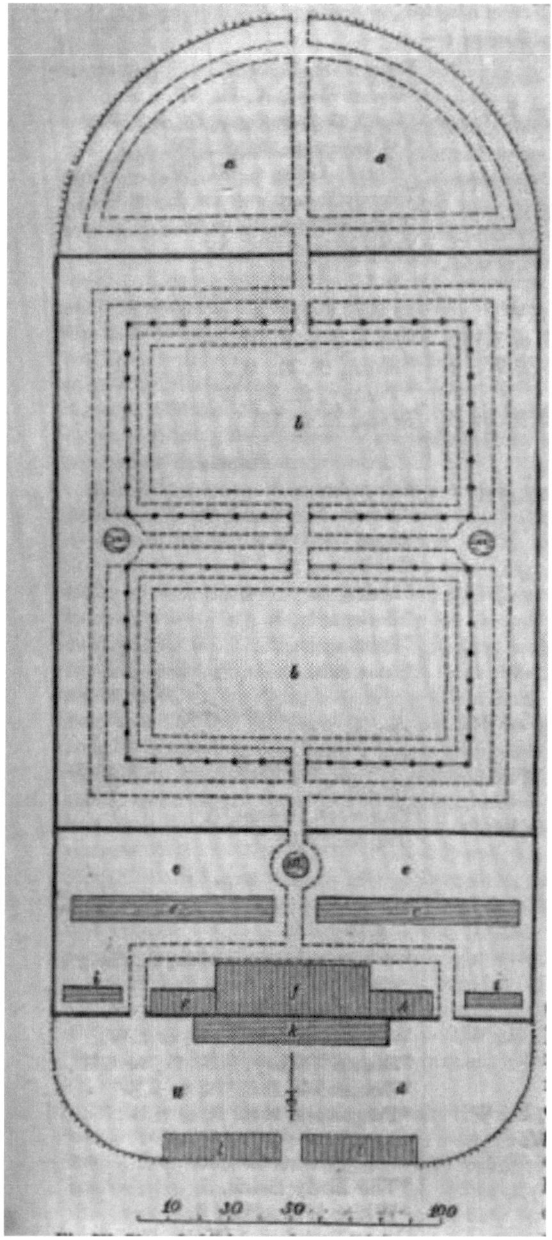

Fig. 331 Plan of a kitchen-garden, containing one acre within the walls, and three quarters of an acre in the slips at the two ends.

impervious subsoil, and deeper still on soil that they will readily penetrate. The following are references: –

THE DISTRIBUTION OF FRUIT-TREES IN A KITCHEN-GARDEN.

CROPPING AND GENERAL MANAGEMENT OF A KITCHEN-GARDEN.

The herbaceous vegetables grown in kitchen-gardens are of two kinds: perennials which remain several years in the ground, such as asparagus, seakale, rhubarb, horse-radish, artichokes, and perennial sweet herbs, and strawberries. The first of these crops remains on the same piece of ground seldom less than ten or twelve years, and the others are renewed generally about half that period or oftener. The other and by far the more numerous crops are annuals or biennials, and many of them only remain on the ground during a part of the year. The proportion of the perennials being fixed on, little more trouble is required with them; but the annuals being numerous and of short duration, the proportionate quantities that require to be sown or planted to supply the demands of the kitchen, and yet to be in due proportion to the extent of the ground to be cropped, and the kinds of crops which ought to succeed each other, require the constant exercise of the gardener's judgment. The first point is to determine the proportion of different crops, and the next is their succession: though the proportions will depend to a certain extent on the peculiar taste or wants of the family, and whether they reside on the spot or at a distance – whether they have a farm for growing the winter supply of potatoes, &c., yet some rules or hints may be devised which are generally applicable.

General proportions of crops. – The greatest breadth of surface in almost every garden requires to be sown with peas; but as this crop only lasts at an average about six months, a second crop may be planted on the same ground in the same year. The cabbage tribe, including cauliflower, broccoli, savoys, Brussels sprouts and borecoles, occupy the next greatest space in most gardens, and they very generally succeed the crops of peas. Turnips are perhaps the next most extensive crop, unless indeed the main summer crops of potatoes are grown in the kitchen-garden, which is not desirable where they can be grown on the farm; the potato being a crop that, for some reason or other which we do not pretend to explain, is seldom found so mealy and high-flavoured when grown in a garden as when grown in a field. There are next several crops, each of which have nearly an equal claim for space, viz. – carrots, onions, beans, kidney-beans, celery, and winter spinach. Jerusalem artichokes and red-beet crops may come next in the order of space required; and then leeks, garlic and shallots, salsafy and scorzonera. Lettuce, endive, radish, cress, mustard, chervil, parsley, and other summer salading, garnishings or herbs, may in general be grown among other crops, or in the front margin of wall-borders.

In determining the extent of each crop, the nature of the produce must be taken as a guide. It would be of little use to have a less quantity of any crop than would not at a single gathering produce a dish sufficient for a family of several persons. This for such articles as asparagus and peas requires considerable breadth of ground; but this breadth once planted and in bearing, will afford several or perhaps many gatherings during the time it is in season. On the other hand, where a succession of crops of turnips or carrots is wanted, if only two or three square yards were sown each time, that space would afford one or two dishes. For such articles as salsafy and scorzonera, which in most English families may perhaps not be asked for above two or three times in a season, a very small surface will be sufficient. When a gardener enters on a new place, before he determines on the extent of particular crops, he ought to consult the cook or housekeeper as to the style of cookery, the ordinary amount of company, and the seasons when extraordinary supplies are wanted, with the periods when vegetables and fruits require to be sent to a distance, with other particulars bearing upon the kind of crops to be grown. Having formed general ideas on the extent of each crop, he will next be able to determine on a system of succession, or, as it is called, rotation.

Editorial Endnotes

1 *The Modern Gardener: Or, Universal Kalendar. Containing Monthly Directions for All the Operations of Gardening. Selected from the Diary of the Late Mr. Hitt.* by James Meader (London, 1771).

11

VEGETABLES AND HOW TO GROW THEM

(Excerpts)

Elizabeth Watts

Source: Elizabeth Watts, 'Preface and Dedication', ch. 1'Economy of Making Gardens Pay', ch. 10 'Roots Less General in Use', *Vegetables and How to Grow Them* (1866), pp. 1–5, 47–51

PREFACE AND DEDICATION.

THERE are many hundreds in the world, even among owners of little gardens, that have never eaten any but bought vegetables in their lives. These vegetables have undergone usual market treatment, involving a certain degree of heating and fermentation, engendering the early stage of decomposition. If all who have a few square yards of ground will test the difference between them and garden produce brought in straight from the garden and cooked at once, by growing as many crops as they can possibly make room for, they will soon create in themselves, and in those who eat at their tables, a taste for wholesome freshness in vegetables. I dedicate this little volume to all who feel a like pleasure with myself in Horticulture, and recommend those who have a little bit of ground at their disposal to devote a portion of it to useful purposes.

John Bull has been accused of living too exclusively on animal food; and certainly the masses of meat we often sit down to, with perhaps only one vegetable (besides potatoes), might be improved both in delicacy and sanitary value by a plentiful addition of garden produce; and a test of the merits of plenty of good fresh-gathered vegetables is especially worthy of attention, now that meat may possibly become dear and scarce among us. Every owner of a garden can aid in this useful experiment.

E. W.

Jersey, 1865.

ECONOMY OF MAKING THE GARDEN PAY.

"CABBAGES at a shilling each, and lettuces costing their weight in silver at least," is the accusation which meets almost every amateur, especially when he *first* turns his attention to the cultivation of a garden, that most healthful, most cheering,

 DOI: 10.4324/9780429198809-14

most delightful of pursuits. Now, none who have bought their own experience, and paid for it in specie and failure, will attempt to deny that gardening may be expensive without limit; but it has this great advantage, that it may be made to suit alike millionnaires who want to get rid of their money, ladies or gentlemen of limited means seeking an inexpensive recreation, and those who wish to make their pleasure profitable. We may hire plenty of gardeners, and place no check upon the introduction of rare plants and other appliances; we may keep a garden in nice order at small outlay in labour and purchases; or we may turn attention to selling produce, and thus make the garden a source of profit, or at any rate self-supporting. Persons who need a shilling manual, a little work in which to find just the information they require, unburdened by elaboration and technicalities, are generally not above the rank in which economy in pleasure and in the every-day pursuits of life is a desideratum: in this little work, therefore, it will not be out of place to consider how moderate economy and good gardening may go hand in hand.

Economy in labour is the first item of economy to consider, since on it hinges economy in hiring. Beginners in gardening, as in other things, often find it very difficult to come to a decision as to their arrangements or alterations. From this cause I have sometimes known a gravel-path and other portions of the garden make most eccentric changes year by year, giving much useless work. It is highly advisable to consider well the intended plan of the garden before the laying out is commenced, to attempt nothing for which the locality is not fitted, and to have the work done in such a manner as may avoid the necessity of having the same thing done over and over again. Always begin at the right end of the work, from laying out a garden to tidying a bed. If a new garden has to be laid out, or an old one remodelled, consider what kind of draining is required before the paths are laid down or the beds shaped; and when you set a man to dig and plant a border, do not let him begin so that he will have to walk over the part he has done to get to a piece afterwards to be taken in hand. In all things manage the work so that the work in hand shall not upset or interfere with that which is done and out of hand. To the same end consider well the best way of doing a job before it is begun; and then, if you are lucky enough to get a tolerably good gardener, or happen to be a good one yourself and will undertake the cultivation of your own garden, all is pretty sure to go well.

Economy in hiring is not difficult for those to manage who themselves take a delight in a garden; for there is much work which ladies and gentlemen, the young ladies of the family, and even the children can do, and which, if they are natty and industrious (as all ladies, gentlemen, young ladies, and children should be), they can do with more neatness than a regular gardener. No gentleman who loves his fruit-trees as we do should leave the pruning of them to a gardener; no gardener's fingers can prick out young plants of cabbages, celery, lettuces, and such like, with the same chance of their doing well that they will have if young ladies' delicate soft little dexterous hands undertake the job; and careful active industrious children are firstrate weeders. In any family where the expense of a gardener is an item worth considering, all the sowing, planting, hoeing, raking, pruning, and weeding may be done by the family; and where is there so pleasant a sight as the members of one family all working together, united in the pursuit of health, enjoyment, and more enjoyable garden produce than money can get? Then it will only be necessary to hire for digging, mowing, and such jobs as ladies and gentlemen would not like to undertake. Poor Richard says the master's eye does more work than both his hands; if economy is to be considered, before hiring a man arrange the work he will have to do. This may be done whether the man be engaged only occasionally, just when there happens to be work for a gardener, or regularly once, twice, or thrice a week. Even where there is entire work for a man or for several, it may be so arranged as to incur less loss of time than there must be in many gardens, to judge by the work done in proportion to the hands employed.

Economy of manure is another thing which ought to be considered, as its purchase is an expensive item in a highly cultivated garden. Crowded localities, such as the immediate neighbourhood of towns, have a great advantage over country places in procuring manure easily, as there, except just at the season when hotbeds are made, opportunities often occur of buying clearings out of accumulations from stables, &c., at a moderate price. These chances should be utilized, that rich fertilizers may be at hand when wanted, thus probably avoiding buying at a high price when necessity presses. In the country, where almost every proprietor can use all the manure from his own homestead, vigilance in watching for chances of making a manure heap is even more necessary. In addition to manure from the stable, and other fertilizers of the kind, there is scarcely any refuse from a house which may not be made valuable to the compost heap, a fact which must never be lost sight of by those who wish to have a productive garden without great outlay of money. To speak further on this subject now would be superfluous, as I think of devoting a chapter to it; so I will only advise that that excellent proverb "Waste not, want not," be brought to bear upon it. Waste no chance of laying up any lot of manure which may be got cheap, and let no liquid or substance from the house which may help the compost heap be wasted. Sweepings of poultry-houses, rabbit-hutches, and pigeon-lofts are more valuable in proportion, if put to good use, than the more extensive accumulations from the stable, cow-house, and pigsty: let no fertilizer be wasted.

Economy in buying trees, plants, and seed is another form in which this useful virtue may be brought to bear upon the cost of the garden. It may be the most satisfactory plan to go to a first-class nurseryman or seedsman, and buy largely of whatever we may want or wish for to add to the beauty and luxuriance of the garden, and this may bring about a good result more quickly than buying more sparingly would do, and so be desirable in the case of those who have no need to consider frugality; but those who have may arrive at the same end with some little delay by purchasing one really good specimen of tree, shrub, or plant, and increasing that by the various modes of propagation suited to it. There is no more delightful branch of the gardener's pleasant labours than the propagation of favourite trees or plants, nor none which makes a more satisfactory return for his labour. Seed is a thing in which economy may be practised with advantage, for it is a mischievous mistake to sow too thickly, and one which many gardeners fall into. The waste of the seed is the least of its evils, for no crop can be thinned without doing harm by disturbing more or less the roots which are left. In the purchase of seeds, get the best that can be got, and buy sparingly, except in the case of those which improve with age. There is no item in the management of a garden on which a wise economy may not be brought to bear; but avoiding the outlay of money is not always nor necessarily the best thrift, and we must be careful not to allow frugality to take the form of false economy. The work of the garden should be done thoroughly, and in the best possible manner, by efficient hands, whether those hands work for their owners or for hire; and it should also be done at the right time, as well as in the best manner. In buying, great economy may be used with advantage; but never buy an indifferent tree, plant, packet of seed, or garden tool, because it costs little; and never keep a bad workman, however small his wages may be; for, depend on it, indifferent things are dear at any price.

ROOTS LESS IN GENERAL USE.

THE Jerusalem artichoke, *Helianthus tuberosus*, is nourishing and wholesome, it imparts a delicious flavour to stews and hashes, and is very good boiled as a vegetable, although a little watery. It is so nice, and so different from other vegetables, that it seems quite to deserve more general cultivation than it gets, especially as any spare corner will do for it, and it requires little care in the cultivation. It will thrive almost anywhere, but best in a rich light soil and open situation. Middle-sized tubers may be planted whole the end of March, or even in February. Some plant in October and leave them to stand the winter. The tubers are put in with the dibble, not quite so deep as potatoes are planted, and three feet apart every way. The ground about them must be hoed from time to time, and some people shorten the stems in August to give strength to the roots.

They will be fit for use in September, when they may be dug as wanted. Later in the autumn, when the stems wither down, the whole crop may be dug and stored in sand for winter, or a few at a time only, for frosty weather, may be taken up, and the remainder left in the ground. In digging them care should be taken to clear the ground of them entirely, or bits which are left in will be cropping up continually. A little frost will not hurt them. As they grow from seven to ten feet high and very thick, care must be taken not to plant them where the shade will do mischief in the garden. No doubt the name is derived from the Italian for sunflower, *girasole*, as it is a helianthus. It came from the Brazils, and produces its small yellow flowers in September, but scarcely ever flowers with us. If they are left to grow year after year in the same spot, coming up from any stray roots which happen to be left in, it is no wonder they are poor; but give them fair cultural advantages like other crops, and they will be fine. When they are pared for cooking each root should be dropped into water as it is done; or a better plan is to peel them quickly after they are cooked. Attention to these little items would make the Jerusalem artichoke better thought of than it is by all who have a bit of ground to spare it, as in truth it well deserves to be. It is said to be good winter food for pheasants.

Salsafy, Scorzonera, Chicory, Rampions, and Skirrets are old-fashioned root vegetables, seldom met with in our gardens now, but still used on the Continent. They are wholesome, make a change with other vegetables, and are eaten boiled tender, and covered with white sauce, or with various vegetables in stews, fricasées, and other made dishes.

The Salsafy, *Tragopogon porrifolius*, thrives in an open situation and a light and rather fertile soil, and requires the culture of all deep-growing root crops, *i. e.* deep working of the earth, and the manure inserted deep down. In March or April the seed may be sown in shallow drills, nine inches apart, and covered with half an inch of earth. When the plants are two or three inches high, weed the bed and remove chance stragglers, so that the plants may remain that distance apart. Water plentifully, sometimes with guano water (half an ounce of guano to a gallon of water), and by September or October the roots will be ready to eat. When frost comes, a few at a time had better be dug and stored for use when the ground may be too hard to dig, but the crop may stay in, as frost does not hurt it. Roots left in the ground will grow in the spring; when the little shoots are a few inches high, young, and tender, they are said to be very good dressed as asparagus. Another way of cooking them is to boil and mash them, shape them into cakes, and brown them in a frying-pan. A root or two may be left for seed, or seed may be bought 6*d.* or 9*d.* an ounce of Messrs. Butler and McCulloch, Covent Garden, and Messrs. Carter and Co., of High Holborn, London.

Scorzonera, viper's grass, or Spanish salsify, *Scorzonera Hispanica*, is a parsnip-like root, with the disadvantage of an unsightly black skin, which should be carefully scraped off before it is seen at table. It may be grown with exactly the same treatment as salsify, only drawing the drills and thinning the roots to a few inches further apart. Roots which are left growing, although they may flower, increase in size and goodness again afterwards. A plant or two may stand for seed, or the seed may be bought 6*d.* an ounce, or the finest kind at 1*s.* It should be used at one year old.

The root of Chicory, succory, or wild endive, *Cichorium intybus*, is eaten in France, and there is no doubt but it is very wholesome. That old-fashioned writer, Culpepper, attributes to it beneficial action on the liver, and other vital organs, and mentions it as a remedy in ague. It has a little bitter flavour, but it affords considerable nourishment, and yields a good crop. It requires an open situation, and a rich, light, and rather moist soil. When it is grown for the sake of the root, the seed should be sown late in the spring, that the plants may not run up to seed. Sow rather more than half an inch deep, in drills a foot apart; thin out the young plants to a foot apart, and keep them weeded. Some roots may be left for seed; they will need no protection, run to seed about May, and the intense blue of the flower is very fine.

The Rampion, *Campanula rapunculus*, is a radish-like root which does best in a moist shady border, where the earth is rich and very light. The seed should be sown in March, April, and May, in drills six inches apart, very thinly, and covered lightly. The seed is small, and therefore care must be taken to put it in very thinly. Thin the young roots to six inches apart – evening is the best time for doing it;

and then the little plants which are thinned out may be planted, if they are wanted, although those not removed will do better. The early sown crops will be apt to run up to seed, but they will come into use for the end of summer and autumn, and the last sowing will do for winter use, and from then as long as any remain. Constant watering throughout the growth will do good. The root must be used while still young and tender; it may be eaten like radishes, and it has a pleasant nutty flavour; or it may be sliced up, green and all, in salads, or boiled until soft, and eaten with melted butter like asparagus. Leave a plant or two standing for seed; they will flower in July or August, and ripen the seed in the autumn, when it must be gathered before it scatters, and properly dried. A threepenny packet of seed may be bought of Messrs. Carter or Messrs. Butler and McCulloch.

Skirrets, *Sium sisarum*, grow something like dahlias. The seed may be sown in drills twelve inches apart, and only just covered. When the young plants are large enough, thin them to twelve inches apart, and the roots will be fit to use in autumn. A better mode of propagation, however, is to divide the roots of the year before in the spring, taking off the offsets, and planting them in rows a foot apart. The earth for them is light rich loam, with manure laid in with the bottom spit. An old root or two may be left to ripen seed in autumn, or the seed may be bought.

The Chinese yam, *Dioscorea batata*, made a great sensation at the time when men seemed driven by necessity to find a substitute for potatoes, and this root was set forth at that time as likely to supply their place, and to grow well in our climate. Potatoes, however, got better, and Chinese yams were cultivated by a few, not by the many; but I believe it has been more extensively grown in France. It must have a warm situation, and the ground must be duly prepared by trenching to a depth of a yard, and thoroughly mixed with rotten manure. The roots are cut into sets two inches long, and planted about three inches deep. April or May is the time for planting, and in November the long narrow tubers will be ready for use, when they may be dug and stored like potatoes. Some recommend starting the plants in pots plunged in a hot-bed in April, and planting them out when all fear of frost is over. Watering is quite necessary.

The Oca, or *Oxalis deppei*, is another root which there has been much talk about cultivating for a vegetable, and some persons like it very much. The way of cooking it is to peel the roots, clean out the hollow centres, and boil them in stock or gravy until tender, when they are served with sauce made of a little of the stock in which they have been boiled, browned and thickened. About ten roots make a dish, and they come in from early in October to the end of December. In Belgium the leaves are used as sorrel, and the flowers are put into salads. The roots may be planted singly in pots early in March, and kept in a cold pit or in the house until there is no longer danger of frost, when they may be put out in a light soil, on a south border, twelve inches apart every way. The ground should be prepared as for most root crops by working it deep, and putting manure deep; and if it be thought less troublesome than starting them in pots, the roots can be planted in it in April. Keep the crop free from weeds, and water plentifully in dry weather; if the roots get dry they will split when wet comes. They must be protected from early frost by mats.

12

ORCHARD AND FRUIT GARDEN

(Excerpts)

Elizabeth Watts

Source: Elizabeth Watts, 'Preface', ch. 1, 'The Orchard', *Orchard and Fruit Garden* (1867), pp. 1–5

PREFACE.

WE have so many excellent large and expensive works on fruit culture, that I should never have thought of writing this one, if its peculiar place had not stood vacant, waiting to be filled. It is the first *cheap work* on the Orchard and Fruit Garden. Those who, like myself, have had through life some space of land, between a pole and an acre, at command, do not need to be puzzled with lists of hundreds in the selection of the trees they want; their need is one which I have made it my endeavour to supply, *i.e.* particulars of a *few* good sorts of fruit trees of all kinds which any careful cultivator can manage, and which may be obtained at any good nursery. The sorts named are good, those I most recommend I have tested, and the directions given respecting their cultivation will be found plain, concise, and practical. I can only hope my little book will be found useful to all who delight in fruit culture, but who cannot command access to voluminous and expensive works upon the subject.

E. WATTS.

Jersey, May, 1867.

THE ORCHARD.

IN the treatment of our fruit trees, sins of omission are more general than sins of commission. Whereas many, in the present advanced state of Horticulture, select with precision, plant with care, prune with skill, manure sufficiently, and accommodate the soil to the development and improvement of the fruit, there are yet also many others who seem to think a tree a tree, and one about as good as another, make a hole in which to bury the root, together with all reasonable hope of success, and leave the rest to nature and to chance. This negligent gardening, although still too prevalent, is much less so than formerly. Excellent treatises and well-conducted periodicals now aid the beginner with advice and instruction, and

DOI: 10.4324/9780429198809-15

offer a theatre for the discussion of all moot points. These valuable works are, however, not so universally within the reach of all fruit growers and lovers of a garden as to render a cheap work on the management of fruit trees superfluous.

Fruit fresh gathered and good in kind, is so pleasant and so useful an addition to diet, and so especially valuable in the case of children and young people, that I can scarcely comprehend why there are so many gardens, even of the smallest dimensions, entirely without fruit trees, and why ground, where they would grow, thrive, and produce, is often devoted to trees which are not more ornamental than they, and which entirely lack their utility. I scarcely like to see even an ornamental flowering shrub or tree occupying a fine wall which might be utilized in the production of good stone fruit or pears, with very little loss on the score of decoration; but this is merely a matter of fancy.

"Be aye sticking in a tree,"[1] was the Scottish Laird's instruction to his heir, "it will be growing, Jock, when ye're sleeping," and "be aye putting in a fruit tree," I echo, after trying the plan with five different residences, situated in various localities, soils, and climates, occupied and tested during the last quarter of a century.

An outcry against a man for *selfishness*, because he shows an unwillingness to plant trees in ground not his own, is hardly fair, since few wish to practise unsolicited bestowal of property on unknown recipients; but as a matter affecting personal remuneration, it is not necessary or prudent to avoid planting because we may have no permanent or lengthened hold on the land we occupy. If our tenure expire soon or unexpectedly, we have, I believe, the right by law to remove trees within three years of the time of planting them, and when they have been planted more than that time, those which have been chosen, planted, and managed with judgment, will already have made a fair return on outlay and work. A well-managed tree will within a *very* few years of the time of planting prove a capital investment, and one the interest of which will continually increase for many years; I therefore very strongly recommend all who own or rent a bit of ground, to plant trees; if the tenure be given up soon, the trees can be removed, and the removal will do them no harm; and if they are held for from three to a dozen years, they will in that time make a much better return than any other crop which can be grown.

If the tenure be short or uncertain, fruits which will make an early return may be planted, – such as apples, gooseberries, currants, raspberries, strawberries, and perhaps some well-chosen stone fruit trees. If the tenure be pretty secure, and of unlimited duration, then plant without reservation pears, and all other choice fruits, only taking care to plant so far apart as to leave ample space for the increase of every tree that is planted.

The best situation for an orchard or fruit garden is a gentle slope towards the South, South-east, or East. A North aspect should be avoided, and a West aspect is too much open to tearing winds, and too burning in the afternoon heat to be beneficial. A low, damp situation should be avoided, as no tree will do well and produce fine fruit if its roots are occasionally in water. At the same time a low, sheltered spot, if dry, is better than exposed high land, being less open to cold, and violent wind.

With regard to soil, good corn land is said to do well for fruit; good loam is best. The soil should be neither too light and dry, nor too heavy and stubborn; if its nature require such treatment, it should be trenched and drained, and its surface should have a sufficient thickness of fine, fertile earth.

An orchard may be stocked with apples, pears, cherries, or plums. In the fruit garden we may have these and all other fruits; the vinery, orchard house and frame will give choice fruit earlier than it can be produced out of doors, and every good wall should be made available for apricots, peaches, and nectarines, and the choicest kinds of pears, plums, and cherries.

If grass land be used as an orchard, it must not be imagined that all may be taken from the land and nothing restored to it. The earth pays double dues in fruit and grass, and therefore it cannot be expected to do its duty and produce fine crops and fine fruit too, unless it be doubly manured: manured on the score of the fruit trees, and again on that of the green or other crop which is taken off it.

The same may be said of fruit gardens which are also used to grow vegetables. If the vegetables are dug and removed without mercy on the fruit trees, their roots will be injured and impoverished, and their fruit spoiled. The digging must be managed with reference to the roots of the trees, and no crop must be put in so close to a tree as to interfere with it either while it grows or in its removal. The manure, too, must not be given to the vegetables without regard to the prosperity of the trees; it must on no account be given close round the roots at a season when forcing them would encourage rampant growth to the detriment of a fine crop of fruit, nor withheld from them when mulching is requisite or beneficial.

Fruit and vegetables, or fruit and orchard crops, may do well in concert if all are fairly treated, and fairly enriched, but for the production of choice fruit it should be made the *specialité* in either the orchard or the garden.

For fencing, there is nothing that may be compared with fine, smooth, high walls, as available for the training of the best kinds of fruit, and a safeguard against the depredations of thieves. Next to these in merit, though far inferior, stand well-made, close wooden fences, and these are about the only methods of enclosure to be used with advantage for orchards or fruit gardens.

A wall should be at least eight feet high, and as much higher as can be made convenient. If not more than eight feet, it may be nine inches in thickness. From eight to fourteen feet high requires thirteen inches and a half in thickness, and above that height eighteen inches. A wall, if high and many feet in length, must be strengthened with buttresses, and the foundation of all walls should be a brick and a half thick, even where the wall itself is only one brick thick.

Flued or hot walls are built hollow, the interstices being used as flues for the circulation of artificial heat. It is necessary for these walls to have hooks along, near the top, under the coping, or other apparatus for arranging protection over the trees when necessary.

A wall like this will require a fire to every forty or fifty feet of wall. Stone walls are good, but they almost require a lattice over them to which to train trees.

All the walls of a fruit garden may be made valuable, even that with the North aspect being needed for shelter, and available for Morello cherries, and some other fruits.

A well-formed, well-kept bank, with a hedge at the top regularly pruned and thickened, is a good fence to an orchard in the country, especially if the locality be cheap in labour, and dear in bricks and other building commodities.

For hedges to a fruit garden, evergreens only can be tolerated, and they are so inferior in utility to walls of any kind that they should never be used unless as a positive matter of necessity.

A nursery for rearing and testing young trees is a valuable adjunct to every establishment where fruit is much thought of: there we can bring forward young trees for future use, increase the choice sorts, try the merits of all which are unknown, and provide stocks for any valuable varieties that may be met with.

Editorial Endnote

1 "Be aye sticking in a tree", from *The Heart of Midlothian* (1818), the seventh of Walter Scott's Waverley novels.

13

BOUVARD AND PÉCUCHET

(Experiments in Agriculture)

Gustave Flaubert

Source: Gustave Flaubert, ch. 2 (Experiments in Agriculture), *Bouvard and Pécuchet* (1881), trans. by D. F. Hannigan (1896), [29–72] [excerpts] pp. 29, 43–48, 56–60

HOW happy they felt when they awoke next morning! Bouvard smoked a pipe, and Pécuchet took a pinch of snuff, which they declared to be the best they had ever had in their whole lives. Then they went to the window to observe the landscape.

In front of them lay the fields, with a barn and the church-bell at the right, and a screen of poplars at the left.

Two principal walks, forming a cross, divided the garden into four parts. The vegetables were contained in the plat-bands, where, at different spots, arose dwarf cypresses and trees cut in distaff fashion. On one side, an arbour just touched an artificial hillock; while, on the other, the espaliers were supported against a wall; and at the end, a railed opening gave a glimpse of the country outside. Beyond the wall there was an orchard, and, next to a hedge of elm trees, a thicket; and behind the railed opening there was a narrow road

. . .

Moreover, he was plagued with the garden. All his time would not have sufficed to keep it in order. Bouvard was occupied with the farm. They took counsel, and decided on this arrangement.

The first point was to have good hot-beds. Pécuchet got one made of brick. He painted the frames himself; and, being afraid of too much sunlight, he smeared over all the bell-glasses with chalk. He took care to cut off the tops of the leaves for slippings. Next he devoted attention to the layers. He attempted many sorts of grafting – flute-graft, crown-graft, shield-graft, herbaceous grafting, and whip-grafting. With what care he adjusted the two libers! how he tightened the ligatures! and what a heap of ointment it took to cover them again!

Twice a day he took his watering-pot and swung it over the plants as if he would have shed incense over them. In proportion as they became green under the water which fell in a thin shower, it seemed to him as if he were quenching his own thirst and reviving along with them. Then, yielding to a feeling of intoxication,

 DOI: 10.4324/9780429198809-16

he snatched off the rose of the watering-pot, and poured out the liquid copiously from the open neck.

At the end of the elm hedge, near the female figure in plaster, stood a kind of log hut. Pécuchet locked up his implements there, and spent delightful hours there picking the berries, writing labels, and putting his little pots in order. He sat down to rest himself on a box at the door of the hut, and then planned fresh improvements.

He had put two clumps of geraniums at the end of the front steps. Between the cypresses and the distaff-shaped trees he had planted sunflowers; and as the plots were covered with buttercups, and all the walks with fresh sand, the garden was quite dazzling in its abundance of yellow hues.

But the bed swarmed with larvæ. In spite of the dead leaves placed there to heat the plants, under the painted frames and the whitened bell-glasses, only a stunted crop made its appearance. He failed with the broccoli, the mad-apples, the turnips, and the watercress, which he had tried to raise in a tub. After the thaw all the artichokes were ruined. The cabbages gave him some consolation. One of them especially excited his hopes. It expanded and shut up quickly, but ended by becoming prodigious and absolutely uneatable. No matter – Pécuchet was content with being the possessor of a monstrosity!

Then he tried his hand at what he regarded as the *summum* of art – the growing of melons.

He sowed many varieties of seed in plates filled with vegetable mould, which he deposited in the soil of the bed. Then he raised another bed, and when it had put forth its virgin buddings he transplanted the best of them, putting bell-glasses over them. He made all the cuttings in accordance with the precepts of "The Good Gardener."[1] He treated the flowers tenderly; he let the fruits grow in a tangle, and then selected one on either arm, removed the others, and, as soon as they were as large as nuts, he slipped a little board around their rind to prevent them from rotting by contact with dung. He heated them, gave them air, swept off the mist from the bell-glasses with his pocket-handkerchief, and, if he saw lowering clouds, he quickly brought out straw mattings to protect them.

He did not sleep at night on account of them. Many times he even got up out of bed, and, putting on his boots without stockings, shivering in his shirt, he traversed the entire garden to throw his own counterpane over the hot-bed frames.

The melons ripened. Bouvard grinned when he saw the first of them. The second was no better; neither was the third. For each of them Pécuchet found a fresh excuse, down to the very last, which he threw out of the window, declaring that he could not understand it at all.

The fact was that he planted some things beside others of a different species; and so the sweet melons got mixed up with the kitchen-garden melons, the big Portugal with the Grand Mogul variety; and this anarchy was completed by the proximity of the tomatoes – the result being abominable hybrids that had the taste of pumpkins.

Then Pécuchet devoted his attention to the flowers. He wrote to Dumouchel to get shrubs with seeds for him, purchased a stock of heath soil, and set to work resolutely.

But he planted passion-flowers in the shade and pansies in the sun, covered the hyacinths with dung, watered the lilies near their blossoms, tried to stimulate the fuchsias with glue, and actually roasted a pomegranate by exposing it to the heat of the kitchen fire.

When the weather got cold, he screened the eglantines under domes of strong paper which had been lubricated with a candle. They looked like sugarloafs held up by sticks.

The dahlias had enormous props; and between these straight lines could be seen the winding branches of a Sophora Japonica, which remained motionless, without either perishing or growing.

However, since even the rarest trees flourish in the gardens of the capital, they must needs grow successfully at Chavignolles; and Pécuchet provided himself with the Indian lilac, the Chinese rose, and the eucalyptus, then in the beginning of its fame. But all his experiments failed; and at each successive failure he was vastly astonished.

Bouvard, like him, met with obstacles. They mutually consulted one another, opened a book, then passed on to another, and did not know what to resolve upon when there was so much divergence of opinion.

Thus Puvis[2] recommends marl, while the Roret Manual[3] is opposed to it. As for plaster, in spite of the example of Franklin,[4] Riefel[5] and M. Rigaud[6] did not appear to be in raptures about it.

According to Bouvard, fallow lands were a Gothic prejudice. However, Leclerc[7] has noted cases in which they are almost indispensable. Gasparin[8] mentions a native of Lyons who cultivated cereals in the same field for half a century: this upsets the theory as to the variation of crops. Tull[9] extols tillage to the prejudice of rich pasture; and there is Major Beetson,[10] who by means of tillage would abolish pasture altogether.

In order to understand the indications of the weather, they studied the clouds according to the classification of Luke Howard.[11] They contemplated those which spread out like manes, those which resemble islands, and those which might be taken for mountains of snow – trying to distinguish the nimbus from the cirrus and the stratus from the cumulus. The shapes had altered even before they had discovered the names.

The barometer deceived them; the thermometer taught them nothing; and they had recourse to the device invented in Louis XIV.'s time by a priest from Touraine. A leech in a glass bottle was to rise up in the event of rain, to stick to the bottom in settled weather, and to move about if a storm were threatening. But nearly always the atmosphere contradicted the leech. Three others were put in along with it. The entire four behaved differently. . .

. . .

Pécuchet was disgusted with gardening, and a few days later he remarked:

"We ought to give ourselves up exclusively to tree culture – not for pleasure, but as a speculation. A pear which is the product of three soils is sometimes sold in the capital for five or six francs. Gardeners make out of apricots twenty-five thousand livres in the year! At St. Petersburg, during the winter, grapes are sold at a napoleon per grape. It is a beautiful industry, you must admit! And what does it cost? Attention, manuring, and a fresh touch of the pruning-knife."

It excited Bouvard's imagination so much that they sought immediately in their books for a nomenclature for purchasable plants, and, having selected names which appeared to them wonderful, they applied to a nurseryman from Falaise, who busied himself in supplying them with three hundred stalks for which he had not found a sale. They got a locksmith for the props, an iron-worker for the fasteners, and a carpenter for the rests. The forms of the trees were designed beforehand. Pieces of lath on the wall represented candelabra. Two posts at the ends of the plat-bands supported steel threads in a horizontal position; and in the orchard, hoops indicated the structure of vases, cone-shaped switches that of pyramids, so well that, in arriving in the midst of them, you imagined you saw pieces of some unknown machinery or the framework of a pyrotechnic apparatus.

The holes having been dug, they cut the ends of all the roots, good or bad, and buried them in a compost. Six months after the plants were dead. Fresh orders to the nurseryman, and fresh plantations in still deeper holes. But the rain softening the soil, the grafts buried themselves in the ground of their own accord, and the trees sprouted out.

When spring had come, Pécuchet set about the pruning of pear trees. He did not cut down the shoots, spared the superfluous side branches, and, persisting in trying to lay the "duchesses" out in a square when they ought to go in a string on one side, he broke them or tore them down invariably. As for the peach trees, he got mixed up with over-mother branches, under-mother branches, and second-under-mother branches. The empty and the full always presented themselves when they were not wanted, and it was impossible to obtain on an espalier a perfect rectangle, with six branches to the right and six to the left, not including the two principal ones, the whole forming a fine bit of herring-bone work.

Bouvard tried to manage the apricot trees, but they rebelled. He lowered their stems nearly to a level with the ground; none of them shot up again. The cherry trees, in which he had made notches, produced gum.

At first, they cut very long, which destroyed the principal buds, and then very short, which led to excessive branching; and they often hesitated, not knowing how to distinguish between buds of trees and buds of flowers. They were delighted to have flowers, but when they recognised their mistake, they tore off the three-fourths of them to strengthen the remainder.

Incessantly they kept talking about "sap" and "cambium," "paling up," "breaking down," and "blinding of an eye." In the middle of their dining-room they had in a frame the list of their young growths, as if they were pupils, with a number which was repeated in the garden on a little piece of wood, at the foot of the tree. Out of bed at dawn, they kept working till nightfall with their twigs carried in their belts. In the cold mornings of spring, Bouvard wore his knitted vest under his blouse, and Pécuchet his old frock-coat under his packcloth wrapper; and the people passing by the open fence heard them coughing in the damp atmosphere.

Sometimes Pécuchet drew forth his manual from his pocket, and he studied a paragraph of it standing up with his grafting-tool near him in the attitude of the gardener who decorated the frontispiece of the book. This resemblance flattered him exceedingly, and made him entertain more esteem for the author.

Bouvard was continually perched on a high ladder before the pyramids. One day he was seized with dizziness, and, not daring to come down farther, he called on Pécuchet to come to his aid.

At length pears made their appearance, and there were plums in the orchard. Then they made use of all the devices which had been recommended to them against the birds. But the bits of glass made dazzling reflections, the clapper of the windmill woke them during the night, and the sparrows perched on the lay figure. They made a second, and even a third, varying the dress, but without any useful result.

However, they could hope for some fruit. Pécuchet had just given an intimation of the fact to Bouvard, when suddenly the thunder resounded and the rain fell – a heavy and violent downpour. The wind at intervals shook the entire surface of the espalier. The props gave way one after the other, and the unfortunate distaff-shaped trees, while swaying under the storm, dashed their pears against each other.

Pécuchet, surprised by the shower, had taken refuge in the hut. Bouvard stuck to the kitchen. They saw splinters of wood, branches, and slates whirling in front of them; and the sailors' wives who, on the sea-shore ten leagues away, were gazing out at the sea, had not eyes more wistful or hearts more anxious. Then, suddenly, the supports and wooden bars of espaliers facing one another, together with the rail-work, toppled down into the plat-bands.

What a picture when they went to inspect the scene! The cherries and plums covered the grass, amid the dissolving hailstones. The Passe Colmars were destroyed, as well as the Besi des Vétérans and the Triomphes de Jordoigne. There was barely left amongst the apples even a few Bon Papas; and a dozen Tetons de Venus, the entire crop of peaches, rolled into the pools of water by the side of the box trees, which had been torn up by the roots.

Editorial Endnotes

1 Numerous references are made to actual publications apart from *Le Bon Jardinier*.

2 Puvis – Marc-Antoine Puvis de Chavannes (1776–1851), author *De l'emploi de la chaux en agriculture* (1836) (*On the Use of Lime in Agriculture*)

3 The Roret Manual – a reference to one of around 500 manuals (and part of a series of encyclopaedias) published in his lifetime by Nicolas Edme Roret (1797–1860), science editor and publisher, from 1820

4 Franklin – Benjamin Franklin (1706–1790), one of the founding fathers of the US. One of his interests was agriculture. See 'Benjamin Franklin – adventures in agriculture' by Card R. Woodward, *Journal of the Franklin Institute*, vol. 234 (1942), pp. 207–228.
5 Possibly Louis Michel Rigaud (1761–1826), agricultural writer
6 Riefel – Not identified
7 Leclerc – Georges-Louis Leclerc, Comte de Buffon (1717–1788), a major contributor to the science of the Enlightenment with his 50-volume *Histoire Naturelle* (1749–1804)
8 Gasparin – Comte Adrien de Gasparin (1783–1862), author of *Cours d'Agriculture* (6 vols, 1843–1860)
9 Tull – Jethro Tull, *The New Horse-Hoeing Husbandry* (1733), about intensive farming, translated into French in the mid-18th century
10 Major Beetson [sic] – Alexander Beatson (1758–1830), who became a major-general and published in 1820 a book entitled *A new system of cultivation without lime or dung*
11 Luke Howard (1772–1864), chemist, known for his classification of cloud systems, *Essay on the Modification of Clouds* (1803)

14

'CULFORD HALL, SUFFOLK'

J. Douglas

Source: J. Douglas, 'Culford Hall, Suffolk', *Journal of Horticulture* (25 February 1875), pp. 164–166

CULFORD HALL, the seat of the Rev. Edward Richard Benyon, is about four miles N.N.W. from Bury St. Edmunds.[1] The estate includes the five parishes of Timworth, Ingham, Wordwell, West Stow, and Culford, and exceeds in extent 12,000 acres. The land varies in character, but is mostly light, upon a subsoil of gravel or chalk. This manor appears to have been given at an early period to the Abbey of St. Edmund, but afterwards became the property of the Bacon family. The Hall is a large handsome mansion, and was built in 1591 by Sir Nicholas Bacon, and passed with the estates by marriage into the noble family of Cornwallis. It was purchased about the year 1823 from the executors of the last Marquis by Richard Benyon de Beauvoir, Esq., uncle of the present proprietor.[2] It is surrounded by a park of some five hundred acres in extent, which is remarkably well wooded, containing fine specimens of ancient Oaks, Elms, Beeches, Horse Chestnuts, &c. It is also embellished by a fine piece of artificial water on the south side of the Hall; this is formed by a tributary of the river Lark, and was designed by the celebrated landscape gardener Repton, and its skilful disposition gives evidence of the talent of its designer. At about a quarter of a mile west from the Hall the water is spanned by an extraordinary massive iron bridge, which was erected at an enormous cost by the first Marquis of Cornwallis on his return from India.[3] The chief approach to the mansion was made to cross the river by way of this bridge. But the principal entrances are now in the village of Culford.

The church stands in the pleasure grounds a short distance from the Hall, and is exceedingly neat and ornamental in appearance, being built of flint, with facings of Caen stone. It is furnished with a square tower of very chaste proportions. It was rebuilt in 1856 on the site of a former church, which was of great antiquity, and contained many ancient and interesting monuments to different members of the Bacon and Cornwallis families, all of which have been carefully preserved. It is worthy of mention that the present proprietor of these estates has, since he came into possession of the property, rebuilt or restored the church of each of his five parishes, and has also built an entirely new one for the benefit

 DOI: 10.4324/9780429198809-17

Fig. 43. – The conservatory and flower garden at Culford Hall.

of the inhabitants of the northern portion of the parish of Culford, which is five miles in length.

The conservatory stands near to the Hall, and has a small flower garden and fountain in front of it; while in the rear of the structure, and communicating with it, is an exotic fernery containing fountains, waterfalls, &c.; and a wilderness or hardy fernery occupies a position near to the gardens.

In the grounds are some good specimens of the Cedar of Lebanon, the Deodar Cedar, Wellingtonia gigantea, Picea Pinsapo, Irish Yews, &c. The gardens are situated on the east side of the Hall, and at a distance of some 500 yards from it. They contain upwards of eleven acres, and slope gently to the south, and are surrounded and intersected by brick walls of various heights. These gardens also contain about 1500 yards of Yew hedges, which vary in height from 4 to 12 feet. The principal walks are 10 feet wide, and the walls are pierced where necessary by circular arches of the same width.

There are a considerable number of glass structures of various kinds – viz., two span-roofed orchard houses each 90 feet long, 20 feet wide, and 12 feet high, chiefly devoted to the culture of Peaches, Nectarines, and Cherries. A lean-to Fig house, 220 feet in length; 200 feet of wall, on which Apricot trees are trained, and dwarf standards planted in front of them at a distance of 5 feet from the wall, and so arranged that a glass roof is placed over them when the trees begin to bloom, and taken off when all danger from frost has passed. By this means a crop of fruit is always secured. The ripening of the fruit of the standard trees succeeds that of the trees trained to the wall, and by this means the season of this useful fruit is greatly prolonged. There are also a greenhouse 30 feet long, a plant stove, Melon house, Cucumber house, and Strawberry house; an early Peach house, two small vineries for early forcing, planted chiefly with the Black Hamburgh and Muscadine varieties – one for late Black Hamburgh Grapes, one for Muscat of Alexandria, one for West's St. Peter's and Black Alicante, one for Lady Downe's and Mrs. Pince, and one house which contains various sorts, together with pits used for the purpose of forcing early Potatoes, Asparagus, &c. We will now add the notes of another correspondent.

Being in the neighbourhood of Bury St. Edmunds, I took advantage of the opportunity to spend a few hours with Mr. Grieve, the head gardener to the Rev. E. R. Benyon, so well known to the horticultural public by his skill in crossing Pelargoniums of the Zonal section, especially the Tricolors. On first entering the kitchen garden the visitor is struck by the way it is laid out, and the method of training the Gooseberry bushes. Very stout poles are inserted in the ground at the base of each bush, to which the trees are trained in the pyramid fashion. A whole quarter is devoted to the bushes, and ample space allowed between the rows to allow of gathering the fruit with facility. The quarters of the garden are divided and subdivided by Yew hedges; these are regularly furnished throughout, and have been clipped and trained with mathematical precision. This is certainly a very pleasing arrangement so far as effect is concerned, but its advantages are doubtful. The hedges may shelter some crops, but they occupy much space, and would not do where ground is valuable.

There are likewise extensive ranges of glass houses in the kitchen garden. The first range we enter is a Muscat house and Peach house. Some excellent Muscats were still hanging on the Vines; the bunches were small, but the berries possessed that high golden colour indicative of good cultivation

Mr. Grieve cuts his spurs back close to the main stem, not caring to cut to a good eye, which soon causes unsightly spurs to form, but which is the sure way to get large bunches. Most interesting to me were the two fine orchard houses. There is a bed in the centre, which is planted with one row of Peach trees 6 feet apart. The side borders are planted with Cherries, all the trees being trained in the bush or pyramid form. From some cause Cherries do not succeed in the open air at Culford, but excellent crops are obtained in the same house with the Peaches. Some cultivators experience difficulty in getting Cherry trees to set their fruit; but this need not be, if only the atmosphere can be dry and air admitted night and day. Black Hamburgh and Royal Muscadine Grapes are grown over the passages from Vines trained to wires near the glass; these two sorts are the very best that can be planted in such position. Another orchard house is devoted to Peaches and Nectarines planted out and trained to a trellis overhead. The training was different from anything I had previously seen. It may be styled cordon training: main branches are trained either horizontally or vertically, and the side growths roughly spurred back after the manner of Vine rods. A large number of varieties are grown. I inquired which were the best Peaches, and they are Noblesse, Belle de la Croix, and Grosse Mignonne; and for early forcing, Royal George. Nectarines – Elruge,

Imperatrice, and Pitmaston Orange. Mr. Grieve does not grow Pine Apple, but it is certainly superior to all of the orange-fleshed sorts.

Decorative plants are trained to the back walls in some of the houses, notably Habrothamnus elegans, which was plentifully covered with deep rose-coloured flowers. Begonia insignis, one of the very best for winter flowering, is much grown. Amongst the bedding plants wintered in these lean-to vineries were a large number of standard Zonale Pelargoniums, single and double-flowered. The plants are trained to a clear stem of 3 or 4 feet high; the heads are trained to a circular wire, and the plants supported by a stout stick when they are planted-out in the beds. These plants, many of them ten years old, are very useful for planting in the centres of beds and as single specimens on a lawn. Of bedding Pelargoniums the best white-flowered variety is a hybrid raised at Culford named Dolly Varden; it is a cross between a bronze Zonal and P. peltatum elegans. Bronze Harry Hieover is the best for small beds. Culford Rose and Culford Pink are both very useful varieties for vases, especially the former. Bridal Wreath, another hybrid Ivyleaf, is a favourite here.

The vineries contain the Vine described and figured in the Journal from which the bunch of Golden Champion was cut from a cane of Trebbiano. I did not see the bunch hanging, but those who did see it had no doubt whatever but that it was Golden Champion. It has also been found that Muscat Hamburgh succeeds remarkably well worked on the Frontignan stock. In the Lady Downe's house, where a very large crop was hanging, were a few rods of Mrs. Pince's Black Muscat, which is much liked for its good flavour, but it does not keep nearly so well as Lady Downe's. The Vines are trained on the double-rod system. A large house is devoted to Fig culture, the trees being trained to a trellis fixed under the glass and also against the back wall, the varieties most approved of being Brown Turkey and Raby Castle.

In the stove the beautifully-marked leaves of many different varieties of Coleus had an excellent effect. A fine specimen of Adiantum farleyense was noticeable. The old Centradenia rosea floribunda was flowering very profusely, and an excellent plant it is for decorative purposes during the winter months. In the greenhouse Coronilla glauca is largely represented; it will thrive under ordinary treatment, and is very useful either for cutting or decorative purposes. I also noticed a very handsome specimen of what seemed to be Clianthus puniceus, the Glory Pea of New Zealand, but it was of a much more robust growth than is commonly seen. It is evidently a variety of the species, and has been named Magnificent. The Clianthus is a good pot plant, but to see it in all its glory it should be planted in a conservatory border. The coolest treatment suits it, and if kept free from red spider by daily syringing when it is making its growths it will soon make a large plant or bush 12 feet high, which when loaded with clusters of its large crimson Pea-like flowers is an object of great beauty. I noticed in the conservatory a number of well-flowered specimens of Salvia splendens; their numerous spikes of brilliant scarlet flowers were very attractive. There was also a good stock of healthy plants of the double white Primula sinensis. This is rather a difficult plant to keep through the winter, but the flowers are useful for cutting, and they also remain in good condition on the plant for a very long time. Attached to the conservatory is a fernery under glass; it is formed of artificial rockwork, the Ferns being planted in nooks in prominent positions, where a sufficient depth of mould can be obtained; the groundwork is principally composed of Lycopodium denticulatum. I noticed the following Ferns: – Adiantum scutum and A. cuneatum, Nephrolepis exaltata; Lygodium scandens, the climbing Fern; Cibotium spectabile, &c.

There is a neat flower garden in front of the conservatory, but the principal part of the bedding-out is done at the kitchen garden. A series of beds have been laid-out in front of the gardener's cottage, which has been erected about the centre of the hothouses. The plants had all been cleared from the beds at the time of my visit, but under the management of such a master as Mr. Grieve they are no doubt worthy of Culford. – J. Douglas.

Editorial Endnotes

1 Rev. Edward Richard Benyon (1802–1883).

2 The history of the estate, through the Bacon family and the Cornwallis family exemplifies the build-up of wealth through acquisition of land and, in the case of the Benyon and de Beauvoir families, through colonial activity in India during the 18th century.

3 The first Marquess of Cornwallis (1738–1805) was primarily a military man who was appointed Governor-General of India (1786–1793) and introduced the 'Cornwallis Code'.

15

WOOD AND GARDEN

(Excerpt)

Gertrude Jekyll

Source: Gertrude Jekyll, ch. 14, 'Large and Small Gardens', *Wood and Garden Notes and Thoughts, Practical and Critical, of a Working Amateur* (1899), [171–188] excerpt pp. 179–183

Beyond this again would come the kitchen garden, and after that the stables and the home farm. If the kitchen garden had a high wall, and might be entered on this side by handsome wrought-iron gates, I would approach it from the parterre by a broad grass walk bounded by large Bay trees at equal intervals to right and left. Through these to the right would be seen the free garden of hardy flowers.

For the kitchen garden a space of two acres would serve a large country house with all that is usually grown within walls, but there should always be a good space outside for the rougher vegetables, as well as a roomy yard for compost, pits and frames, and rubbish.

And here I wish to plead on behalf of the gardener that he should have all reasonable comforts and conveniences. Nothing is more frequent, even in good places, than to find the potting and tool sheds screwed away into some awkward corner, badly lighted, much too small, and altogether inadequate, and the pits and frames scattered about and difficult to get at. Nothing is more wasteful of time, labour, or temper. The working parts of a large garden form a complicated organisation, and if the parts of the mechanism do not fit and work well, and are not properly eased and oiled, still more, if any are missing, there must be disastrous friction and damage and loss of power. In designing garden buildings, I always strongly urge in connection with the heating system a warmed potting shed and a comfortable messroom for the men, and over this a perfectly dry loft for drying and storing such matters as shading material, nets, mats, ropes, and sacks. If this can be warmed, so much the better. There must also be a convenient and quite frost-proof place for winter storing of vegetable roots and such plants as Dahlias, Cannas, and Gladiolus; and also a well-lighted and warmed workshop for all the innumerable jobs put aside for wet weather, of which the chief will be repainting and glazing of lights, repairing implements, and grinding and setting tools. This shop should have a carpenter's bench and screw, and a smith's anvil, and a proper assortment of tools. Such arrangements, well planned and thought out, will save

DOI: 10.4324/9780429198809-18

much time and loss of produce, besides helping to make all the people employed more comfortable and happy.

I think that a garden should never be large enough to be tiring, that if a large space has to be dealt with, a great part had better be laid out in wood. Woodland is always charming and restful and enduringly beautiful, and then there is an intermediate kind of woodland that should be made more of – woodland of the orchard type. Why is the orchard put out of the way, as it generally is, in some remote region beyond the kitchen garden and stables? I should like the lawn, or the hardy flower garden, or both, to pass directly into it on one side, and to plant a space of several acres, not necessarily in the usual way, with orchard standards twenty-five feet apart in straight rows (though in many places the straight rows might be best), but to have groups and even groves of such things as Medlars and Quinces, Siberian and Chinese Crabs, Damsons, Prunes, Service trees, and Mountain Ash, besides Apples, Pears, and Cherries, in both standard and bush forms. Then alleys of Filbert and Cob-nut, and in the opener spaces tangles or brakes of the many beautiful bushy things allied to the Apple and Plum tribe – *Cydonia* and *Prunus triloba* and *Cratægus* of many kinds (some of them are tall bushes or small trees with beautiful fruits); and the wild Blackthorn, which, though a plum, is so nearly related to pear that pears may be grafted on it. And then brakes of Blackberries, especially of the Parsley-leaved kind, so free of growth and so generous of fruit. How is it that this fine native plant is almost invariably sold in nurseries as an American bramble? If I am mistaken in this I should be glad to be corrected, but I believe it to be only the cut-leaved variety of the native *Rubus affinis*.

I have tried the best of the American kinds, and with the exception of one year, when I had a few fine fruits from Kittatinny, they had been a failure, whereas invariably when people have told me that their American Blackberries have fruited well, I have found them to be the Parsley-leaved.

Some members of the large Rose-Apple-Plum tribe grow to be large forest trees, and in my wild orchard they would go in the farther parts. The Bird-cherry (*Prunus padus*) grows into a tree of the largest size. A Mountain Ash will sometimes have a trunk two feet in diameter, and a head of a size to suit. The American kind, its near relation, but with larger leaves and still grander masses of berries, is a noble small tree; and the native white Beam should not be forgotten, and choice places should be given to Amelanchier and the lovely double Japan Apple (*Pyrus malus floribunda*). To give due space and effect to all these good things my orchard garden would run into a good many acres, but every year it would be growing into beauty and profit. The grass should be left rough, and plentifully planted with Daffodils, and with Cowslips if the soil is strong. The grass would be mown and made into hay in June, and perhaps mown once more towards the end of September. Under the nut-trees would be Primroses and the garden kinds of wood Hyacinths and Dogtooth Violets and Lily of the Valley, and perhaps Snowdrops, or any of the smaller bulbs that most commended themselves to the taste of the master.

Such an orchard garden, well-composed and beautifully grouped, always with that indispensable quality of good "drawing," would not only be a source of unending pleasure to those who lived in the place, but a valuable lesson to all who saw it; for it would show the value of the simple and sensible ways of using a certain class of related trees and bushes, and of using them with a deliberate intention of making the best of them, instead of the usual meaningless-nohow way of planting.

16

THE WASTED ORCHARDS OF ENGLAND

(Excerpts)

'The Gardeners' Magazine'

Source: George Gordon, 'Introduction'; Anon, 'Kent'; Anon, 'Staffordshire', *The Gardeners' Magazine; The Wasted Orchards of England* (1896), pp. v–viii, 1–5, 84–88

INTRODUCTION.

THE prominent position which the question of the fruit supply of the United Kingdom has occupied during the past twenty years, and is still occupying, is in a large measure due to the vigorous and continued efforts that have been made by the GARDENERS' MAGAZINE.

All aspects of fruit culture have been fully dealt with, but our most strenuous efforts have been in the direction of increasing the productiveness of the fruit garden and orchard, improving the quality of the fruit and the methods of storage and marketing, and bringing into prominence the importance of fruit culture as a national industry. For several years past we have devoted special attention to farm orchards, which represent the most unsatisfactory feature of English fruit culture. Through the apathy and ignorance of the owners and occupiers, these orchards have in innumerable instances been allowed to fall into such wretched plight as to have long since ceased to yield either pleasure or profit. It has, of course, been known to those specially interested in fruit culture that worn-out or otherwise worthless orchards exist. But we are fully persuaded that few even of the most observant of the experts have an adequate idea of their numbers, or of the wretched condition in which so many of them are in. Knowing from continuous observation in widely separated districts how terrible is the state of neglect in English orchards, and recognising the urgent necessity for a drastic reform in their management, we determined, as the best possible means of arousing public attention to their deplorable condition, to investigate those in the principal counties and publish the results. This important task has now been accomplished, and it is eminently satisfactory to us to have full justification for the belief that the reports of our Special Commissioner, which have involved much labour and a large expenditure of money, will greatly aid in effecting the much-desired reform.

 DOI: 10.4324/9780429198809-19

These reports, which are limited to a plain statement of facts, and the expression of opinion founded upon a thorough acquaintance with all phases of fruit production, have, as they have appeared in the columns of the GARDENERS' MAGAZINE from week to week, created an immense amount of interest, far beyond the boundaries of the counties whose orchards have been investigated. We have received many hundreds of letters having reference to them, and they have been freely commented upon in the provincial press, but in no instance have the statements of our Special Commissioner been called into question. No better proof of the fairness with which the investigations have been made could be desired, and we would here state that our object has not been to attract attention by sensational reports, but to quicken interest in the orchards, and suggest the means by which those in a wasted state might be improved. Many requests have been made to us that on the completion of the series we would publish the reports in a handy form for reference, and in compliance therewith we have decided to issue them as inexpensively as possible, that those who appreciate their value may not only be able to place the brochure on their book-shelves for purposes of reference, but may without any serious tax to themselves hand copies to those of their friends who have orchards, but have no adequate idea of their value for the supply of fruit for market or home consumption. There can be no question that it would be an immense advantage to the rural districts in which fruit growing can be carried on without any special difficulty, were the lessons in which the reports abound taken to heart and have practical expression given to them.

To us it is a deplorable state of things that, with the depression which has for so long overshadowed the agricultural interest in consequence of the low price of cereals, farmers should have almost completely neglected so important a source of income. We do not, of course, suggest that fruit culture should take the place of the growth of wheat or other of the staple crops, but we submit that it might be enormously increased with advantage to the owners and occupiers of land, and to those who have to obtain their supplies of fruit from the market. There are few English counties in which districts more or less favourable to hardy fruits cannot be found, and as so clearly shown by our Special Commissioner in his reports, there are many counties specially adapted to fruit culture for profit. For example, Kent, Devon, Worcester, Hereford, and Gloucester have a soil and climate so well suited to hardy fruits that they are unsurpassed in any part of the world. The large acreage of the orchards in these counties testifies to their being well adapted to apples, pears, and plums, and the excellent fruit obtained from many orchards that receive the scantiest attention justifies the assertion that from these counties tens of thousands of tons of apples should be annually sent to the markets. Could the owners of land in these counties be aroused from their apathy and fully recognise the possibilities of fruit culture in them, our markets would in a comparatively short time be made almost, if not wholly, independent of the hundreds of thousands of barrels that are annually sent to us from America. In Berkshire, Oxfordshire, Bedfordshire, Sussex, Hampshire, Nottinghamshire, and, indeed, in all the counties reported upon, there is room for the extension as well as the improvement of the orchards.

It is undoubtedly to the improvement of existing that orchards attention should be first directed, for even with a moderate change the value of the English orchards could be increased very largely, and by carrying out the reforms suggested in our Special Commissioner's reports their value would be at least doubled. Trees that have become so old and decrepit as to be past recovery should be swept away, for they serve no useful purpose, but simply become a harbour for insect pests of various kinds. All trees that possess a moderate degree of vigour might be greatly improved by having their heads moderately, and judiciously thinned, and their roots dressed with either natural, or artificial fertilisers, good farmyard manure being the best of the former, and superphosphate of lime, muriate of potash, or Kainit and nitrate of soda the best of the artificials. If the varieties are worthless, as in the majority of cases they are, and the trees are healthy the orchards can be converted into a profitable state at a small outlay by grafting them with the finer cooking or dessert varieties suitable for the country according to the suggestions in the report relating to it. No apple or pear tree in a thrifty state should be cut down because the variety is indifferent, but should be grafted, as such trees, when grafted rather high up to admit of a score of grafts being put in, form large and fruitful heads much more quickly than do newly-planted trees.

There is yet another aspect from which farm orchards may be regarded, and that is as a source of supply of wholesome fruit for the households of the owners. They might in most cases be made to supply the household with fruit of high quality for a period of from four to six months. But instead of their doing so, the supply is indifferent and limited to a few weeks only, and the deficiency made good by the purchase of fruit, American in all probability, from the neighbouring town. This is not a wise policy, and we trust the farmer who grows for his household as well as he who engages in its production for market, will, as the results of our efforts, be induced to take more interest in his orchards, and endeavour to make them profitable instead of being, as in innumerable instances is now the case, wasted and a blot upon the homestead.

THE EDITOR OF THE "GARDENERS' MAGAZINE."

KENT.

THE Editor of the GARDENERS' MAGAZINE rightly points the moral that, given cultural skill and intelligent methods, apples and pears may be grown in this country equal, if not superior, to the finest products of trans-oceanic orchards. Our soil and climate admit of possibilities which might well strike dismay into the heart of the foreign producer, and convince him how lucky a man he has been to gain command of our markets. It is not English soil or English air which have given the case of the home grower away; it is apathy and inertia, aided by slovenly cultural methods.

Are the weak points amended now? Are the wonderful examples of British-grown fruit which have aroused the astonishment and admiration of the most experienced growers and experts the sign and token of a new order of things, wherein intelligence, energy, and foresight have taken the place of slothfulness and neglect? Alas! the battle of progress is still only half won, and side by side with the trim and well-managed gardens we find orchards full of stricken and failing trees, a standing disgrace to British husbandry.

Although it is impossible to arouse an active and widespread expression of public opinion on a question such as this in the same way as it may be aroused on the condition of the army or the navy, it is nevertheless a national misfortune that English soil should still be cumbered as it is by thousands of acres of worthless fruit trees. And it was with a deep sense of the loss and discredit reflected on a great industry that I was instructed to make my way from village to village in some of the most important fruit growing counties, and to give a plain, unvarnished tale of what I saw in my wanderings. My duty lay clear before me. It was to give a report, gilded with no rhetorical embellishment, and based on the evidence of my own eyes alone, on the condition of our English orchards, and, where circumstances demanded it, to frame an indictment against those senile and turgid cultivators who refuse to read the signs of the times. If these lessons of the lanes and byways do anything to remedy a condition of things that forms a standing menace to the prosperity of the nation, and darkens the cloud of want which hangs over the rural labourer, their performance will remain a pleasant instead of a troubled memory.

The train that took me into the "Garden of England" was enlivened by the cheerful conversation and tender melodies of a few hundred hoppers. They boarded it in force at St. Paul's, and simultaneously lifted up their dulcet voices in fearful and mysterious harmony. But they were not unwelcome, for I wanted some information of the road, and your seasoned hopper, albeit he only goes down once a year for a month or so, knows almost every stone. I alighted at Shoreham, proposing to make my way round to Halstead, Knockholt, and Cudham. This is a great fruit district, and some of the hoppers had worked it for fruit earlier in the season. An orchard of standard apples on the right of a long hill soon gave me material for a survey. The field was one of some twenty acres, and the trees seemed to be about fifteen years old. I was struck by their miserable appearance. In the whole field there was hardly one that looked fresh and healthy. With but few exceptions they were dying of canker, being covered with great gaping wounds and leafless branches. Fruit, it is true, they bore, or some of them at least, but in appearance and quality it was of the most wretched description. It will go into the market of course, stunted, specked, and discoloured as it is, and there it will come into competition with large, clean, fresh-coloured American produce, which will infallibly thrust it into the background.

What, it may be asked, was the cause of what ought to be an exceedingly valuable crop being practically worthless? Simply this. In defiance or ignorance of the most common cultural laws, the planter of these apples had chosen delicate varieties, and had planted them in deep holes in a heavy and damp soil. For more than half the year the holes were simply water-butts, in which moisture collecting and turning stagnant had killed the young roots and deprived the trees of nourishment.

For years they had been going backwards instead of forward, dying, in fact, of starvation. The capital invested in them has been wasted, and the fruit they bear is a mere caricature of imported produce.

I worked my way to the farm buildings, and found, as I expected, that they were in a mournful state of barrenness. I ran my eye over stables, barns, and cartsheds. There was at least two thousand square feet of wall space, not an inch of which was utilised. On that north aspect there are broad expanses suitable for Morello cherries and cordon gooseberries and currants. On just such a west wall on an Oxfordshire cottage I have seen fine apricot trees, the produce of which rarely fails to pay the rent; but there are no apricots here. The wall of the farmer's home garden is a brick erection some four feet high, and the border at the base is smothered with vigorous weeds. From soil that an allotment-holder complained was too heavy to grow anything, I once pulled scores of thistles four feet high, and I could do it again here. Yet, against just such a wall, I know of tomatos laden from top to bottom with ruddy fruit, each plant averaging eight to ten pounds of it, but occupying only two feet of lateral space. It has been a grand tomato year, but there was not a plant to be seen about this farmstead, any more than there were apples or pears, currants or gooseberries, apricots or peaches, cherries or plums.

It was a natural concomitant to this state of affairs to find, when I made my way to the village for a midday meal, that wholesome vegetables and fruit were impossible to obtain. I called at three inns, one after the other, and the burden of the story was the same, though given in varying degrees of politeness. One worthy landlady listened to my requirements with an ill-concealed contempt. Another was considerate, but equally hopeless. There were no vegetables to be had in the village, she said; no fruit, no tomatos. Questioned as to the possibility of fresh fruit being welcomed by many if it could be got in the hot weather, she admitted that it was quite likely, but anyway, there wasn't any. The idea of steps being taken to provide a supply evidently never occurred to her. After some weeks in the different villages, I have found this state of affairs nearly universal. Fresh vegetables, notwithstanding the plenitude of gardens, can rarely be had. Fruit is scarcer still. Yet, in passing cottage homes and allotments, I saw peas drying on the haulm, kidney beans yellowing, marrows developing to a monstrous size, and plums falling from the trees to rot. There is want on one side, waste on the other, and all from lack of a system of distribution. It is a sad story, but a true one, as I shall give further evidence to prove.

There is an extensive fruit-growing district between Canterbury and Sandwich. The route to follow from the Cathedral city, and a most pleasant one it is, is through the villages of Littlebourne, Wingham, and Ash. Not far from the former place I came upon as instructive an example of the noble principle of how-not-to-do-it in fruit-growing as could be found in the whole kingdom. A peculiar whiteness on the trees, as though they were silvery with a fall of summer snow, attracted my attention, and on getting amongst them I found, as I expected, that they were attacked by American blight. But what an attack! The sight of those trees, thick with heavy masses of the loathsome pest on trunk, limb, branches and spurs, will not quickly fade from my memory. I have heard it hazarded that American blight sometimes leads to canker. There is not the faintest doubt about it. Next to bad planting it is probably the most fertile cause of all, and when you get a combination of both, as there was in the case under notice, but one result can follow, and that is, bad fruit in the market, loss to the grower, damage to the credit of British fruit.

I got into conversation with a friendly and intelligent grower in the district about the orchard, and ventured on a little guess as to the way in which it had been planted, of which the summary is as follows: A blunt spade to trim in the roots, a hole about large enough for the head of the planter (which ought to have been stuck in it for an hour or two to teach him better), a twisting of the roots to get them in, and a final application of the huge foot of the heaviest labourer. The grower began to smile long before I had finished, and told me I had hit it to a hair. He had seen the trees put in with his own eyes nearly fifteen years ago, and had strenuously warned the planter that he was going the wrong way to work, for there was only the remotest chance of the trees succeeding. However, the farmer was one of the old school, and was set on getting the trees in as quickly and cheaply as possible. If he could have drilled them in with a machine he would have done it. They never had any more care. A man who would not listen to common-sense suggestions about planting his trees was hardly likely to go to any trouble or expense in order to keep them clean. Every noisome pest that came along was allowed to have its little fling, and one after another they bit and sucked to their heart's content. But the American blight came to stay. When it had got tired of one patch it tried another, leaving the canker to take its place, and right nobly the latter strove to finish off what its fellow-worker had begun. The trees were covered with huge rugged wounds, and hundreds of large branches thrust their naked and withering stems into view. The trees had endeavoured to bear a crop, but half of it lay on the ground, rotten to the core through the operations of grubs of the Codlin Moth, and the rest would have been dear at nine-pence a bushel.

We have heard the old story about plums rotting on the trees again this year. We shall hear it again. It is a hardy perennial, and it will spring up regularly until – well, until some of the hundreds of thousands of trees, the produce of which a wise public will hardly buy at any price, are swept away. There are plenty of them about Ash and Sandwich, and occasionally the crop pays for the picking, but not often. I said "swept away"; I had better add, "or grafted with sorts that the people want." I have never yet heard of Rivers' Early being left to rot on the trees, although it is common enough with the Small Black that comes with a fearful glut about mid-season. Rivers' comes in at the time when plums are most wanted, and these are the things which pay. To it might be added Monarch, a good paying plum as a rule. I saw neither of these about Ash, but I saw acres of the common Black, which is very little better than a good damson in appearance, and has not a particle of the brisk, enlivening flavour of that modest but useful fruit.

It was in the Sandwich district that I dropped upon such a specimen of a small holder as I had long wished to see. He had about ten acres, devoted mainly to strawberries and bush fruit, but including some vegetables. He was not an old hand at market work, which, perhaps, accounted for his freely confessing that he still had something to learn, and also for the still more remarkable fact that he had nothing to say about Free Trade, railway rates, and land charges. He simply pumped me for all he was worth, and told me that he had benefited greatly by County Council technical lectures. The ten-acre man sent very little stuff to London; it paid him better to dispose of it in the seaside towns thereabouts, such as Deal, Ramsgate, and Margate. He was paying his way, he told me. He had a small cottage on his plot, and did nearly all the work himself. He found his strawberries answer pretty well; indeed, was so satisfied with them that he was going to take a few acres more ground to increase his plants, when he hoped to get a comfortable living out of the little place. There is this to be said about his chances. Although – or, perhaps, I should say because – he has not been trained to the business, he does his strawberries much better than his bigger neighbours do theirs. Market men have told me that they do not count Sandwich strawberries as Kentish ones at all, which is rather an unkind remark, although it may be deserved.

On the Ashford side of Maidstone I saw a good many gooseberries, but they were far inferior, taken as a whole, to those which are to be found over the hills at Sittingbourne. The favourite method of gooseberry culture with many of the farmers hereabouts appears to be to divide a worn-out old stool, and plant the divisions for new stock. I came across one field of about ten acres, in which every bush was three parts composed of suckers that had sprung up from the base, and a more melancholy lot I have not often seen.

Gooseberries are not usually a particularly profitable crop, but they appear to be a sure and useful one, doing a little more than paying expenses, besides keeping the pickers going until they are wanted for the great bulk of the fruit crop later on. Sensational prices are heard of now and then, but in a general way a modest balance on the right side about represents the return, and in these times it is entitled to respect; the gold-mine days of fruit-growing have gone by. There is this one great danger about a steady-going crop like gooseberries – growers are apt to think that it will manage itself, that the bushes want no treatment, and that the picking is a humdrum affair which may be left to the foreman. Hence it is that we see those masses of suckers, incapable, on the face of them, of giving fine fruit, and a source of slow and disagreeable work to the pickers. Fruit growers of the take-it-easy school are unaware that it is as cheap on the first cost to raise a good gooseberry bush as a bad one, and immeasurably cheaper in the end; and, what is more, they do not want anybody to tell them, invariably falling back on the tawdry, threadbare argument of "practical experience." I have seen enough in my rambles among the orchards of Kent to make my temper rise at those words. As a rule they are the stand-by of the ignorant, the obstinate, and the incompetent.

As there are cultural so there are market points about small fruits such as this. I have frequently seen growers setting-to at their bushes before the crop is half ready, and rushing baskets of pea-like fruit off to market in red-hot haste. I suppose they have something in the way of return to justify them, but, all the same,

I have the evidence of one of the foremost and most successful growers in the country that it pays him better to develop his fruit to its best size before letting it go. It takes a large number of gooseberries to fill a sieve when they are about the size of haricot beans, and a good many bushes have to be stripped to make much of a show; but the developed produce of thinly-trained, clean, and well-fed bushes tells quite a different story.

STAFFORDSHIRE.

NOTWITHSTANDING the high natural fertility of a large portion of the land in Staffordshire, very little is devoted to market gardening and fruit culture. This seems very regrettable, because there are thousands of acres well adapted for either or both purposes. Fruit culture, in the best sense of the term, is practically non-existent, and there is, in my opinion, a splendid chance here for men with the necessary skill, capital, and enterprise. There are a few acres of fruit trees about Tamworth, Lichfield, and Denstone; the remainder consists chiefly of old and worn-out orchards – of the usual dimensions – attached to farm-houses. I am fully persuaded that landlords would act very wisely in their own interests, as well as in the interests of their tenants, if they were to encourage more attention to fruit culture where the soil and situation are adapted to it. I am aware that landlords generally cannot afford a great outlay on their estates, but it is quite as much to their benefit in the end to keep up a good orchard on a farm as to maintain good barns. In other words, it is as essential to the permanent welfare of an estate, and as profitable an outlay, to repair old orchards with young and good varieties, or form new ones entirely, as it is to repair buildings or construct new ones.

In the course of my inspection of the Staffordshire orchards, I found that there are large numbers of worthless varieties of apples in them, just as there are in orchards in other counties which have been reported upon. The small and inferior fruit, samples of which I saw in Uttoxeter market, produced by these trees, cannot be profitable. So far from this being the case, the trees only cumber the ground. I may fairly be asked by the owners of orchards in Staffordshire, If you condemn some of our varieties of apples, what varieties would you substitute for them? My reply is, plant a large number of trees of the varieties of sterling merit which are known to succeed in various parts of the county; and they are Keswick Codlin, Lord Suffield, Dumelow's Seedling (Wellington), Warner's King, Ecklinville Seedling, Bess Pool, Cellini, King of the Pippins, Worcester Pearmain, Annie Elizabeth, and Northern Greening. To these I would add White Transparent, a splendid early and free-cropping variety, Stirling Castle, Lane's Prince Albert, Bramley's Seedling, Tom Putt, and Pott's Seedling.

In commencing my inspection of the orchards of Staffordshire I made Birmingham my starting point and Uttoxeter my centre. Leaving New Street Station by the Midland Railway for Tamworth I soon passed on the left the immense sewage farm at Saltley belonging to the Corporation of Birmingham. From this farm many tons of vegetables are sent into Birmingham, and not entirely to the benefit of the market gardeners of Tamworth and Lichfield. From Saltley a purely agricultural district, watered mainly by the river Tame, is reached. This part of the county forms a broad valley, mainly in grass, and widens out until Burton is reached, where Needwood Forest (Crown property) intervenes, and the valley there divides into two, the one to the right developing beyond Burton-on-Trent into the fertile valley of the Dove, and the one to the left following the course of the Trent, in the smaller and earlier stages of that beautiful river, by Lichfield, Rugeley, and Stafford, up to and beyond Stoke-on-Trent. These two valleys comprise the greater part of the most important agricultural districts of the county.

The orchards between Birmingham and Tamworth consist of the usual type of farm-house orchards. That is to say the trees are few in number, mostly old in years, starved and neglected, with a sprinkling of young trees planted among them, to become prematurely old in their turn.

At Tamworth I found, as might have been expected, the fruit trees in better condition, but fruit culture is not carried out on a large scale there. Some four or five hundred acres of land are devoted mainly to vegetable culture, the chief crops being celery, broccoli, onions, cauliflower, and cabbage. There are several good orchards, and these are in the occupation of Messrs. Jenders, Pickering, Rowley, and others. Mr. Pickering grows Keswick Codlin Apple in large quantities, having gathered several thousand bushels in 1895. This variety does well generally in the county of Stafford and usually commands a paying price. He also speaks well of Dumelow's. His favourite pear for market is Marie Louise, followed by Williams' Bon Chretein; he has obtained as much as 25s. per bushel for the latter when the sample was fine and pears scarce. Catillac bears most abundantly with Mr. Pickering. I saw one tree carrying an immense crop on an old orchard standard. Mr. Rowley finds Jargonelle a good paying pear, having this year obtained 12s. per bushel for it as against 6s. for Williams' Bon Chretien. He speaks most highly also of Louise Bonne of Jersey and White Lammas; the latter are grand trees. Of plums Victoria is the favourite around Tamworth for market purposes and the most reliable, but Pond's Seedling, Coe's Golden Drop, and Jefferson succeed in the same district.

Proceeding on my way towards Burton-on-Trent and Uttoxeter I pass Elford, where the varieties of fruit just mentioned succeed equally as well as at Tamworth, and where a variety of apple is grown that is not common. It is a good bearer in that part of Staffordshire and in North Worcestershire, and is locally known under the name of Low Bough, but its proper name is supposed to be Grey Pippin. It is not only a prolific bearer but it is rather a handsome apple, good for cooking or dessert, and keeps well until February. The junction of the river Tame with the Trent is soon reached, and passing by Barton on the left and Walton on the right, I found the whole district dotted with farm homesteads and small orchards, the latter with but few exceptions being unsatisfactory and containing a few good varieties and many worthless ones. The best varieties are substantially the same as those mentioned previously and the soil and subsoil being the same generally, viz., a fair depth of loam with gravel below, and decidedly suitable for fruit culture.

Leaving Burton-on-Trent with its gigantic breweries and immense pyramids of empty barrels, behind, I entered the valley of the Dove, where the landscape gradually assumes a richer and more fertile appearance, and where practically all is old pasture devoted to grazing and meadow land for hay. Hardly an arable field is to be seen for miles; so few are they that it appears as though there were none, and this continues for some sixteen miles by Tutbury, Sudbury, Uttoxeter, and Rocester, until the Dove and the Churnet unite, and the most hilly and beautiful part of Staffordshire is reached.

There are the usual orchards around Tutbury, but nothing of special interest in that way, the majority being old and dilapidated. At Sudbury there is a large dairy established a few years ago by Lord Vernon for the benefit of the farmers of the district. This was the first dairy of the kind established in this country, and, of course, it came in for the usual amount of adverse criticism. Would that some landowners in the valley of the Dove had the same amount of courage to promote fruit culture in their districts; not necessarily upon their richest grazing pastures, but upon the lower slopes about Hanbury, Marchington (where some fine apples are grown), Doveridge, Eaton, and Rocester! Fruit culture would undoubtedly prove as great a success as in any part of the kingdom, greatly to the benefit of the landlords and the population generally. Properly planted and intelligently managed, the orchards of the Valley of the Dove would compare very favourably with the best orchards of the Valley of the Thames or the Vale of Evesham, and prove equally as profitable.

Uttoxeter is reached five miles beyond Sudbury, and is an important market town of about 6,000 inhabitants. There is no fruit culture here, except such as is found in private gardens, although all kinds of fruit succeed in the neighbourhood. The air is pure, the water abundant and of excellent quality, and the soil is rich and warm, resting mainly upon gravel, and there can be no doubt that orchards properly planted would prove profitable. Of apples, Keswick, Lord Suffield, Wellington, Tom Putt, Cellini, Warner's King, Blenheim, Worcester Pearmain, Court Pendu Plat, and Yellow Ingestre all do well; good pears are Williams' Bon Chretien, Marie Louise, Louise Bonne de Jersey, Jargonelle, and Doyenne du Comice; favourite plums are Orleans, Victoria, Cox's Emperor (locally known as "Dove Bank"), Jefferson, and White Magnum Bonum.

Walking from Uttoxeter towards Rocester, I passed through the hamlets of Spath and Combridge, where there are a few small orchards mainly devoted to apples. I may here remark that practically all the Staffordshire orchards are on turf, and the grass eaten off by stock; also, that except the manure obtained from the stock in this way, the trees here, as in other counties under the same circumstances, do not receive any assistance, and therefore it is not surprising that so large a proportion should present a starved, worn out and unprofitable appearance.

In the neighbourhood of Rocester are some old-established orchards of apples and damsons; these are principally in the hamlets of Denstone and Stubwood. The damsons once formed a very important crop, and were a great source of revenue to the farmers and cottagers of the district. Buyers from Lancashire secured the fruit year after year for the purpose of obtaining dyes therefrom, and paid from fifteen to twenty-five shillings per "strike," or bushel in average years; and when damsons were very plentiful they seldom fell below ten shillings per bushel. This having been the case, it will be readily understood that several good free-bearing damson trees were a great help to a cottager. Aniline dyes have spoiled this source of income. Proceeding on my way through Denstone and its damson orchards, I crossed the river Churnet, which enters the Dove near Rocester, and passing through the Quixall Lodge, we enter the famous demesne of the Earl of Shrewsbury – Alton Towers. Here I called upon Mr. Gilman, the head gardener, and in discussing the question of fruit culture with him, he told me that the varieties of apples which best succeed there are Lord Suffield, Warner's King, Bess Pool, Ecklinville Seedling, Bramley's Seedling, Newton Wonder, and Worcester Pearmain; and he places Victoria plum first, then Kirke's and Green Gage. Pears: Williams' Bon Chretien and Louise Bonne of Jersey. From Alton up to Leek the country becomes very high and cold, and increases in picturesqueness, but it becomes less fruitful, and fruit crops become more scarce and orchards few and far between.

Returning to Uttoxeter, my next journey was up the valley of the Blythe, where there are fruitful orchards within four miles of Uttoxeter. These are at Bromshall, Loxley, Highfields, and Leigh, and are attached to farms. Apples do well, especially in those orchards situate in miniature valleys between the low hills. I next reached Cresswell, Blythe Bridge, and Normacot, which are just within the smoke radius of the Potteries. As the Potteries are approached vegetation becomes less luxuriant, and orchards become scarcer until they cease to exist until Tunstall is reached, when they again begin to appear.

Turning towards Birmingham again, we soon have Keele in view, a charming little village belonging to the Sneyd family, of Keele Hall. Here hardy fruits succeed fairly well, especially where they receive a little shelter. Newcastle-under-Lyme is soon reached, and from thence to Trentham. Hereabouts the local orchards, considering the deleterious influence of the very impure atmosphere, look remarkably well. The apples, pears, and plums appear healthy and fruitful, and judicious planting of young trees is kept up from year to year as circumstances require. Stone is soon reached from Trentham, and is more famous for its beer than its hardy fruit. Orchards in the neighbourhood are in an unsatisfactory condition, and some effort should be made to improve them. The orchards between Stone and Stafford are mostly old and worn out, like too many in the county, and the unproductive trees met with are living examples of the apathy and ignorance that so generally obtains in this and so many other of the English counties. Weston, Ingestre, and Tixall are a short distance from Stafford on the east side, and Gnosall and Eccleshall on the west. The orchards look tolerably well at Weston, Gnosall, and Eccleshall, but elsewhere they are unsatisfactory. Journeying down the Trent valley to Lichfield, I found that, while market gardening is largely carried on, not much ground is devoted to orchard fruit. The small area is, I am pleased to say, occupied chiefly with young and healthy trees of approved varieties. The most popular apples hereabout are King of the Pippins, Irish Peach, Blenheim Pippin, Bramley's Seedling. Normanton Wonder, Annie Elizabeth, and Northern Greening, Victoria Plum is the favourite here as everywhere else in the county, and Rivers' Early Prolific is winning its way into favour. Williams' Pear holds its own here against all comers. The orchards between Stafford and Wolverhampton are of the usual character, or, in other words, they, with few exceptions, consist of old trees that have become exhausted and unprofitable.

On the whole, the orchards of Staffordshire are in an unsatisfactory condition. A few are good, and giving a fair return to the owners, but many are so bad that they ought to be destroyed forthwith, and replaced by new ones formed in accordance with the praetice of the more advanced fruit growers. Staffordshire is not, generally speaking, so well adapted for fruit culture as the highly favoured county of Kent; but there are large areas in which the industry might be carried on with profit to the landlords, the tenants, and the workers in the district. The general community also would derive considerable benefit from having the markets supplied with an abundance of fresh fruit of high quality produced within the county.

c)

GREENHOUSES AND CONSERVATORIES

Editorial Headnotes

17. J. C. Loudon, 'Preface', 'Introduction', 'Of the placing or arrangement of the Plants in a Green-house', *The Green-House Companion* (1824), [v–viii] excerpt, pp. v–vii, 1–5, 135–142
18. John Dillwyn Llewelyn, 'Some Account of an Orchideous House constructed at Penllegare, South Wales', *Journal of the Horticultural Society of London*, vol. 1 (1846), pp. 5–6
19. Edouard André, 'Conservatories in the Natural Style' and 'A Cool House', *The Garden* (20 January 1872, 17 February 1872), pp. 181–184, 288–291
20. Shirley Hibberd, 'Introduction', 'The Conservatory and Winter Garden', *The Amateur's Greenhouse and Conservatory* (1873), pp. 1–3, 215–219
21. Rhoda Broughton, *Not wisely, but Too Well* (1867) (excerpt), pp. 107–108
22. Frances Jane Hope, 'Sweet-scented Greenhouse Plants', *Gardeners' Chronicle* (12 May 1877), pp. 590–591, 'Interesting Stove Plants' (22 March 1879), pp. 366–367
23. H. Harland, 'Magic Grottoes and Show Ferneries', *Gardeners' Magazine* (23 December 1876), pp. 699–700

The greenhouse was intended for protection and the conservatory for the more permanent display of plants. Greenhouses were generally separated from the house while the conservatory might be attached to it, although the latter might serve both functions. These sources represent variations in scale and, apart from the 'Orchideous House' at Penllergare, generally focus on the 'temperate' or 'cool' houses (conservatories). For the evolution of construction, glass, metal, wood, and heating systems, see Volume 4. The sources share tropes of glasshouses as spaces that are compatible for the education and pleasure of all family members and as recreations of nature and of the 'other' landscapes in British gardens, a phenomenon that is amplified by some of the illustrations published alongside the texts. While the practice of naming authors was becoming more widely established in the press from the 1880s, the naming of the 'artist' of listed illustrations was rare (with the exception of botanical plates), although some are signed with initials. Further identification, however, has not been possible in most cases.

J. C. Loudon's *Greenhouse Companion* (1824), published anonymously, is compact and has few illustrations; it went through three editions (1824, 1825, 1832). It was intended for a broader readership than his earlier, more expensive publication for the landed gentry, *Remarks on the Construction and Hothouses . . .* (1817). Loudon's claim for the greenhouse as 'an appendage to every villa' needs to be understood in the context of the price of exclusive suburban housing at this date and of the cost of a greenhouse as noted in the introduction. Note that a 'commercial establishment in the King's Road' refers to that founded by Joseph Knight (1778–1855) from 1808, later known as the Exotic Nursery, which traded for almost half a century and was later sold to James Veitch.

The fluid meanings of the 'natural' and of 'nature' are evident in these sources. Loudon argues that the greenhouse is 'entirely a work of art' because it is an artificial or man-made environment, as opposed to the 'natural' environment from which the plants in the greenhouse originated. He urges readers, however, to plant according to the 'Natural' system, in the sense of scientific, botanical orders of Antoine Jussieu defined by their visual resemblance (the Linnean system is the most artificial). In contrast, the approach of John Dillwyn Llewelyn in 'Orchideous House, constructed at Penllergare, South Wales', *Journal of the Horticultural Society of London* (1845/1846) over two decades later is to recreate the natural conditions and foliage of South America. The lush environment, based on the accounts of Robert Schomburgk (see Volume 3), is indicated by the illustration (signed with the monogram 'H'). Dillwyn Llewelyn (1810–1882) inherited wealth and was a pioneering photographer and botanist whose estate was located near Swansea, South Wales. His photographic samples (as opposed to dried plants or drawings) of orchids that he sent to Kew for identification were among the first of their kind.

A series of articles on 'Conservatories in the natural style', *The Garden* (1872), opens with a reference to Llewelyn's orchideous house to evoke the natural environment of the 'other' and to consider the distribution of plants in their place of origin (the lists of plants featured in the articles and the ground plans are not included here). The high tropical planting of the well-known stoves and large-scale glasshouses at Chatsworth and Kew that had predominated up to the 1850s entailed expensive heating bills. The emphasis on temperate planting by the author of the articles, Edouard François André (1840–1911), a French horticulturalist and leading landscape architect of the late 19th century, follows the trend set at the Crystal Palace (1851/1854), Enville Hall (1854), and the Temperate House at Kew (1862). Intermittent references to those of 'moderate means' and possible economies are somewhat undermined by the scale of the plans of conservatories of 60 feet in length; *The Garden* was keen to attract a mixed audience of professionals and wealthy amateurs in its first year of publication. Hibberd's chapter on conservatories in *The Amateur's Greenhouse and Conservatory* (1873) proposes a modest-scale version of this luxury addition to the home.

The sense of the erotic stimulated by the plants and perfumes of 'other' landscapes is suggested by the novelist Rhoda Broughton in a short extract from her second published novel *Not Wisely but Too Well* (1867). The scene is set for a kiss in the conservatory between the heroine Kate Chester and the self-centred Dare Stamer. Broughton (1840–1920), was one of the late 19th century's best-selling novelists. Her early works were associated with the controversial genre of 'sensation' fiction. Frances Jane Hope (1822–1880) supplied the names, perhaps, for the plants of Broughton's fiction. Hope's article 'Sweet-scented Greenhouse Plants' first appeared in *Gardeners' Chronicle* (1877) and was published posthumously in a collection of her writing, *Notes and Thoughts on Gardens and Woodlands* (1881). The granddaughter of a professor of botany at Edinburgh University, she was a botanist and garden journalist in her own right. Her access to the greenhouse (as opposed to the scientific institution or laboratory) fits with John Loudon's early 19th century recommendation of domestic glasshouses as liminal spaces suitable for women. The relative obscurity of Hope and the generally unacknowledged scientific and aesthetic contribution made by women to horticulture in the period are largely due to their operating in a domestic environment.

An alternative 'other' landscape is evoked by the illustration (artist unknown) for the transformation of a greenhouse in a practical article 'Magic Grottoes and Show Ferneries' (1876) by Henry Harland (n/d). An appropriate title for the Christmas edition of the *Gardener's Magazine* (1865–1916), the conjunction of magic, fairies, and ferneries was part of later 19th-century popular culture. Hibberd was editor at this date; 'Songs of the Fairies' was the title of a poem in an early collection of his poetry, *Song of Songs*, published in 1851.

17

THE GREEN-HOUSE COMPANION

(Excerpts)

J. C. Loudon

Source: J. C. Loudon, 'Preface', 'Introduction', 'Of the placing or arrangement of the Plants in a Green-house', *The Green-House Companion* (1824), [v–viii] excerpt, pp. v–vii, 1–5, 135–142

PREFACE.

A GREEN-HOUSE, which fifty years ago was a luxury not often to be met with, is now become an appendage to every villa, and to many town residences; – not indeed one of the first necessity, but one which is felt to be appropriate and highly desirable, and which mankind recognise as a mark of elegant and refined enjoyment.

The taste for these exotic gardens, indeed, has increased much more rapidly than the skill requisite to manage them to the best advantage, – for the progress of imitation is more rapid than that of knowledge; and hence it is much more common to see a green-house, than to see one filled with a proper selection of plants in high health and beauty.

The management of plants in a green-house requires a higher degree of knowledge, than is called for in the management of the open garden; and though this knowledge is fast extending among the rising generation of gardeners, it is not yet in such abundance as to be general.

The object of the GREEN-HOUSE COMPANION is to supply what is wanting in this respect, not only to gardeners, but to their employers. In composing it we have had in view the twofold object of informing such gardeners as may not have attended much to this branch of their profession; and of enlightening their employers so as they may not only be aware when their servant neglects his duty, or performs it properly or improperly; but may, in the case of employing common labourers in their gardens, be able to direct all the operations themselves. The directions we have given, and the reasons for them, are so ample, and so plain, that no lady or gentleman can be at a loss fully to comprehend them, or discern when they are properly acted on.

Though we have chiefly had in view, in preparing these directions, the villa and town green-house; yet we have treated the subject of management so amply, and given so complete a catalogue of green-house plants, with all that it can be desirable to know as to their culture and management, that we have no hesitation

DOI: 10.4324/9780429198809-21

in presenting our work as a treatise on green-house culture and management as complete as the present state of things admits of. We have composed it in conformity to the best written authorities, and according to what we have seen in our extensive observation and communications with botanical cultivators and nursery-men, in the neighbourhood of London and elsewhere. The work has also had the advantage of revisal by a nurseryman who is extensively engaged in the culture of green-house and hot-house plants at his commercial establishment in the King's Road. This gentleman's chief business is with exotics, and he supplies and manages town green-houses, and executes London contracts for plants in pots, to a considerable extent: he is therefore peculiarly fitted for the task he so obligingly undertook at the request of the publisher. To mention his name he considers would look like a partial advertisement.

There is one advantage which this work possesses over every other of the kind, which, as it may not strike at first sight, we shall here point out. It is the arrangement of all the green-house plants of this country under the natural orders of the system of Jussieu.

INTRODUCTION.

THE most refined enjoyments of society have gradually arisen from desires more simple, and even from wants. Man is fond of living beings, and after assembling those plants around him which he found necessary for food, he would select such as were agreeable to the eye, or fragrant to the smell. A flower in the open parterre, though beautiful and gay, has yet something less endearing, and is less capable of receiving especial regard, than a plant in a pot, which thus acquires a sort of locomotion; and becomes, as it were, thoroughly domesticated. After choice things were planted in pots, things rare would be planted in them; and from things rare to things rare, foreign and tender, the transition would be natural and easy. Tender rare plants in pots would be taken into the house for shelter, and set near the window for light, and hence the origin of the Green-house.

In what age of the world, and in what country a green-house first appeared, it is impossible to determine; it is sufficient to have shown that a taste for this appendage to a dwelling is natural to man; to experience that it adds to his enjoyments; and to feel that it bestows a certain claim to distinction on its possessor. A green-house is in a peculiar degree the care of the female part of a family, and forms an interesting scene of care and recreation to a mother and her daughters, at a season of the year when there is but little inducement to walk in the kitchen-garden, and nothing to do in the parterre or the shrubbery. The progress of vegetation, interesting in all scenes, and at all seasons, is more especially so in a green-house during winter. There the objects are of limited number, brought near the eye by their position, and rendered striking by their contrast with the cold, naked, and dreary scenes which are shut out: – then it is that the genial climate, the life and growth, the deep tone of verdure, and the prevailing stillness of repose within, cause this winter garden to be felt as a luxuriant consecration to man.

But all green-houses do not yield the enjoyments which a green-house is calculated to produce; because all are not well contrived, or judiciously managed for that purpose. Some do not know what a green-house will, and what it will not afford; and others expect all its peculiar enjoyments without their accompanying cares. Some erect a green-house of such a form and position, that the plants within can never prosper; others in such a situation, relatively to the house, that if they prosper they can never be enjoyed; and not a few think they have done every thing when they have completed the construction, stocked it with plants, and committed it to the future care of a house servant or mere out-door labourer.

But a green-house, if it is to be worth any thing, must not be trifled with in this manner. It is entirely a work of art: the plants inclosed are in the most artificial situation in which they can be placed, and require constant and unremitting attention to counteract the tendency of that artificial state to destroy them. It is a common notion that a plant in a pot is in a safer state than a plant in the open ground; but this is a most erroneous notion, and directly the reverse of the fact. Placing plants in pots is often more convenient for the cultivator; but it always checks and counteracts the natural habits of the plant: it checks the extension of the roots, and, by consequence, of the shoots; and it subjects these roots to be alternately deluged by water, and dried up for want of it, – and all this under the best management. Under a careless gardener, if the pots are not properly drained, and this drainage kept in repair, the soil will be kept soaked in water till the roots are rotted; or by neglecting to shift the plants to pots of a larger size at proper seasons, the roots will get matted so as to derive no benefit from the soil, preclude the water from entering, and thus first stint and then kill the plant. But plants in a green-house are not only in an artificial and injurious state as to the soil, but also as to their climate, and especially as to heat, light, and air. These requisites to vegetation require also to be particularly attended to, so as partly to imitate nature, and partly to effect particular purposes of art on natural principles. Finally, it must be obvious, that where there is so much art, there must be a greater tendency to disease and accident than in ordinary nature; and consequently, that no small degree of vigilance is required in this respect.

Let none however feel alarmed at these obstacles, or imagine that by a moderate degree of care and attention all these evils may not be avoided, and the enjoyments of the green-house fully obtained. It is the object of this little work to enter into the details of the subject in such a way, as may enable any lady or gentleman, with the assistance of a footman or common labourer, to manage their own green-house as completely as if they employed a regularly bred and skilful gardener.

Of the placing or arrangement of the Plants in the Green-house.

The stock for the green-house being got together, the next thing is to dispose of them on the stage in the best manner. One common rule is, to place the smaller plants in front, and the larger behind; this is most generally proper, because it brings the smallest plants nearest the light. In some cases however, where there is a front wall which reaches higher than the lowest shelf of the stage, or which throws a shadow on it, it is better to place the smallest plants three or four shelves high on the stage, and plants of a larger description on the bottom shelves. This more effectually attains the object of getting the smallest plants nearest the light, and also brings them near the eye.

A second general rule is, to mix all the different plants as thoroughly as possible, so that no two of a species, or even if possible of a genus, may be seen together. The object of this mixture is to produce variety; but a little reflection will convince any one, that instead of variety it produces a sameness of mixture exactly the reverse. Variety requires a certain degree of distinctness of character or feature, on which the eye can repose itself before, and then proceeding to another: but where every thing is indiscriminately mixed together, there can be no features, nothing on which the eye can dwell with satisfaction; all is confounded, and reduced to a mere chaos of forms and colours. Instead of this mode of mixture, we recommend, as much as is practicable, that each genus and species be kept by itself; and where a number of genera form a striking natural order, as the *Ericeæ, Geraniaceæ, &c.*, that the whole be grouped together; not in a formal manner, but so as to show a sort of relationship or connexion, and at the same time to blend in with surrounding genera. It may be alleged, that where some of the species of a genus are large plants, and others small, this could not well be done consistently with showing off the plants to advantage: but such an objection arises from taking too formal a view of the subject. By keeping plants together, it is not meant to keep them in contact in a compact clump, but to place them in visible connexion in irregular groups, which is quite consistent with placing the tallest plants of the group on the upper part of the stage, and the lesser plants nearer the spectator. It is sufficient that the connexion of the species be recognised by the eye, and that there appear in the green-house, what there always is in natural scenery, something like a natural gradation, and blending of character, in shrubs, trees, and plants growing together.

Besides some attention to keeping together genera and species, or natural orders where the case admits of it, it is desirable also to see that they harmonize in growth with the adjoining genera or natural orders. For example, after the group of Camellias, it would be too violent a contrast to commence with Succulents; but the Myrtle tribe, or *Proteaceæ*, might very well adjoin; then *Leguminoseæ, Geraniaceæ*, and next the Succulents. In short, the most perfect mode of arranging the plants on a green-house stage would be to follow the natural system of Jussieu: but as to do that completely is impracticable, the next thing is to follow it as far as we can – and the third best thing, to keep the genera together. Neglect of the last principle as a general guide is quite inexcusable even in the smallest green-house, – attention to it will be amply repaid by the effect produced even on a chamber flower-stage: as a proof of this, we have only to refer to the green-houses of nurserymen who are extensive growers of green-house plants, and especially to the well-arranged collections of Messrs. Loddiges in their garden at Hackney.

Let it not be thought, however, because we recommend every genus to be kept by itself, that we carry this so far, or hold the principle in so arbitrary and absolute a manner, as to exclude all or any good effects that may result from a deviation from it. No man of sound sense ever holds any principle in so absolute a manner. That good effects may and do arise from a deviation from general principles, we readily allow, and on all fitting occasions would gladly avail ourselves of them. For example, in the midst of a broad group of heaths, it may produce a striking effect to place a tall handsome camellia, or to distribute two or three hyacinths and polyanthuses in flower: a tall plant of tree mignionette may have a fine effect among the succulents, and so on. The effect produced always carries its own argument, and justifies the deviation from the general principle; for extraordinary effects, in fact, only become such by being beyond the reach of ordinary rules.

The third common rule in placing plants on a stage is, to adjust their heights so accurately as to make them dress off from the floor to the top of the back wall in one even surface of verdure, like a shorn hedge. This produces a striking effect at the first glance, but is unfavourable to a prolonged interest, by a more detailed examination, of whatever species the collection may be composed; only their tops are to be seen, and no flowers but what issue from the points of the shoots. What we would recommend is, to make the sloping surface much more irregular; by which means the eye of the spectator would examine the sides of many of the plants as well as their tops, and the plants themselves would receive more benefit from the light and air. Any mode in which plants can be placed on a stage has a tendency to draw them up to slender forms, naked below; but when they are crowded, and so nicely adjusted to an even slope as is commonly aimed at, the deformity is greatly aggravated. This evil is beginning to be felt by the most eminent exotic gardeners, who now place their plants much thinner than formerly, and so adjusted in regard to size, that the direct rays of the sun, in many cases, repose on the earth in the pots. The green-house of the Comtesse de Vandes at Bayswater, affords a fine example of the beneficial effects of this sparse mode of placing the pots. The plants there, are generally so well clothed with foliage from the pot upwards, that almost any one of them taken at random, and placed on a pedestal, would form a fine single object. This, indeed, is the true test of success in the culture of plants in pots.

With respect to the arrangement of plants in a conservatory, the general rules are the same as for the green-house; viz. to place the tallest-growing sorts behind, or towards the back wall or north side of the house, so as to present a sloping surface to the sun. This surface is also best adapted for meeting the eye of the spectator in the front path, while in the back path he is under the shade of the higher trees and shrubs.

But it very frequently happens that a conservatory is placed south and north with a span roof, a walk round the whole, and one up the middle. In this case the higher-growing plants are placed along each side of the middle walk, and lesser ones sloping towards the side walks. The creepers are planted on the props which support the span roof, and which form a sort of groined arcade over the middle walk: this is the situation in which creepers do leastinjury; and in short this plan, all circumstances considered, is the best for a large conservatory.

It is a desirable circumstance in a conservatory, to mix evergreens and deciduous sorts together (though of the latter, only the most beautiful flowerers should be introduced); and when this is done, the evergreen shrub should always be contrived to cover the surface under the deciduous tree: at any rate, the contrary practice should never be adopted; for nothing can be more opposite to natural principles, than for a deciduous tree to grow in the shade or under the drip of an evergreen. By trees, of course, we are to be considered as referring to such woody plants as come up with single erect stems, as the orange, in opposition to such as come up with bushy stems, or numerous suckers, as the myrtle, rose, fuchsia, &c.

Wherever there is a tolerable collection of green-house plants, it is desirable that they should be named. To make use of the proper and correct names of objects, is an important part of common conversation, and nothing can lead to a correct application of botanical names, but affixing them by some means to the plant, or pot containing it. It is also very desirable that plants should be named, with a view to children, as the perusal and recollection of these names will aid in strengthening their memory, enlarge their powers of pronunciation, and create a taste for natural history.

The most common mode of naming plants in pots, is to take a flat slip of wood, sharpen one end; rub a little white lead on an inch or two of its smoothest surface at the opposite end, write the name with a black lead pencil on this lead, and then insert the stick in the pot. This is the mode adopted in the nurseries, and will last several years. Another suitable mode for a private green-house, is to put the name in black letters of oil colour on a neat small stick painted white: but the most recent and pleasing mode is that of using earthenware as a substitute for wood, and either writing the name with common ink, pencilling it with black or common lead, or painting it in oil. Such naming instruments are now made in abundance at the Potteries, and are to be had at Spode's, Wedgwood's, or any of the principal earthenware shops in the Metropolis and large towns. They are in use at Messrs. Loddiges, Sion-house, and in the hot-houses of the Horticultural Society.

A small book containing a list of the names should be prepared, and occasionally the stock examined by it, to ascertain what sorts may have been lost, &c. A very agreeable and instructive task for a young person studying plants, would be to turn this list into a descriptive history, with a graphic sketch of every plant appended.

18

'SOME ACCOUNT OF AN ORCHIDEOUS HOUSE . . . '

John Dillwyn Llewelyn

Source: John Dillwyn Llewelyn, 'Some Account of an Orchideous House constructed, South Wales', *Journal of the Horticultural Society of London*, vol. 1 (1846), pp. 5–6

[Mr. Llewelyn having mentioned to the Vice-Secretary that he had constructed an epiphyte house, through which a waterfall had been directed so as to dash over rocks, and finally to flow into a basin forming the floor of the house, that gentleman was solicited to favour the Society with some account of it, which he has done in the following interesting communication, accompanied by an interior view of the house, which forms the frontispiece of the present volume.]

I enclose with this the ground-plan and section of the stove, which I promised to send. These will show the size and shape of the building, and the arrangement of its pipes and heating apparatus, and the manner also in which the water for the supply of the cascade is conducted to the top of the house by means of a pipe communicating with a pond at a higher level. This pipe is warmed by passing with a single coil through the boiler, and terminates at the top of the rock-work, where it pours a constant supply of water over three projecting irregular steps of rough stone, each of which catches the falling stream, dividing it into many smaller rills and increasing the quantity of misty spray. At the bottom the whole of the water is received into the pool which occupies the centre of the floor of the stove, where it widens out into an aquarium ornamented with a little island overgrown like the rock-work with Orchideæ, ferns, and lycopods.

The disposition of the stones in the rock-work would depend much on the geological strata you have to work with: in my case they lie flat and evenly bedded, and thus the portions of the rock-work are placed in more regular courses than would be necessary in many other formations. In limestone or granite countries, designs much more ornamental than mine might, I think, be easily contrived.

DOI: 10.4324/9780429198809-22

Interior of Orchideous House at Penllergare.

The account of the splendid vegetation which borders the cataracts of tropical rivers, as described by Schomburgk, gave me the first idea of trying this experiment. I read in the 'Sertum Orchidaceum' his graphic description of the falls of the Berbice and Essequibo, on the occasion of his first discovery of Huntleya violacea. I was delighted with the beautiful picture which his words convey, and thought that it might be better represented than is usual in the stoves of this country.

With this view I began to work, and added the rock-work which I describe to a house already in use for the cultivation of Orchideous plants. I found no difficulty in re-arranging it for its new design, and after a trial now of about two years can say that it has entirely answered the ends I had in view.

The moist stones were speedily covered with a thick carpet of seedling ferns, and the creeping stems of tropical lycopods, among the fronds of which many species of orchideæ delighted to root themselves.

Huntleya violacea was one of the first epiphytes that I planted, and it flowered and throve in its new situation, as I hoped and expected. The East Indian genera, however, of Vanda, Saccolabium, Aerides, and other caulescent sorts, similar in habit and growth, were the most vigorous of all, and many of these in a very short time only required the use of the pruning-knife to prevent their overgrowing smaller and more delicate species.

Plants that are grown in this manner have a wild luxuriance about them that is unknown to the specimens cultivated in the ordinary manner, and to myself they are exceedingly attractive, more resembling what one fancies them in their native forests – true air-plants, depending for their subsistence on the humid atmosphere alone.

Different species thus intermingle together in a beautiful confusion, Dendrobium, and Camarotis, and Renanthera, side by side, with wreaths of flowers and leaves interlacing one another, and sending their long roots to drink from the mist of the fall, or even from the water of the pool below.

Many species are cultivated upon the rocks themselves, others upon blocks of wood, or baskets suspended from the roof, and thus sufficient room is secured for a great number of plants. At the same time the general effect is beautiful, and the constant humidity kept up by the stream of falling water suits the constitution of many species in a degree that might be expected from a consideration of their native habits; and I would strongly recommend the adoption of this or some similar plan to all who have the means of diverting a stream of water from a level higher than the top of their stove.

This, I think, in most situations might be easily contrived. My own house lies on high ground, and the water is brought from a considerable distance, but yet I found very little of difficulty or of expense in its construction; for it must be borne in mind that a small quantity of water is sufficient, and that this, after passing through the stove, might be conveniently used for garden purposes.

It must be remembered also that this plan may be added to any existing stove, and that the sole expense will be for the pipe to conduct the stream, and for the labour of the carriage and arrangement of the rock-work.

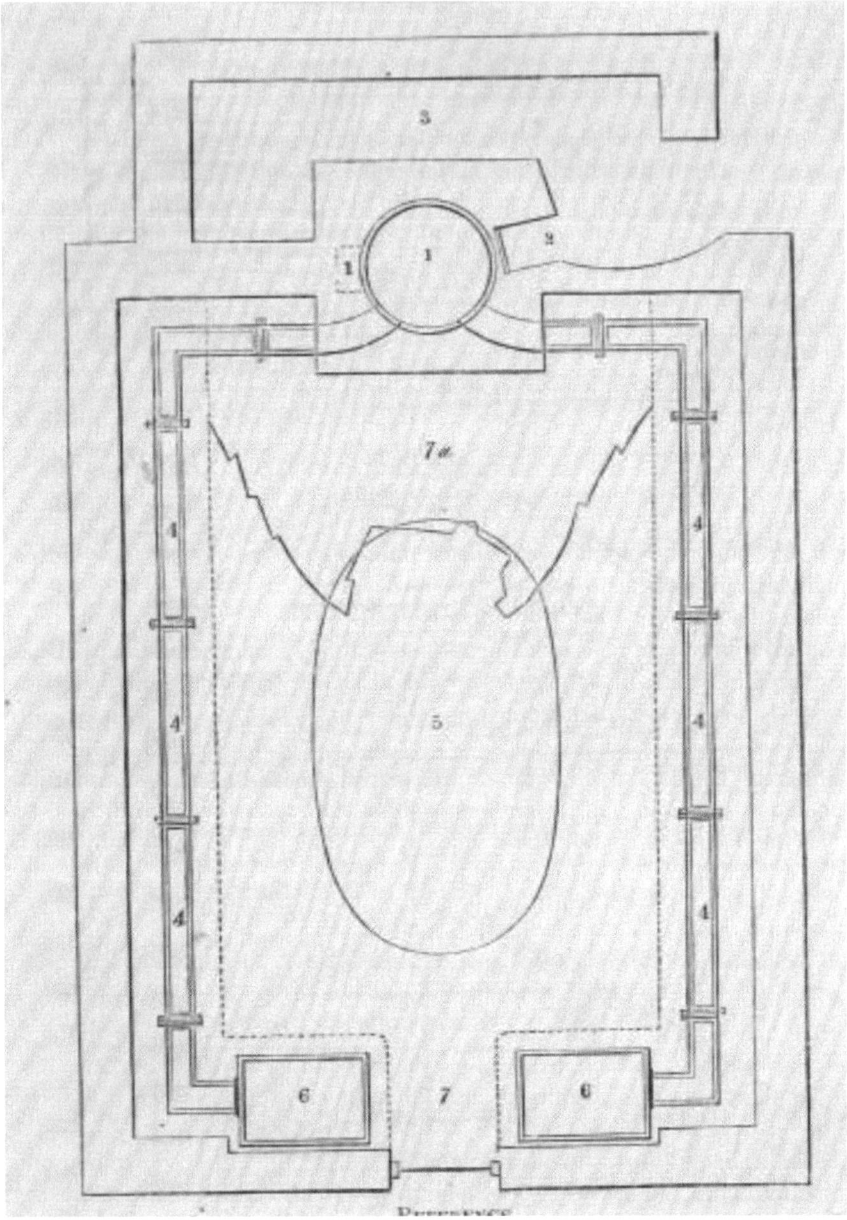

Reference.

1. Boiler and flue. 2. Fireplace. 3. Coalshed. 4. Hot-water pipes. 5. Water tank. 6. Hot-water cisterns. 7. Doorway. 7*a*. Rock-work. 11. Pipe with hot water to fall over the rock-work. 12. Cold-water pipe. 13. Stop-cock. 14. Boiler. 15. Ash-pit.

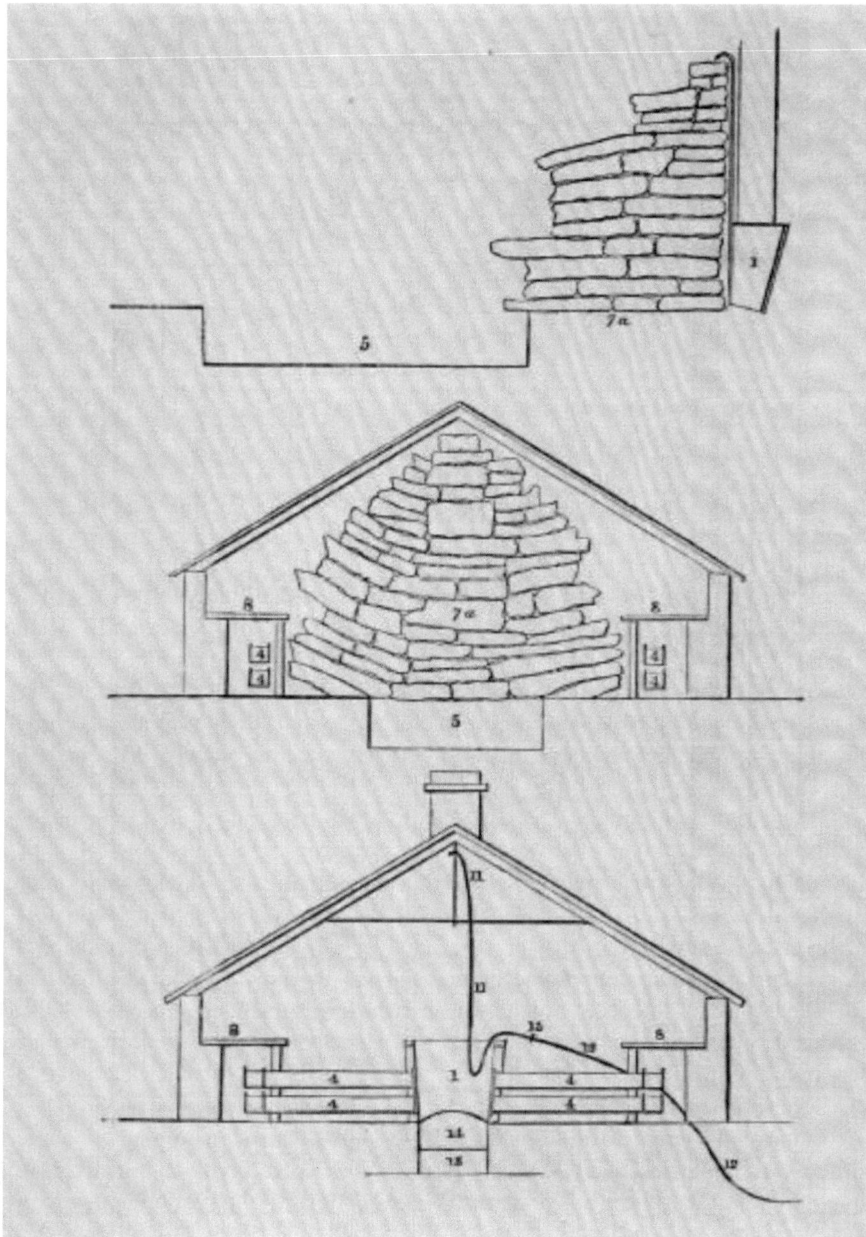
1
7a
5
8
4

19

'CONSERVATORIES IN THE NATURAL STYLE', 'A COOL HOUSE'

Edouard André

Source: Edouard André, 'Conservatories in the Natural Style', and 'Conservatories in the Natural Style. A Cool House', *The Garden* (20 January 1872, 17 February 1872), pp. 181–184, 288–291

CONSERVATORIES IN THE NATURAL STYLE.

ALTHOUGH considerable progress has been made during the last few years in the art of grouping plants in the open air in harmony with their natural affinities, the same cannot be said of their disposition under glass. In England more especially, where amateur gardening is so highly developed, and where more attention is concentrated upon the beauty of individual plants than general effect, next to nothing has been done towards developing a more picturesque style in conservatories. Even our largest establishments leave much to be desired in this respect. Of course, in a mercantile or a small private establishment where the plants are either awaiting sale or where they are frequently removed, they will be arranged for convenience, like the bottles in a chemist's shop. But it is surprising that in planting the noble glass palaces of the rich, such as we find at Sion House, or spacious palm houses, like that of Kew, so little regard has been paid to general effect by their originators.

The cause is simply this, that practical cultivation alone has been considered in England, and in all other European countries, and that the knowledge of plants has hitherto been limited to experimental culture. Gardeners have ignored the teachings of plant distribution in so far as it relates to their natural characteristics and the effects resulting therefrom. Information on this subject can only be gained by studying the narratives of travellers, and these unfortunately are often incomplete and uninstructive on this point.

To supply in some measure this defect, we shall attempt to show how a winter garden in the picturesque style should be treated. Before doing so, however, we should mention that there are some notable exceptions to the monotonous repetitions which we so heartily deplore. In Mr. Llewellyn's garden at Penllergare, in Wales, might have been seen a few years ago an orchid-house aquarium where these lovely denizens of the tropics were planted in the admirable disorder which they present in their native forests. True, it was only on a small scale, but the effect produced was most pleasing. Again, at Paris, the Gardens of Acclimatisation in the Bois de Boulogne possess a large and beautiful house, which is designed and

 DOI: 10.4324/9780429198809-23

planted in the most attractive manner, representing a tropical scene, terminated by a background of rockwork with a cascade falling into a pool at the base and continued onwards in a rivulet meandering through a valley of Selaginellas.

There are also other places known to us, though too few in number, where there are plants of sufficient development to permit of similar disposition, and which might form the main features of such a tropical semi-wild garden as we have in view. There would be little difficulty in thus beautifying the Palm House at Kew, the large Pavilion of the Museum at Paris, the Horticultural Society's fine conservatory at South Kensington, a portion of the Crystal Palace at Sydenham, the large houses at Sion and Chatsworth, &c. When in Belgium last year we visited the seat of M. Varocqué, at Mariemont, where a new winter garden, between fifty and sixty feet high, had been erected on the site of an old orangery, in which a Belgian landscape architect of some note (M. Fuchs) had essayed to introduce the style that we advocate. Magnificent palms, tree ferns, cycads, &c., formed the principal features in this splendid building, with an undulating carpet of Selaginellas beneath them. Immediately facing the grand entrance door was an artistic group of rockwork flanked by a water basin, and planted with evident care. But, in our opinion, the result was not commensurate with the pains bestowed. The walks are too winding, the surface too uneven, the rockery too imposing to appear natural, and the plantations confused, the sides being naked and the centre excessively dense. The artist aimed at the picturesque, but he has failed to produce it in consequence of too great profusion, to the detriment of the general effect and the loss of harmony in the details. We quote this example to show that it is easy to err in carrying out the best of plans, and that exaggeration in the intended natural style is more to be deprecated than crowding in the inevitable uniformity of houses with stages and benches.

Even in winter gardens where the aim has been to produce picturesque effect, there is always something in the structure to remind us of the artificial surroundings. Thus, walks are indispensable, and they should be of sufficient breadth to admit of free and agreeable promenading. To try to imitate the forests of Brazil by compelling the spectator to scramble over the rotten remains of trunks of fallen trees, rough stones, and withered fern fronds, would be the height of absurdity. And those interminable winding tortuous walks are equally opposed to good taste, with the rectangular paths which remind us of the system and order of a purely botanic garden. There is thus a *mezzo-termine*, a happy medium, to be studied; and the *nil nimis* of Horace, which neither admits of too much nor too little – the great rule that reigns supreme in all matters of taste and judgment – is equally applicable in the case of our ideal garden.

An outer circular or slightly devious walk near the well-clothed side-walls or lights should surround the central area, where the eye rests upon choice specimens standing out distinctly upon a carpeting of Selaginella denticulata, trailing Commelynaceæ, Lippia repens, Spergula pilifera, and other plants which readily form a close and compact verdure. To be brief, the space between the walks and the walls should be filled with dense masses of foliage effectually concealing the stems of the plants, and rising gradually from the walk outwards; and the centre should show isolated trees and little groups upon an open lawn of creeping plants.

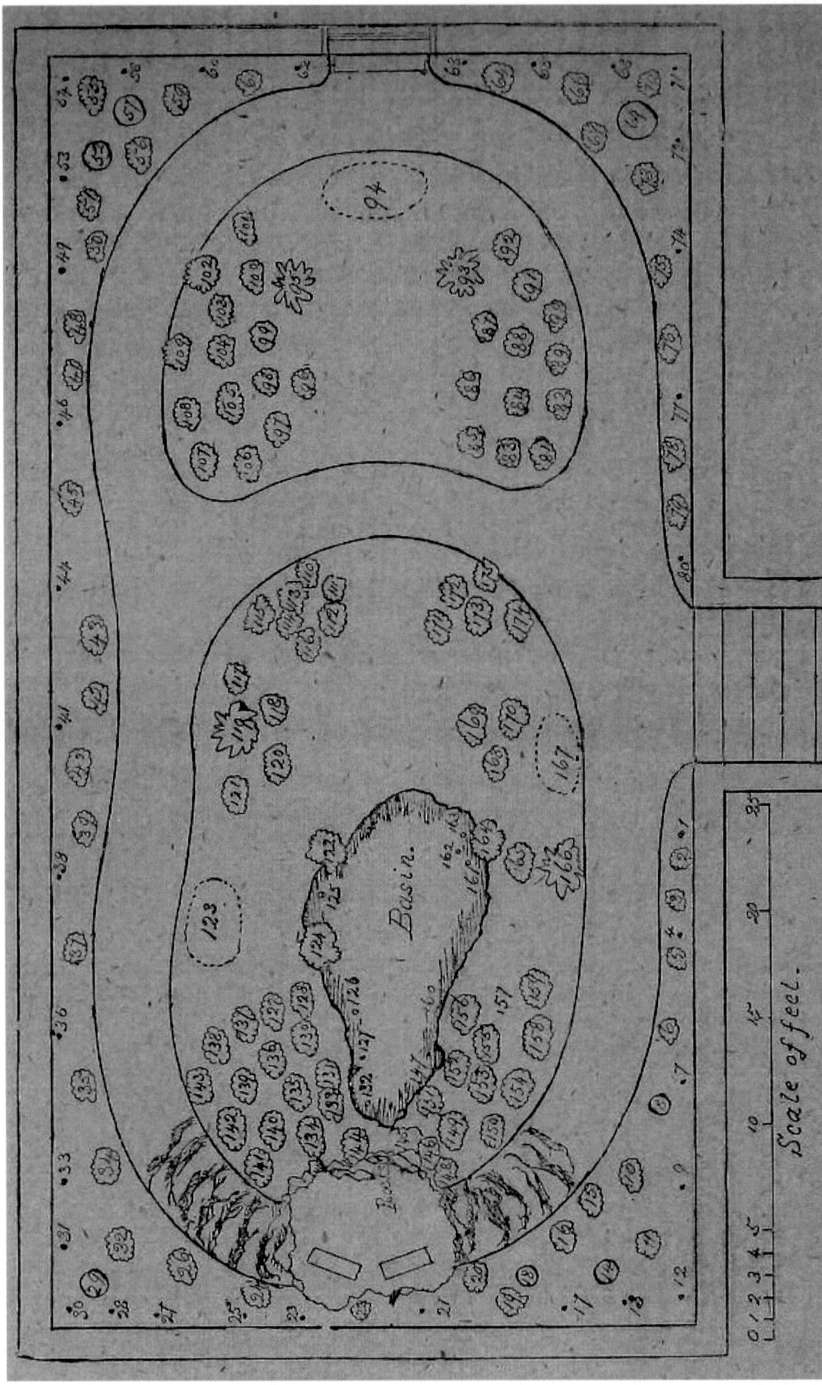

Ground Plan of Conservatory in the Natural Style.

Such is the arrangement indicated upon the plan (see p. 184), though the detached specimens must not be planted indiscriminately in the central area, which is surrounded and intersected by the walks. They should be grouped in combinations or planted singly, according to size and foliage, and in such a way that the view between their trunks is uninterrupted; and at the same time their heads should harmonise together in colouring and outline.

Instead of undulating the surface in an infinity of insignificant hillocks, as in M. Varocqué's winter garden, we recommend limiting it to two intercepting dells. A longitudinal one from the rockery, terminated by a bower, under which are placed a table and seats, and ending, at the side entrance, in a single hollow, of which the pool or basin is the lowest point. From this pool to the flower-bed No. 94, the ground rises slightly, and the cross walk curves towards the middle in agreement with the lower ground. The boundary walk is of the same level throughout, except towards the rockery, where it rises and terminates in four or five rustic stone steps leading to the alcove. The beds Nos. 94, 123, and 167, should be elevated about two feet above the walk; and each of the isolated trees should be planted on a scarcely perceptible mound of earth, with the exception of the groups on either side of the rockwork, which are on an abrupt slope, and the large clump on each side of the bed No. 94, which should be raised about nine inches above the walk and gradually sloped off.

It now remains for us to enter into the planting, which is certainly the most important part of a winter garden. This may be considered from several points of view, according to the class of house, whether cold, temperate, or hot, or for orchids or an aquarium. For the present we will content ourselves with treating of a warm-temperate winter garden with a mean temperature of from 65° to 68° Fahr., in which we could place palms and similar house plants, that would thrive all the better for the increased warmth.

We shall take it for granted that, previous to planting, due care has been given to the drainage and the composition and preparation of the soil. As bottom heat is the main thing to produce luxuriant vegetation in plants, we will assume that the hot-water pipes are properly arranged and covered over with flag-stones to prevent the plants on the walls from being scorched, instead of being around the outside, and exposed, as usually is the case. As we consider the question of foliage-clothed walls of paramount inportance we must be explicit. If uncovered pipes are considered absolutely essential along the base of the walls, a brick or other partition of some sort should be placed between them and the plants, to prevent too great an accession of direct heat from the pipes. But a still better method is to conduct the heat through underground brick channels to points with open gratings in the walks to allow the heat to rise. The soil, well drained at the bottom, and crossed here and there by the pipes we have mentioned, will materially induce the luxuriant vegetation we so much desire around the walls of the house. The middle will have less need of bottom heat. A good drainage of broken bricks and mortar rubbish will suffice for the subsoil, leaving a depth of about three feet from the mould intended for the plants. Two drains running below this through the house, with an outlet outside, will carry off the surplus water. The composition of the soil may be varied according to the class of plants it is intended for. But, as a general rule, for strong growing plants and large palms it will be found advantageous to prepare the mass in the following manner: – Upon the layer of broken bricks and mortar rubbish a layer of reversed turfs about eighteen inches thick, should be placed, and upon that a second formed of –

Argillaceous earth	3	parts.
Ordinary garden soil	3	”
River, or white sand	2	”
Coarse pieces of earth and brickbats	1	”
Leaf mould	1	”
	10	

Mix this well together and apply it about a foot thick. Then for the surface add a layer about five or six inches thick of peat, loam, and leaf-mould from hedgerows, or rotten willow, oak, or chestnut trunks, if attainable, with a tenth part of sand. The peat should be but slightly broken up except just at the top, and the roots left in it. Prepared in this way, it will be ready to receive the Selaginella, which will thrive admirably in it and speedily clothe it with verdure.

It will be understood that these three layers, forming a total thickness of three feet, will not be of uniform thickness all through the house, because the surface will be undulated. The measurement of one yard is merely given as a basis or unit, for the total depth of subsoil in different parts of the house will vary, having, for instance, a depth of only eighteen inches near the basin in the centre of the little lawn, and from four to five feet for the clumps abutting on the flower-bed No. 94. But the top-dressing will be equal all over.

The majority of large tropical plants can be grown in such a compost. And those which may require special treatment may be potted in suitable soil and plunged; the pots being plunged deep enough to conceal them. Allow the Selaginella to spread over their surface. Liquid manure, if used in moderation and judgment, will accelerate the growth of most plants, including ferns. So much for the preparations for planting. But the key of the question, the secret of success, depends above all upon the choice and grouping of the plants. This selection may vary to infinity, such are the riches of exotic flowers at our command. It is difficult to lay down rules on the mode of arranging plants according to the colour and character of their foliage. We may, however, remark that in houses, no matter how splendidly constructed, the effects of uniformity are bad, and that masses of one species or of one genus of plants should be avoided. The harmony and grandeur of unity in composition, attained with difficulty even in large parks, is here impossible. The object to be sought, then, is contrast in the foliage and habit of the plants employed. Two species of massive foliage, for example, should not stand side by side, such as a Musa and a Ravenala, or a Coccoloba pubescens and a Theophrasta. But a large tuft of a Strelitzia beneath the shade of a Cocos plumosa is very effective, and a fine contrast is presented to the eye by backing up the grand foliage and yellow spikes of Hedychium Gardnerianum with clumps of ferns, bamboos, or feathery conifers. The rigid foliage of rhododendrons and camellias should be excluded, these beautiful plants being reserved for a separate house, where they will better display their charms in a collection.

In dense masses of foliage, like those adjoining the rockery in our plan, the arrangement should be in gradual rising ranks, thick, heavy foliage forming the basis, surmounted by lighter and more graceful forms of palms and tree ferns, whose slender plumed columns break through the sombre undergrowth.

Limit the use of "flowers" to the borders and special beds, with the exception of here and there one on the rockwork, beyond those belonging to the plants themselves. We mean by "flowers" such plants as are grown in pots in special houses and taken to the winter garden for temporary decoration, as – primulas, cyclamens, Van Thol tulips, hyacinths, heaths, crocuses, &c.

The side walls, or sashes, should be provided with wire trellis-work, or wooden lattice-work against a dead wall, to support climbing plants all round the house. For covering the surface of the soil nothing is better than Selaginella denticulata; and this should be planted or re-planted in autumn, or in spring, as the dry heat of summer is unfavourable to the success of this operation. Small fragments, about three inches long, planted four or five inches apart, will soon cover the ground. For edging the walks use slight castings in imitation of rustic woodwork or archlets such as may be found in the squares of Paris. These will keep the border even, and prevent the feet of visitors from straying upon them and the loose earth from crumbling away. Above all a good system of shading must be ensured to protect the plants from the direct rays of the sun in summer.

The paths may be formed of fine gravel, or, better still, asphalted or paved. The construction of the rockery should be of the simplest kind, composed of a few stones naturally disposed and projecting slightly from the earth so as to be discovered rather than seen. Monumental rockwork should be avoided, and, above all, reject the so-called pretty stones. Geological or mineralogical toys may be all very well in the cabinets of the learned, but they are altogether out of place in a garden where the object is to reproduce natural beauties in their native simplicity. If a small pond be added to the rockwork it should neither be absolutely round nor sinuous. The golden rule – simplicity without excess – should everywhere prevail.

In order to secure a warm, humid atmosphere, so essential to the well-being of plants, we recommend having the pipe that feeds the cascade and replenishes the basins made so as to pass through the boiler, which will sufficiently heat the water in its transit to cause it to give off a portion by evaporation when discharged. In this way the atmosphere will be thoroughly saturated with moisture, adding thereby greatly to the healthy appearance of the plants. We would direct especial attention to this contrivance for ensuring a congenial atmosphere.

As regards choice of plants, as has already been observed, it may be varied indefinitely. But instead of treating of this subject under the numerous aspects it presents we prefer giving the arrangement we suggest for the house represented in the accompanying plan. This arrangement is, of course, applicable to a far larger structure than the one under consideration. But we have selected this example

View in a Conservatory Planted in the Natural Style.

because it comes within the means of a greater number of amateurs. Furthermore our design is open to every conceivable modification.

. . .

Such are the prominent distinctive features of this mode of grouping the principal furniture in this palace of flowers. As we have already observed, these combinations are capable of endless variations, and numerous small species may be inserted in empty spaces between the larger plants; and we may add baskets of ferns, Bromeliaceæ, orchids, &c., suspended from the rafters on slender wire. These may be planted in the baskets or the pots placed in them, the interspaces being filled with growing moss. The rockery will be adorned with all sorts of plants that will flourish in the interstices between the stones, including a complete collection of Selaginellas creeping amongst innumerable ferns.

Lastly, dead trunks of trees, fixed in the ground by means of iron stakes to prevent them from falling in decomposition, should be placed at the spots Nos. 8 14, 18, 29, 52, 57 and 69 and covered with tropical creepers and epiphytes, and a whole collection of Bromeliaceæ, especially Tillandsia usneoides and orchids. A short time will suffice for these to assume all the picturesque appearance they present in the tropics. Here ends this rough sketch of our winter garden. That this rude model may be the means of bringing cleverer minds to bear upon the subject is our most ardent wish.

[In justice to M. André's admirable suggestions for the carrying out of this most desirable improvement which he advocates, we may state that the "view" prepared by our artist from the section kindly supplied by M. André fails to show the abundant vegetation by which our correspondent rightly wishes the walls to be hidden. His plan, however, puts this clear. We have only to add that the greatest charm we can give to our large conservatories would be to plant them in the natural manner, after M. André's fashion.]

CONSERVATORIES IN THE NATURAL STYLE. A COOL HOUSE.

In a preceding article we have given consideration to the laying out of a winter garden or conservatory for plants belonging to the warmer regions of the world, and have grouped together such representatives of tropical vegetation as can exist in the same atmosphere. The only exceptions which we have made to this rule is in the case of some greenhouse kinds which acquire larger dimensions under the influence of a more intense heat than they would in a cooler atmosphere, such as certain Palms, Aralias, &c. For instance, the Rice-paper plant, which grows in the open air in the south of France, acquires gigantic dimensions in a hothouse. A specimen of it, thus treated, measured thirteen feet two inches high, and threw out magnificent leaves, exceeding six feet six inches in length, including their stalks.

In most cases, nevertheless, the plants named for a greenhouse suffer under too high a temperature. Besides, the cool winter garden is still more valuable than that which we have already placed before our readers, inasmuch as it is as rich in ornamental specimens, and more within the reach of moderate means. Very little artificial heat is enough to keep the temperature in winter at a minimum of three degrees above the freezing point, which is quite sufficient for the period of repose which is required for many of the plants from Australia, China, Japan, New Zealand, and mountainous tropical regions, &c. One can hardly believe what numbers of plants there are, often supposed to belong to tropical climates, with which a cool conservatory can be furnished. Numbers of our beautiful Palms would yield to cool treatment; and hundreds of Ferns require no better situation than the shelter of glass. The Dracænas, Agaves, Acacias, Dasylirions, Ficus, Aralias, Banksias, tender conifers like the Norfolk Island Pine, Yuccas, Grevilleas, Rhopalas, and the Cactuses, would certainly submit to the same treatment, without mentioning the smaller kinds, which only thrive under a low winter temperature.

The experience acquired of the natural style of arrangement in conservatories during fifteen years in Europe, enables us to recommend it with confidence. The treatment carried out in reference to some tropical species has often no relation to the altitude at which the plants naturally grow. Should a plant arrive from Mexico, it would naturally be placed in a warm conservatory. But as regards the Ferns of that country, they are found between 3,600 and 6,000 feet above the sea level, that is to say, at the limit where begin the Pines and Heaths of the sub-alpine region; and it is at this height that the magnificent Alsophilas spring up.

More than thirty Palms now flourish in our cool houses. A great number grow in the cold regions of tropical mountains, such as the Ceroxylon andicola, which is found at 10,000 feet and upwards. The Oreodoxa frigida, and several kinds of Chamædorea, rise up to the Pine region; the Areca humilis reaches to 8,000 feet in Java; the Chamærops Martiana to 7,800 feet in Nepal; the Phœnix humilis to 6,000 feet; without reckoning the Chamærops excelsa of China, the Rhapis flabelliformis of Japan, Corypha australis, &c.

. . .

Our large illustration will convey a faithful picture of the general aspect of vegetation that may be obtained in a large, slightly-heated structure.

COOL CONSERVATORY IN THE NATURAL STYLE.

We speak of such only as have been proved to succeed; and if the altitudes at which they grow spontaneously are compared, it will be seen that a conservatory with a minimum of 38° Fahr. is all that they require. It is known that for every six hundred feet of altitude above the sea, the mean temperature decreases one degree. But, if it be admitted, as is generally the case, that the mean temperature of the tropics is 80° at the sea level, at six thousand feet it would be only 65°, and at ten thousand feet, 50°, that being equal to the mean temperature of southern England. It is not, therefore, very extraordinary that the Palms of these high regions, as in the case of the Chamerops excelsa, can live out of doors in those climates where they have nothing to fear but exceptional winters (such as the intense cold of December 1871). If we have not certain data as to this mode of culture, it is because Palm trees have hitherto been too costly to risk their sacrifice in the open air.

GENERAL TREATMENT.

The Palms for a winter garden ought to be kept in pots up to the period when their leaves divide and show their character and their stems become at their base as thick as the arm. They must not be put in the ground before this, nor until they shall have been frequently repotted and have been kept as much as possible in a warm greenhouse where the pots have been plunged in tan. They should be repotted twice a year, in spring and summer, when their growth is rapid, without cutting the roots, and in pots deep and narrow. A quiet and warm atmosphere, somewhat shady, but without stagnant moisture, is best suited to Palms when young. Growing Ferns have nothing to fear from the open air or the sun; it is only the stemless kinds which flourish in the shade and under other plants, their roots requiring nourishment. The Alsophila australis may be placed outside in the full sun without injury; if it be watered from time to time with liquid manure it will acquire considerable dimensions in a short time and be of unsurpassable beauty.

A great number of the plants named will remain uninjured if protected from the frost; but it is better, as has been already said, to keep up the winter temperature a little over the freezing point; and even when the sun strikes upon the glass, raising the temperature, it will not be necessary to open the house at all during the winter. After February however, when vegetation is getting active, it will be necessary to give air gradually and to water in the evening. In March you must begin to shade with some light material up to the time that you can uncover the greater part of the conservatory, and at last place some of the plants in pots or boxes in the open air. As to the great Palms and Tree Ferns, Dracænas, Aralias, &c., they will be better slightly shaded throughout the year, taking care to give plenty of air. Where it could be easily done, it would be desirable to remove the roof and allow the contents to be refreshed by the summer rains. Thus managed, with plenty of water and a proper amount of shade, it is very possible to develop splendid vegetation in such a structure.

Here, then, is another example of a winter garden in the natural style. Should the readers of THE GARDEN take some interest in the matter, we may, at some future period, show how it may be suitably varied by tropical plants, hothouse fruits, and medicinal plants. ED. ANDRÉ.

20

AMATEUR'S GREENHOUSE AND CONSERVATORY

(Excerpts)

Shirley Hibberd

Source: Shirley Hibberd, 'Introduction', ch. 17, 'The Conservatory and Winter Garden', *The Amateur's Greenhouse and Conservatory* (1873), pp. 1–3, 215–219

INTRODUCTION.

"Who loves a garden loves a greenhouse too.
Unconscious of a less propitious clime,
There blooms exotic beauty, warm and snug,
While the winds whistle and the snows descend.
*　　*　　*　　*　　*
All plants, of every leaf, that can endure
The winter's frown, if screened from his shrewd bite,
Live there, and prosper."

COWPER.[1]

THE briefness of the British summer and the frequent interruptions to out-door enjoyment, occasioned by adverse weather, render the well-kept plant house a place of most agreeable resort at every season of the year. The capabilities of our climate are truly wonderful, for we may, with a considerable degree of safety, enrich our gardens with plants that are natives of subtropical and arctic climes; and yet we find our varying selection of subjects seriously restricted unless we are aided by plant houses of some kind or other, to ensure for our favourites better conditions than the unprotected soil and free atmosphere would afford them. Hence, for pleasure and utility alike, the various structures that have been adopted as aids in plant-culture are of the utmost importance to the horticultural amateur, for they bring within the range of his observation and practice the vegetable products of climes that differ greatly in conditions from our own, and they may be so managed as to provide the best possible imitation of a perpetual spring, or of "summer all the year round."

 DOI: 10.4324/9780429198809-24

The object of this little work is to afford some useful information, systematically arranged, on those departments of plant-house construction and plant cultivation which may be properly considered under the general heads of the "greenhouse" and "conservatory," in which a temperate climate is maintained, as distinguished from the "stove," the "orchid-house," and other structures in which a tropical climate is required for the advantage of tropical plants. Considering the treacherous nature of our climate and the length of the winter season, it cannot be said we have as yet attained to a full knowledge of the value of glass in horticulture. Nevertheless immense progress has been made since glass and bricks and timber were rendered free of duty, and the vast number of patented plant houses and protective pits that are now in the market sufficiently prove that the demand for such has increased and is increasing. An amateur who purposes to provide some kind of glass structure to enlarge the uses and enhance the enjoyments of the garden may well be perplexed at first as to the best mode of procedure. The horticultural papers teem with advertisements of "portable," "imperishable," and "multum in parvo" plant houses, and with an almost endless variety of apparatus for heating. It will be found, however, on careful inspection, that in plan and material these do not greatly differ, and that in certain leading particulars they very nearly agree all round, so that a blind man could scarcely go wrong, except, perhaps, as to price, in making choice amongst them. But it is not every one who desires the latest patented invention turned out complete as from a bandbox. In one case an amateur may elect to be his own builder; in another there may be need to give a builder directions, and, perhaps, to watch over the work. One important condition of success, whatever be the mode of procedure, is that the amateur should have a clear idea of the sort of house required to suit the plants he intends to shelter in it.

The great point is to begin well, and the business of the writer of this will be to point out as clearly as possible how the desired end is to be secured, at the least possible expense, and with the greatest possible promise of a successful issue. Matters of fact will concern us chiefly, but matters of opinion will, perhaps, occasionally intrude and be governed always with good intentions, and seasoned with some little knowledge of the subject – sufficient, let us hope, for the purpose.

A disquisition on the nice distinctions that might be drawn between the meanings of such terms as "stove," "hothouse," "greenhouse," and "conservatory," would serve no useful purpose. They are all plant houses, and depend for their distinctions quite as much on the furnishing and the management as upon structure and fittings. A house may be heated to 80° or 90° to-day for the comfort of orchids, and be called a stove. If we remove the orchids and put pelargoniums in their place, and lower the temperature to 40° or 50°, it becomes a greenhouse. We have but to enlarge it, and introduce camellias and acacias, and give the whole affair a somewhat elegant aspect, and it becomes a conservatory. These several terms are convenient because they refer to different things. But there is no occasion to define them precisely, and we might indeed go wrong were we to attempt the definition.

THE CONSERVATORY AND WINTER GARDEN.

It is too much the custom to devote the conservatory to ignoble purposes which necessitate labour and anxiety in proportion to their worthlessness. To grow soft-wooded plants in a lofty, airy, roomy edifice is always a mistake, for while they are unsuitable in character, the conditions are unfavorable to their prosperity, and they neither grow nor flower as they ought to justify the keeping of them. The noblest conservatory plants require far less skill and entail less expense to do them perfect justice than the ephemeral flowering plants that look so fresh and bright when thriving in a warm, damp, low-roofed house, when there is no fear of their being "lost" in a large perspective. The first step towards a proper recognition of the kind of embellishment required is to remember that a conservatory is not a stove, or a greenhouse, or a pit, or a hand-light, consequently it should not be used as any one of these things, or as all of these things combined. We employ the several structures enumerated, the conservatory alone excepted, for production simply; and, therefore, although a tasteful arrangement of plants is everywhere or everyhow to be desired, yet where *production* is the primary proposal, mere display is of secondary importance. On the other hand, we may, and do, and should employ the conservatory for production, but display is the matter of first importance, and therefore the conservatory claims the first labours of the "hand of taste." If it be necessary to tie out a plant to a forest of rude stakes, it may be done in the greenhouse; if it be necessary even to suspend a plant head downwards, and make the pot containing it the principal attraction to the eye, it may be done in the greenhouse. But nothing of the kind should be allowed in the conservatory. The glass structures that are devoted to horticultural production are related to manufactures; the conservatory is related to the drawing-room, and, in the broad scheme of horticultural work, it is not a workshop or a museum, but a *garden under glass*.

To render this covered garden attractive and interesting at all seasons should be the principal object of its possessor. It may indeed be gay with geraniums and calceolarias in the months of June and July, and a damp, cold, cheerless den during the remaining ten months of the year. It should be *always* attractive, and should at all seasons offer for admiration something fresh and good, and at all seasons it should be as destitute as possible of any exhibitions of manufacturing processes. For ensuring permanent attractions, permanent features are required; mere display is above all things to be avoided, for the mind needs food of a better sort than colour simply, however strong in tone and perfect in combinations. Noble plants that have a history, that endure for many years, that acquire local renown, that present striking outlines, that exhibit distinctive stages of development, are much to be desired, and the more of such that we can find of a constitution very nearly adapted to the peculiarities of our climate, the better for the tasteful furnishing of the conservatory. Above all things, the conservatory should be attractive in the dull months of the year, and then its occupants should not be such as require a high temperature, for it is not well to take a vapour bath every time the mind desires refreshment in the midst of vegetable beauty.

It will be understood, of course, that we regard as proper conservatory plants Dracænas, Palms, Agaves, Yuccas, Camellias, Acacias, Beaucarneas, Dasylirions, Azaleas, Oranges, Tree ferns, Bonaparteas, Lomatias, Rhopalas, and others of like character, notable for distinctness and elegance of outline, rather than for high colouring, although amongst them we shall find some gay subjects. But we are no advocates for tameness and sameness in a conservatory. Given a sufficient breadth and variety of green furniture, and flowers may be employed with singular advantage. The stove, the greenhouse, and the pit will severally contribute of their productions to enrich the covered garden. The most tender plants, when in flower, will bear without harm a temperature many degrees lower than that in which they have attained their full development, provided they are carefully prepared for it by removal in the first instance to an intermediate temperature, and during the whole time of their stay in the comparatively cool conservatory are supplied with less moisture than they had to promote growth in a higher temperature. Every season should supply new flowers to the conservatory. In the spring, potted bulbs will make a gay beginning, and if orchids are grown in the stoves the bulbs will be followed closely by some of the most resplendent of the family. As the season advances the greenhouse will supply pelargoniums, heaths, herbaceous calceolarias, and specimen petunias, and in autumn the pits will prove their usefulness by providing a glorious display of chrysanthemums. As a rule, however, the less we see of bedding plants in the conservatory the better, for we see enough of them in the open garden during the summer, and it is simply a tax on one's patience – that is, on the patience of one who believes in eclectic horticulture – to pass from a blaze of geraniums in the parterre to another blaze of geraniums in the conservatory. It is neither our business nor our pleasure to denounce people who, in their horticultural enterprises, are content with some half-dozen genera of plants; but we are bound to say, in defence of plants in general, that there are many fine things adapted to the conservatory which many who profess to love plants have hitherto not made acquaintance with. A rabbit cooked a hundred different ways is tiresome, and the cooking must be very tiresome to the rabbit.

Above all things that contribute to make a gay conservatory, the best of the greenhouse climbers should be thought of. As a rule, it is a folly to grow any of these plants in pots; they should be planted out in borders of sufficient breadth and depth to encourage a free growth, and consistently with the aerial space at command for training them; the number should be few rather than many, that each may display its character fully, and a succession of distinct and decisive features be produced rather than a mere confusion of vegetable tracery. It may suit the purposes of a botanical and experimental cultivator to plant in a conservatory as many kinds of climbers as there are rafters to afford them support, but a beautiful scene cannot be obtained by such practice.

There are many fine plants, showy, hardy, easy of culture, that are but ill adapted for the conservatory, and therefore the selection is not a matter of taste solely. We object to bedding plants, because they belong properly to other scenes, but first-class specimens are admissible while in their prime. None of the fast-growing and free-flowering of soft-wooded plants that are most esteemed as bedders are adapted for permanent occupation of the conservatory; for, irrespective of their unfitness in habit and associations, they will not thrive in such a structure. The fuchsia is not strictly a bedding plant, for it loves not the dry soil and the burning sun as the geranium does, and thrives in the subdued light and constantly humid atmosphere of the conservatory. How fortunate! for the free-growing fuchsias make superb pillar plants under glass, and harmonise with whatever other subjects have an equality of claim to shelter with them. Turning in another direction, it may be said that plants of the heath tribe are as unfit for permanent residence in the conservatory as geraniums and calceolarias are. They need more light, more air, less warmth, less humidity, than the more proper inmates of the house, and, therefore, if employed at all, should be as moveable furniture, brought in when perfect, and removed when the flowers begin to fade.

Editorial Endnote

1 "Who loves a garden loves a greenhouse too" – 'The Garden', *The Task, a Poem in Six Books,* II, William Cowper (1731–1800), poet and writer of hymns. Cowper is quoted elsewhere in the collection. His poetry was favoured by garden commentators partly because of the suggestion that the joys of gardens and gardening may be shared across class.

21

NOT WISELY, BUT TOO WELL

(Excerpt)

Rhoda Broughton

Source: Rhoda Broughton, *Not wisely, but Too Well* (1867), ch. 14 [107–112] excerpt, pp. 107–108

OFF they strolled then, as slowly as two people with the right complement of legs apiece could stroll, in the direction of the conservatory; and as it was not much more than a hundred paces off, even they got there before long. And how marvellously pleasant it was when they were fairly inside that 'box where sweets compacted lie;'[1] how almost oppressive, overpowering, the fragrance of the warm damp atmosphere, where a thousand sweet smells strove perpetually for the mastery! There, side by side, gathered from the far east and the far west, blossomed and reigned Nature's most regal flower-daughters. Gorgeous, stately flowers, that had hitherto revealed their passionate hearts, fold after fold, to the fainting air of some cloudless, rainless, brazen tropic sky, now poured forth all their sweets, put on all their brilliant apparel, under our watery, sickly sunbeams. There great dark leaves, moss-green, rose-veined, drooped heavy with their own weight; there crimsons and scarlets burned and flamed, imperial, with a depth and intensity of colour which our dear, pale-faced northern flowers never dreamed of putting on. What of man's devising can be more intoxicating than one of these temples dedicated to rich odours and brave tints? And when there stands in this temple, among these gorgeous flowers, a lovely woman – lovely, with the ripe womanly development of one of Titian's Venuses, not with the emaciated prettiness of modern young ladies – the subjugation of the senses may be supposed to be complete. Kate was in ecstasies. She ran hither and thither, smelling first one, and then another.

'Delicious!' she cried, 'wonderful! I wish I was gardener here. Flowers are one of the very few weak points in my character. Oh, oh!'

The wealth of enjoyment there was in that last 'Oh!' beggars description.

'Well, you certainly are an adept at smelling,' called out Dare at last, from the comfortable position he had taken up on a rustic seat with wooden legs, very ingeniously contorted in a sort of elephantiasis. 'There, you've gone the round of them all now about seven times. Do come and sit down here; I want to talk to you.'

Kate worshipped at the shrine of one more gardenia, and then came and filled the situation indicated.

DOI: 10.4324/9780429198809-25

Dare rested his arm on the much-twisted wooden back behind her, and prepared for a comfortable chat.

'You have not told me yet,' he began, sinking his voice to that low soft key which made tender things sound so infinitely more tender, – 'you have not told me yet how much you have missed me these last few days.'

Kate, very rudely, took no notice of this question. She was much interested in drawing an ingenious design on the pavement, with the point of one of the aforementioned natty little boots. Let us hope she did not hear it. At last the silence became rather a weight to her, so she raised herself up, and asked with great irrelevancy:

'Where's the rose you said you wanted to show me?'

'Oh, hang the rose!' replied Dare, laughing; 'there is not one, that I know of; but come, Kate, you have not answered my question yet. How much did you miss me?'

Editorial Endnote

1 "Box where sweets compacted lie" – George Herbert, 'Virtue' (1633), pp. 1, 10.

22

'SWEET-SCENTED GREENHOUSE PLANTS'; 'INTERESTING STOVE PLANTS'

Frances Jane Hope

Source: Frances Jane Hope, 'Sweet-scented Greenhouse Plants', *Gardeners' Chronicle* (12 May 1877), pp. 590–591, 'Interesting Stove Plants' (22 March 1879), pp. 366–367

SWEET-SCENTED GREENHOUSE PLANTS

It always seems to me a pity that the plants in the mixed greenhouse are limited in variety, and that a show of flowers, even if only of one or two sorts, is the one point aimed at. We all know the look of the medium-sized greenhouse in small places, which is "always gay." In winter it is bright with Chinese Primroses and Dutch bulbs, succeeded in spring by Cinerarias, Azaleas, or Pelargoniums, Calceolarias, and Fuchsias in summer, with Japan Lilies and Chrysanthemums in autumn. If the owner of the greenhouse is omnivorous in his love of plants, the result is disastrous in the overcrowded stage of ill-used and ill-grown plants – Camellias with yellow foliage, Heaths in a dying state, or we might rather say dead, for in the amateur's mixed greenhouse when a Heath is observed to be unhealthy, hopeless mischief has been done. But setting aside Heaths, which are an acknowledged difficult family to grow in mixed houses, look at the wretched state of the so-called hard-wooded plants, and we must own that Epacrises and Cytisus, &c., are uninteresting when past blooming; and it strikes me in these ordinary greenhouses, so cramped for space, that no plant should be admitted that is not for some reason or other always interesting. For instance (1), plants of sweet-scented foliage; (2), those whose habit or varied growth is peculiar; (3), plants whose seed-vessels are interesting or ornamental, thus having two crops of beauty; or such as should be studied from the seed-pan and known in their different stages of growth, as Acacias, certain species of which in their young state could never be known by one unacquainted with the changes they go through; (4), climbing plants, including specimens of Mr. Darwin's divisions, viz., twining plants, leaf climbers, tendril bearers, hook and root climbers. A selection from these four sorts of plants, with a sprinkling of the ordinary flowering occupants of the greenhouse, along with Ferns, Aloes, and succulents, would make the amateurs' greenhouse always

DOI: 10.4324/9780429198809-26

varied and instructive, enabling them to appreciate more highly and learn better from botanic or large establishments which they may be in the habit of visiting. I do not see why we should encourage growth and progress in our plants, and get mentally contracted and stunted ourselves.

SWEET-SCENTED PLANTS.

To begin with plants of sweet-scented foliage, both soft and hard-wooded, as they give an immediate return. Soft are quickest grown and safest to cut. We have more than thirty varieties of Cape Pelargoniums of distinct perfume, but half that number would be a choice selection of the sweetest, leaving out the beautifully-cut Pheasant's-foot and glutinosum section, as the greenfly adheres to them, and they stick to the cut flowers they are mixed with, and stain gloves and dress when worn. Capitatum, or rose-scented, and its variegated variety, Lady Plymouth; quercifolium, or Oak-leaf; Fair Helen, an Oak-leaf – the oldest; Fair Emily, Prince of Orange, and its variegated variety; lemon-scented, nutmeg-scented, Radula or Stag's-horn, citriodorum and its minimum variety, Lady Scarborough (of the citriodorum section), lobatum or Peppermint, abrotanifolium or Tansy-leaved, vitifolium or balm-scented, Laurencianum or serrulatum. There is great confusion with even these unsatisfactory names, but I have selected those by which they are most generally known.

Balm of Gilead (Cedronella triphylla) one seldom meets with, and Verbena (Aloysia citriodora) is never grown of sufficient size to cut freely from. The Pineapple scented Salvia? is another pleasant smelling plant; Lavandula Spica (French Lavender), and Prostanthera lasiantha, are sweet and useful for cutting, lasting long in water, as are also Cedronella incana, and a Scutellaria (?) with pale lilac blossoms, both labiate plants of pleasant smell, while Teucrium Marum, not perfectly hardy, must have a place. Humea elegans overpowers all other perfumes, and adheres to whatever it touches, but many are partial to it.

Of hard-wooded plants the sweetest are the Diosmas and Adenandras (heath-like plants), Leptospermum (allied to the Myrtles), and Boronias. Diosma, capitata, fragrans, ciliata, and speciosa are fresh and fragrant, lasting weeks in water, and retaining their leaves and sweetness when dead; but beware of D. crenulata, the intolerable smell of whose foliage is not counterbalanced by its beautiful white starry flowers, in bloom in December.

The Diosmas and Adenandras are so mixed up in nurseries that it is prudent to see and smell the different sorts for one's self, and not vaguely order by a catalogue. Of Leptospermum, bullatum and scoparium are desirable. The latter is covered with its little Myrtle-like flowers, which ripen their seed and come up freely. All these small-leaved plants take up little room, and are not affected by insects. Boronia tetrandra has a very sweet leaf when bruised, not unlike the smell of the flowers of the well-known B. serrulata, and the charming B. megastigma. B. alata, again, with a good-sized lilac flower, has as pertinacious a smell as Diosma crenulata. They are a pair to be avoided; nevertheless that great charm of foliage belongs to them – viz., that unless bruised or drawn through the hand they do not give out their perfume like flowers, and are therefore innocuous. At the same time the perfume, or its opposite, is always there in its true character, unlike a flower which has an evanescent time of perfection, and which requires careful study to watch the exact time of ripeness. But for one individual who looks for sweetness in foliage there will be a hundred who sniff at every flower, and yet never know the real delicious fragrance of a Rose, Carnation, Stock, or Mignonette in its perfection. How annoying it is to the thoughtful observer to have a sweet-smelling flower cut too soon, or when past its best; too early or too late in the day, when it has had too much or too little sun; too damp or too dry. Bees at flowers are like wasps at fruit, the very best judges of perfection; not having their instinct, and not using our own reason and thought, we lose very much enjoyment we might have from sweet-smelling plants. Of greenhouse shrubs the Musk-smelling Aster argophyllus is worthy of a place, both from the silvery upper surface of its leaf, which is brown underneath, as well as for its perfume, for those who like it. A Citron should be grown for its leaves alone, so refreshing; and Myrtles, say four varieties, should certainly be included. Every one likes sprigs of Myrtle, but because it is almost hardy few give a place in the greenhouse for good-sized shapely plants that will bear cutting and bruising *ad libitum*, and with foliage unbrowned by frosts and winds. With the aromatic never-sickening smell of this evergreen I conclude the list of greenhouse plants with sweet foliage. *F. J. Hope, Wardie Lodge.*

INTERESTING STOVE PLANTS.

THE mixed stoves of amateurs are too often, like their greenhouses, filled up with plants of ephemeral beauty, like Coleuses, Caladiums, and Begonias, to the exclusion of those of permanent and growing interest.

In a mixed house, be it hot or cool, there will always be cramming and loss of plants, and much time wasted in shifting pots back and forward, for plants are mightily human in their selfish love of the best place, and resent being removed from the hottest or coolest, driest or dampest spot which suits them, but which is required to revive in turn some other favourite plant. Coleuses are notoriously troublesome to keep over the winter in small establishments, and Caladiums take up much room in summer, and cannot be at all crushed. As cramming must be the order of things, let there be subjects characteristic when of small size, and that always repay watching and observing. Of Coleuses let there be a selection of half a dozen of the very best, not merely the "new ones of the season;" and let the same strict rule be adhered to with both the old Rex section of Begonias and of the tuberous sorts, as also Caladiums, Tydæas, Gloxinias, Achimenes, &c.

Following the order noted for the greenhouse – 1st, plants of sweet-scented foliage; 2d, those of peculiar growth; 3d, ornamental fruit bearers; and 4th, climbers – I can only name under the first head Pogostemon Patchouly and Cinnamomum zeylanicum. The first is very easily struck and grown. I have never seen the spike of small whitish flowers. Those who dislike the tenacious smell need not touch it, and then it would not be known to be in the house. A small plant of the Cinnamon, with its glossy ribbed leaves, of a bright rose colour when young, is ornamental, as well as pleasant to nibble at while working in the stove, and it is refreshing to smell if one has brushed past the patchouly with its overpowering perfume; and coming out in a winter's day from the close stove into snow or bitter winds a bit of the astringent leaf is no bad preservative against sore throat, if one is subject to such. I never throw an old leaf away; it retains the smell and flavour for any time. Xylophylla latifolia and Reidia glaucescens both flower peculiarly; the first on the edge of the so-called leaf, really a flattened out branch; the other hanging in a thin red fringe below the edge of its leaves. Where X. latifolia and X. fraxinifolia are not both grown, the former is surest to flower, and a most attractive plant it is, although one botanically without leaves. Reidia glaucescens gets leggy when old; small specimens are therefore best, and it is quite distinct and graceful withal. Phyllanthus nivea, allied to both Xylophylla and Reidia, should also be grown. The mottled white leaves – occasionally they are pure white – are very pretty. It flowers like Reidia, but sparsely.

Dorstenia is a very singular genus of herbaceous plants. The largest collection I ever saw was Mr. Wilson Saunders' at Reigate. D. maculosa, with long heart-shaped leaves, blotched with white, is useful for cutting; the peculiar fruit or receptacle is a flat shield at the end of a 6-inch stalk; it also lasts long in water, and seeds and increases quickly. Allied to Figs, flowers so understood are of course not to be looked for. D. lasiantha is still more singular. Instead of a flat shield the receptacle is forked, a pair of dark brown outcurved horns 2 inches long. They catch the eye at once, as also the flat buttons at the tip of each branch of D. Contrayerva. This last is a firm little bush of 1½ foot high. There are between 2 and 3 dozen species, but these three are a very fair example of what perhaps, save for microscopic study, is a more curious than beautiful tribe of plants. They are no trouble to grow, and are free naturally of vermin.

Another set of plants I observe attract my young friends, who are always on the look-out for the present of a rat's-tail, but the Peperomias have many of them beautiful foliage. P. arifolia is a charming little plant, the beauty of the family. P. maculosa and P. argyreia have glossy smooth marbled oval leaves, 6 inches long, and the spadix or flower-spike of maculosa (rat's-tails) is 13 inches long and lasts for weeks: seedlings come up in the pots within reach. Then there are tiniest mouse-tails, as in the small growing Saxifrage-like P. pulchella and P. rubella, with red-stalked bright green leaves – all easily grown and taking up little room.

Aspidistra lurida, although so well known a plant, has its curious flower too seldom observed, and one plant should be kept in the stove until that has been watched and known. I have often been told that the great drawback of so useful a room plant is its having no flower, and have simply been thought imposing on my friend's superior wisdom or ignorance, when on lifting the surfacing of moss I have displayed the purplish flower growing as it were out of the soil. The miniature and scarce A. angustifolia variegata is a treasure for a crowded house: there seems no possibility of increasing it. The leaves are 3 inches long, and have much white in them. I got our little plant above twenty years ago from the well-known Mr. Buckley, of Rollisson's, who reluctantly parted with one of his two plants. I am not aware of any soil or treatment that we have not tried with it. We are thankful to have it safe, of exactly the same size, and visions of getting up a stock are at an end. Marica Sabini is an Iris-like plant, and at the end of one or two of the long leaves, which then hang over the pot, is a nice tufty young plant. The flower, which I have never seen, is beautiful and evanescent, lasting, Iris-like, but one day, but the plantlet swinging about in the air at the tip of the plain ribbon leaf, has a curious look. Strictly speaking this ribbon-leaf is a runner and a flower-stalk. Ficus diversifolia has a succession of its yellowish fruits all the year, and plants of even 1 foot high are useful; the branches when cut last long in water, and the thick cuneiform leaf ought to stand, one would think, as well as F. elastica or angustifolia, but we have not hardened it sufficiently or paid attention to that point. I am confident it is our fault, and that it might be a most desirable room plant. F. religiosa is more like a Poplar than a Fig, and its 2 inches of elongated tip of leaf gives it a distinct character. Then the minima variety of F. repens (although it should be placed under head 4, "climbers") is not often enough employed. It is such a temptation to get the back wall quickly covered where nothing will flower (that stumbling-block of the amateur, as if there was no interest in a plant save in its flower) that F. repens usurps every such space: a portion should certainly be left for the tiny F. minima, which also does well in a basket; grown in this way, however, one loses the opportunity of watching its root-climbing properties, which, tiny as it is, are very persistent, and it brings down the lime and plaster when detached from the wall. I presume, therefore, it also, as Mr. Darwin writes of F. repens, emits a "viscid fluid," which, ultimately modified into an unelastic cement, is used by the F. repens to cement its rootlets to any surface which it ascends.

I have frequently heard Aralia Sieboldii called an evergreen Fig, and perhaps it requires as much faith to believe that between such there is no relationship as is needful to credit the statement that F. elastica, F. religiosa, F. repens minima, and the Fig of commerce are of one and the same family. The structure of the fruit is, however, essentially the same in all, great as is the difference in habit and appearance. It is well to grow these half dozen interesting species, if for nothing else than to prevent oneself flying to mistaken conclusions. *F. J. Hope, Wardie Lodge, Edinburgh.*

23

'MAGIC GROTTOES AND SHOW FERNERIES'

By Henry Harland, Esq.

Source: H. Harland, 'Magic Grottoes and Show Ferneries', *Gardeners' Magazine* (23 December 1876), pp. 699–700

THE information you have afforded us on the subject of actinism[1] and the effects of various coloured lights in the growth of plants, interested me so much, that some time ago I resolved to indulge in a little fanciful work, and by degrees this has developed into something more than I originally intended. As writing is not my *forte* I must convey my meaning in a few words, but I do not fear I shall be fully understood. At certain times and seasons I have about me a throng of young people, and I resolved to make some fun for them in the garden. So I constructed a large fernery forming three sides of a square, the two opposite sides being connected by means of a romantic hollow way, which is in places arched in the fashion of a ruin with many ins and outs and parts of buttresses and bastions. It is richly clothed with ferns and

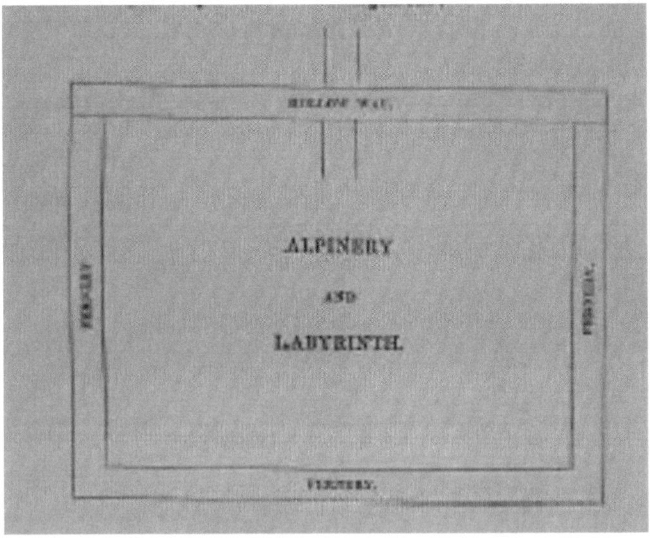

 DOI: 10.4324/9780429198809-27

ivies and other suitable plants. This hollow way forms an irregular walk from either end of the house, so that we must go through it to reach the house, and on quitting the house we come into the hollow way again. The subjoined diagram represents the arrangement: –

The house is divided into a series of bays by suitable banks, arches and abrupt rockeries that reach to the roof, the walks passing round these and opening again into roomy bays that are richly dressed with waterfalls and fountains. The result of this arrangement is that the visitor is completely deceived as to the extent and form of the structure, and it is really difficult for a new comer, even if accustomed to plant-houses, to discover the very simple plan of the arrangement. Indeed, to help the deception, the roof, which is, for the most part a regular span, is in three places varied, so that in the centre of each length we have roomy bays with ridge and furrow-roof with transverse ridge. The result is perfect uniformity and extreme simplicity of plan, but a bewilderment is produced by the interior arrangements. At each of the entrances there is a quiet bay, with a touch of the artistic in the dressing, and it forms a distinct fernery of a very pleasing nature. These compartments are roofed with common colourless glass of good quality. We then proceed through a narrow way arched over and lighted by perforations in the arches with yellow glass, and presently come into a romantic bay covered with green glass. The effect is startling, and when the young people behold how ghastly they look, the fun justifies the fancy. The next compartment is roofed with white glass and clothed with trailing plants only to make a complete change, and then we come to a dead stop. Presently a passage is discovered, and this leads to a bay roofed with rose-coloured glass, and another magical effect is the result. We next have a round-about grotto with yellow glass, and then a little nook like a chapel with white glass, and next a bay with blue glass. These are the more distinctive effects, the exit being in colourless glass, to correspond with the entrance.

It should be observed that under yellow glass plants will not live for any length of time, and therefore it should be used only to make a change and to connect passages and bays well clothed with vegetation. The best growth is under colourless glass and green glass; the blue and red occasion much trouble, but we keep things right by inserting cheap plants here and there in pockets made for the purpose, so that on gala days the place is richly dressed. The green glass suits ferns so well that we have our best displays in these compartments, one of which is represented in the drawing that accompanies this note.

The space enclosed by the constructions is laid out as a labyrinth, but in place of clipped trees we have an imitation of winding passages in the foundations of a ruin, the walls and arches being clothed with alpines and ivies. A distinct maze was marked out first, and then low brick walls were built. On these arose fragments of rubble foundations, broken arches, and occasionally high enclosing walls with occasional open spaces to vary the scheme and make situations for masses of beautiful vegetation.

GROTTO FERNERY.

Editorial Endnote

1 'The property by which light or other electromagnetic radiation causes chemical change, as in photography.' (*OED*)

d)

GARDEN STRUCTURES AND ORNAMENT

Editorial Headnote

24. Humphry Repton, *Observations on the Theory and Practise of Landscape Gardening . . .* (1805) (excerpt), pp. 157–161
25. J. C. Loudon, 'On the Employment of Vases as Receptacles for Plants in Town Gardens with some Remarks on their Use in Garden Scenery in the Country. . . ', *Gardener's Magazine,* vol. 10 (October 1834), pp. 489–495
26. J. C. Loudon, 'Descriptive Notice of the Villa of Mrs Lawrence at Drayton Green. From the "Suburban Gardener" ', *Gardener's Magazine*, vol. 14 (July 1838), [305–322] excerpts, pp. 306–308, 311–319, 321–322
27. Andrew Jackson Downing, 'Embellishments; Architectural, Rustic and Floral', *A Treatise on the Theory and Practice of Landscape Gardening* (1841), pp. 351–357; 383–388
28. Shirley Hibberd, 'Garden scenery and ornaments', *Rustic Adornments for Homes of Taste* (1856) [327–353] excerpts, pp. 327–329; 335–353
29. Frances Jane Hope, 'On Arbours or Bowers', *Gardeners' Chronicle* (6 March 1875), pp. 305–306
30. 'P.', 'Lamport', *Gardeners' Chronicle* (25 September 1897), pp. 209–210

Continuity between the architectural features of house and garden was one of the guiding principles for the display ornament or decoration. Illustrations support and augment the text.

Observations on the Theory and Practice of Landscape Gardening (1805) was a collection of designs for the landed gentry and elite households by Humphry Repton (1752–1818). He was by this time an established designer of garden landscape, and *Observations* further enhanced his reputation (see also Volume 5). In this excerpt a discussion of trelliswork is followed by the distinction he makes between architectural ornament and decoration. J. C. Loudon elaborates on the universal aesthetic sense of humankind from that of the savage to the civilised in his conclusion to 'On the Employment of vases as receptacles', *Gardener's Magazine* (1834). He argues that this aesthetic sense, which can be refined through education, may guide 19th-century consumers through the plethora of garden ornaments and structures increasingly available. He edited an edition of Repton's complete works in 1842 to disseminate landscape design to a wider public.

Unlike Repton, Loudon favours cheaper substitutes for marble and stone to enable the less wealthy to display freestanding ornaments. The variety of scale, form, and material of urns and vases made them desirable for smaller gardens. In contrast, although statuary was becoming more widely available with the beginnings of mass production in the 18th century through the introduction of moulds (that were initially filled with lead because it was weatherproof), the larger size of most statues limited their place to elite or public gardens.

The number of statues situated in the 28-acre site of the garden of Louisa Lawrence was, observes Loudon, more appropriate for a larger site. A month before the article 'Descriptive notice of the villa of Mrs Lawrence at Drayton Green' (1838) was published, Lawrence's husband, William, a celebrated surgeon, had purchased a 100-acre site at Ealing Park. The article takes the reader on a tour of the garden; statuary and ornament serve as a focus or means of directing the eye. Louisa Lawrence (1803–1855) was one of the first women to be allowed title of Fellow of the Horticultural Society of London (later RHS), and she won several gold medals for her plant innovations and collections. Note that the article was first published in the *Suburban Gardener* (1838) intended for amateur readers and reissued in the *Magazine* to include working gardeners (and to manage the costs of a heavily illustrated text).

Andrew Jackson Downing recommends alternative materials in his discussion of ornament that range from the classically styled terrace to the rustic vase (for a cottage-type abode). He includes different kinds of seating and garden houses of more recent lineage for the comfort and entertainment of the garden visitor. His *Treatise on the Theory and Practice of Landscape Gardening* (1841) was his first published book and a commercial success; the small-scale illustrations of ornamental structures are taken from a number of US landscape gardens and nurseries. Shirley Hibberd's *Rustic Adornments for Homes of Taste* (1856), a compact publication with numerous small black and white illustrations (not included here), achieved similar success amongst middle-class readers in Britain. It was reissued a year later (and immensely expanded in the 1870 edition) and launched his career as editor of the *Floral World* (1858–1880). In this excerpt (from the last chapter of the first edition) the range of ornaments is diverse and their manufacture economical. Willow, for example, was produced on an industrial scale in Somerset during the 19th century to meet demand. The filigree metalwork mentioned in relation to 'wire bordering and baskets' was more widely available to consumers because of technological developments. The emotional significance of statuary and ornament is evoked by Dickens in the garden of Mr. Wemmick in *Great Expectations* (see Ia).

Natural arbours are mentioned by Hibberd; Frances Jane Hope (1822–1880), however, was an enthusiast. In her article 'On arbours or bowers', *Gardeners' Chronicle* (1875), she favours the wildlife that a natural arbour attracts. She is a relatively unknown early proponent of planting in a natural style.

Over two decades later, the article 'Lamport', *Gardeners' Chronicle* (1897), by 'P' (not identified) and photographed by E. W. Smith (n/d), features the rockery. The excerpt focuses on the gnomes who inhabit it at the estate of Sir Charles Isham in Northamptonshire. Gnomes were unusual ornaments in late 19th-century British gardens and were more often seen in interior settings as Twigs Way has suggested in *Garden Gnomes* (2009) because of their earlier general status as miniature statues. The elision of gnomes with dwarfs may be attributed to Walter Scott who combined the two in his Waverley novels: *The Monastery* (1820) and *The Talisman* (1825). Isham (1809–1903) was an aristocrat, vegetarian, and spiritualist with an interest in folk law. His interest in gnomes derived from holidays in North Wales where stories of gnomes and dwarves circulated, as they did in the mountainous regions of central Europe. By the 1870s, Germany was the main producer of ceramic gnomes that we might recognise in the 21st century. The bright colours of garden gnomes did not conform to the requirements of taste suggested by contemporary experts. The gnomes of Lamport represent outdoor workers, miners, and builders. The striking miner gnome ('fig. 63') recalls the Miners' Federation of Great Britain founded in Newport, South Wales, in 1888 that co-ordinated miners' unions around the country and showed solidarity and a shared belief in egalitarianism. After Isham's death, most of the gnomes were removed. Gnomes flourished, however, in small gardens. Garden gnomes continue to be banned from the Chelsea Flower Show of the Royal Horticultural Society (an exception was made for the 2013 show's centenary year). For the design, planting, and situation of the rockery in 19th-century gardens, see Volume 6.

Further Reading

Twigs Way, *Garden Gnomes: A History* (2009)

24

OBSERVATIONS ON THE THEORY AND PRACTICE OF LANDSCAPE GARDENING

(Excerpt)

Humphry Repton

Source: Humphry Repton, *Observations on the Theory and Practice of Landscape Gardening . . .* (1805), pp. 157–161

The propensity for imitation, especially where no great trouble or expence is incurred, has made treillage ornaments so common, that some observations concerning them may be expected in this work, especially as I believe I may have contributed originally to their introduction; but I little thought how far this flimsy ornament might be misapplied.

The treillages of Versailles and Fontainbleau were of substantial carpentry, preserving architectural proportions, in which plants were confined and clipped to form a sort of vegetable and architectural *berceau*, or *cabinet de verdure*; these being made of strong wood and painted, were more costly and more durable; and as they only formed a frame for the plants, they might perish, without injuring the forms of these leafy buildings; but the English treillage is made of such slight materials, and so slightly put together, that they can hardly outlive the season for which they are erected; this, however, is no objection where they are used in flower-gardens, or where they are merely to be considered as garden sticks supporting plants, but when added to architectural houses, and made the supporters of a heavy roof, or even a canvass awning, it looks as if the taste of the country were verging to its decline; since shade might be obtained by the same awning supported by iron, if architectural forms and projections are to be despised or discarded.

I should therefore suppose that no treillage ought to be introduced, except in situations where creeping plants may be fastened to the framing, which should be stout in proportion to its height, or its intentions: it is a common mistake to suppose a thing will look light by being slender; if it be not equal to its office by its apparent substance, it will look *weak* not *light*; but the lattice work is supposed to support nothing, and may therefore be of any dimensions, and being always painted, it will be invisible at a distance.

DOI: 10.4324/9780429198809-29

Architectural Ornaments and Decorations.

I could wish, in speaking of architecture, if the use of language would admit of such distinction, to make a difference between the words *Ornament* and *Decoration*. The former should include every enrichment bearing *the semblance of utility*; the latter is supposed to have no relation whatever to the uses or construction of the building; thus for instance, a house may answer all the purposes of habitation without a column, a pilaster, an entablature, a pediment, a dome, an arcade, or a balustrade, which I call the external *ornaments* of Grecian architecture.*

I include under the word *Decorations* – statues, vases, basso-relievos, sculpture, &c. which have no use, but as additional enrichments to the ornaments of architecture; on the contrary, where these *decorations* are applied to plain buildings without *ornaments*, they are marks of bad taste.†

The ornaments of architecture must be correct in design, since no degree of costliness in their materials or their workmanship, can compensate for any defect in proportion, order, or disposition. The eye of good taste will be equally offended with columns too large or too small, too near or too far apart; in short, with every deviation from the established rules of the respective orders, whether such column be composed of marble, of stone, or of plastered brickwork; the costliness of the material makes no difference in the design; but this is not the case with decorations. The cheapness and facility with which good designs may be multiplied in *papier maché*, or putty composition, have encouraged bad taste in the lavish profusion of tawdry embellishment.

This consideration leads me to assert, that every species of enrichment or decoration ought to be costly, either in its materials or in its workmanship: and if we attend to the common opinion of all, except children and savages, we shall find that no real value is attached to any decoration, except upon this principle; on the contrary, it becomes contemptible in proportion as it *affects to seem what it is not*.‡

The idea of costliness in ornament is increased by its rarity, or rather by its being used only where it is most conspicuous, and this sort of œconomy is observable even in the works of nature; for instance, the most beautiful coloured feathers of birds are on the surface, while those for use, rather than for shew, are generally of a dirty brown; it may also be observed, that those butterflies, or moths, whose wings are ornamented on the under side, generally bear them erect; while those which have the upper side most beautiful generally spread them flat. The same remark may be extended to all the vegetable tribe, every flower, and every leaf, has one side more ornamented, more glossy, more vivid, or more highly finished than the other, and this is always the side presented to the eye. Hence we are taught by the example of nature, not to lavish decorations where they cannot generally be seen.§

Source Endnotes

* That these ornaments, although not absolutely necessary, should appear to be *useful*, is evident from the disgust we feel at seeing them improperly applied; as in a column, without an entablature, or an arch supporting nothing, or a pediment without a roof; but

I do not consider columns, or pilasters, as ornaments, when used, as we often see them, to the doors of houses; they may then more properly be called decorations in a bad taste. A column is the most sumptuous ornament of Grecian architecture, and should never be subordinate to any other part of the edifice; it should either belong to the entablature and cornice of the building, or it should be wholly omitted.

† Instances of this often occur in the neighbourhood of large cities and towns, where the taste of a carpenter, and not of an architect, puts balustrades to houses without any entablatures, or perhaps places them in a garret window, while the plain parapet wall is loaded with Mercuries, vases, pine-apples, eagles, acorns, and round balls.

‡ If a lady of high rank were to decorate her person with gauze and gilt paper, with glass beads, and the feathers of common English birds, instead of muslins and gold lace, diamonds, and the feathers of an ostrich, or a bird of paradise; although she might be equally brilliant, and even dispose her dress with grace and fancied taste, we should pronounce it *tromperie*, as affecting to *seem what it is not*.

§ Good taste can only be acquired by leisure and observation; it is not therefore to be expected in men whose time is fully employed in the more important acquirement of wealth or fame; while on certain subjects of taste the most elegant women often excel the most learned men; and although they may not have investigated the causes of the pleasure, they either derive or communicate, yet they are more exquisitely sensible to both. This, if it were necessary, might be used as an apology for occasionally introducing allusions more familiar than the philosophic reader may deem conformable to the nature of a didactic work.

25

'ON THE EMPLOYMENT OF VASES AS RECEPTACLES FOR PLANTS IN TOWN GARDENS . . .'

J. C. Loudon

Source: J. C. Loudon, 'On the Employment of Vases as Receptacles for Plants in Town Gardens with some Remarks on their Use in Garden Scenery in the Country', *Gardener's Magazine,* vol. 10 (October 1834), pp. 489–495

PASSING one day lately Mr. Austin's museum of works in artificial stone in the New Road, London, we were much struck with the effect produced by the vase *fig.* 82. planted with a group of flowers. The flowers were of the commonest kinds. In the centre of the group were purple georginas; around these nasturtiums with their bright orange flowers, white phloxes, and mignonette; and,

82

 DOI: 10.4324/9780429198809-30

hanging over the rim of the vase, were long slender stolones of moneywort. The general effect was exceedingly good; and it immediately occurred to us to recommend this kind of ornament for gardens, especially for those of the town. Nothing, however, ought to be recommended without assigning reasons for doing so. Now, there are several reasons for recommending elevated vases as receptacles for flowers, in order that they may form ornaments for gardens in confined situations; and there are also reasons for recommending vases not to be filled with flowers for gardens in the country. We shall take these reasons in the order of their importance.

In the first place, town gardens, by which we mean chiefly the front gardens of street houses, are generally so confined by walls, that no herbaceous plant will thrive in them longer than one season. The cause is to be found in the want of air, of light, and of fresh soil. The soil, indeed, gets sodden by moisture and want of drainage to such a degree, that no stirring or manuring will do it much good; and,

83

84

therefore, if fine flowers are expected to grow in such gardens, the soil must be renewed annually, or at least every two years: but, even if the soil were renewed frequently, this would not supply the want of a free circulation of air. To attain this object, either the boundary walls of the front garden must be removed, and open iron railings substituted for them, or the beds or borders containing the flowers must be elevated as high as the walls. Sometimes the former mode might be adopted, but the latter will generally be found impracticable; and therefore, as a substitute for it, we would propose elevating the finer flowers in vases fixed on pedestals. The soil contained in these vases, however large they may be (*fig*. 82. is 8 ft. in diameter), could be renewed every year for a few shillings, and a fresh stock of plants supplied for a few shillings more. The vase would cost, perhaps, 3*l*. or 5*l*.; and this is actually less than what the renewal of the soil of the whole garden would cost every year, or at least every other year. If, however, this first cost be thought too great, smaller, and, of course, cheaper, vases might be employed, such as *figs*. 83, 84, 85, 86, and 87. One large vase, or one large and two small vases, would be quite enough for an ordinary street garden: for example, such as

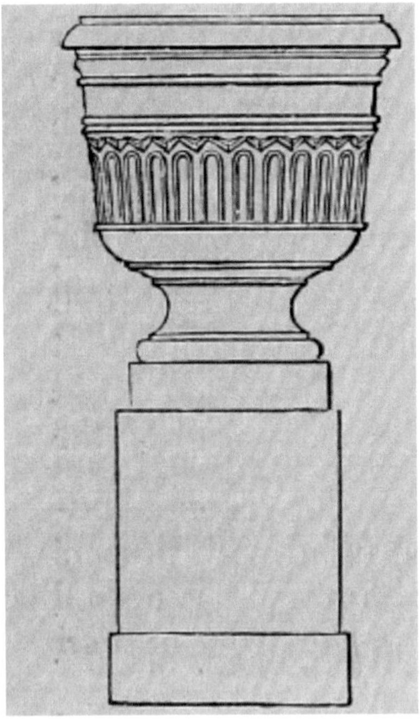

85

86

those along both sides of the New Road. The area of the garden might be of turf, of gravel, or paved, with a narrow dug border along each of the side walls and in front, planted with bulbs and the commoner evergreen herbaceous plants (such as pinks, saxifrages, &c.), with a few trees and shrubs of showy flowering kinds. If these front gardens were either wholly in turf, excepting always the walk from the gate to the door of the house, or wholly paved, with a very narrow marginal dug border, and the vase or vases forming a point or a line in the centre of the turf or pavement, they would look incomparably better than they do at present. They are now generally laid out entirely in dug beds, either on turf or with box edgings, and small narrow gravel paths; in which beds, for the reasons before given, fine plants never thrive, and which are exceedingly difficult to keep orderly and neat, though they never look well unless they are so. By the plan of having vases instead of beds, these gardens may be kept neat all the year round at a very trifling expense. In the summer the vases might be filled with showy flowers, and in winter either left empty or planted with evergreen shrubs and spring bulbs. Thus far as to what may be called the economical reasons for employing vases as receptacles for flowers in town gardens.

The next reason is, that these vases are beautiful objects, either with or without flowers. We are pleased to see flowers in them, because, in addition to the natural beauty of the flowers, there is the important one of their being presented to us in a new and striking situation. The value even of common flowers thus becomes enhanced; and what would hardly be noticed in a bed, or in a border, has a new interest, and a degree of dignity lent to it, by its being placed in a handsome vase elevated on a pedestal. The vase is beautiful as a work of art, whether it is filled with plants or not; but when it serves as a receptacle for fine flowers, in a situation where such flowers are not seen in beds or borders, its importance is increased by the additional beauty which it confers on them, in presenting them in a distinguished situation. There is a certain degree of distinction conferred upon a plant by placing it even in a common flower-pot; but where, instead of the common form and fragile material of the flower-pot, the elegant shape and durable massive appearance of the sculptured vase of stone (natural or artificial) is employed, the superior value of the latter must be felt by every one. A farther reason why the common flower-pot is deficient in dignity arises from its being portable, and generally placed on the ground; while the fixed and elevated vase, from its elevation, magnitude, weight, and solid foundations, has all the dignity and permanence of an architectural structure.

From the influence of fixed and elevated vessels, or other permanent structures, for containing plants, in giving consequence to them, has arisen, not only the employment of stone vases, but even of that description of rustic baskets and vases, for containing flowers, now frequently placed on lawns in extensive pleasure-grounds; particularly suitable for cottage gardens. But the satisfaction which

87

rustic vases give in such situations is far inferior to that produced by fixed stone vases in town gardens; because, in the latter situation, flowers of any kind are comparatively rare and cherished; whereas, in extensive pleasure-grounds in the country, where there are beds of pelargoniums and other flowers without end, it seems almost needless profusion to elevate them in vases. Hence it is, that when stone or pottery vases are introduced into gardens in the country, they are very seldom filled with plants of any kind. They are introduced there as beautiful works of art, to give pleasure, by their contrast, to the beautiful works of nature with which they are surrounded.

Another reason why vases of flowers should be introduced into the little walled gardens of streets is, that they harmonise admirably with the masonry and architectural forms by which they are surrounded. For this reason, also, stone vases should be sparingly introduced in pleasure-grounds in the country, except as appendages, or ornaments to architecture; such as on the parapets of terraces near the house, on the stone borders, balustrades, &c., of architectural flower-gardens, &c. They should never be set down on the naked ground, and, above all, they should always have a fixed and permanent character. Even a common flower-pot, when fixed on a pedestal, assumes a character of dignity, which it never can have in a situation from which the spectator imagines it can be easily removed. Nay, more: a vase, or a common flower-pot, though it be not fixed by mortar or cement so as to become a part of the masonry on which it is placed, yet, if it is only set on a wall, or in any other position from which it cannot be removed without destroying the regularity or symmetry of the composition or whole to which it belongs, acquires a fixed and permanent character. This character is given when a flower-pot or vase is set on a wall, so as to form a termination to piers, pedestals, pilasters, or pillars of any kind: but place it on the coping of the wall, in the interval between such piers, &c., and its character becomes at once temporary and unmeaning: because there is no obvious reason for placing a vase there at all; for limiting the number to one; or for not moving it, either to the right or left, or even for not taking it away altogether.

If it be asked, why use vases in the country at all, unless plants are to be put in them? The answer is, they are employed as appropriate ornaments, as beautiful forms, and as articles of rarity and of value. But it may be farther asked, can an object be beautiful, or confer ornament on another object, when it is in itself of no use? Certainly it can. The noblest ornaments are those which are without use, in the common sense of the word; though, in the sense of art and refinement, everything is of use which adds to the strength of the emotion of that kind of beauty which it is desired to produce. What would be the splendour of the drawing-room, or the effect of the picture-gallery, without gilding? or an assemblage of highly dressed beauties, without those useless articles, diamonds and other precious stones? Nature has planted in the mind of man, from the rudest savage to the most refined philosopher, a capacity for deriving pleasure from beauty. It is necessary that this want should be supplied; and it will be found gratified, to a certain extent, in every stage in the progress of civilisation: partly by objects which are useful in the ordinary sense of the word, as dress, &c.; and partly by such as have no other use than the power of conferring ornament. In the rudest state of society, we have the tattooing and nose-rings of the savage; next the earrings, scarcely less barbarous, of a more civilised period: and, though these must pass away with the introduction of a taste more truly refined, the necklaces and bracelets, which are common to all ages and countries, will probably continue to be so, as long as women dress highly, and large parties are given by artificial light.

Another reason for the introduction of vases, and other architectural ornaments of a similar kind, both into town and country gardens, is, their tendency to create and cultivate a taste for natural beauty in the spectator; since, however paradoxical it may seem, it is, nevertheless, true that we can only be instructed in the beauties of nature through the medium of art. No person can either enjoy, understand, describe, or remember a landscape properly, who has not been instructed in drawing landscapes; any more than he could enjoy, understand, or analyse any particular style of writing, without having been taught grammar and composition. Teaching a person to draw trees, for example, is teaching him to find beauties in trees which he had never seen in them before. Setting before a person beautiful vases, is to familiarise him with beautiful forms, which he cannot help contrasting with ordinary shapes; and, according to his natural capacity, or the suitableness of his organisation for discernment in forms, he will find in the vases a unity of tendency in the lines which constitute their outlines, a symmetry in their general forms, and a richness, an intricacy, and a character of art in their sculptured details, which will induce him to search for these qualities in other objects of art, and to observe every tendency towards them in the works of nature. To such an observer, the productions of architecture and sculpture would assume a new interest; and he would gradually, and almost imperceptibly, acquire a knowledge of, and taste for, the beauty of forms and lines in objects generally. He would thus learn to distinguish symmetry, regularity, unity, variety, and other abstract qualities in works of art, which he would probably have never discovered in the forms of nature without such assistance; though they exist there, and are merely imitated by art in a way which renders them more obvious to untutored man, who sees only that mind in other things which he understands and feels in himself.

The last reason which we shall offer for the introduction of vases into architectural and garden scenery is, the gratification which such objects afford to the man of intelligence and taste. There are, perhaps, few objects, next to the human figure, which afford as many interesting historical associations as the vase. It may truly be said to be the first and the last production of the plastic art. The first utensil formed by man in the dawn of civilisation, in every country, is a vessel or vase for holding water; and that on which the highest resources of art are bestowed, in ages of the greatest refinement, is a vessel or vase for holding wine. In the first case, it is hollowed out of a gourd, or rudely shaped of clay and dried in the sun; and, in the latter case, it is manufactured of costly metals or precious stones; or, if of common materials, such as stone, earthenware, or glass, it is rendered valuable by the taste and skill bestowed on its form or its ornaments. The history of every country may be traced by its vases, no less than by its coins; and the history of all countries is set before us in the vases of all countries. It would be foreign to the object of this Magazine to say more on the subject: a little reflection will bring materials for thinking on it to the mind of every reader; and the young gardener will, we trust, after he has read and considered this paper, look at an ornamental vase, or even a common flower-pot, with a new interest.

26

'DESCRIPTIVE NOTICE OF THE VILLA OF MRS. LAWRENCE AT DRAYTON GREEN . . .'

J. C. Loudon

Source: J. C. Loudon, 'Descriptive Notice of the Villa of Mrs. Lawrence at Drayton Green'. From the "Suburban Gardener" ', *Gardener's Magazine,* vol. 14 (July 1838). [305–322] excerpts, pp. 306–308, 311–319, 321–322

THE *Lawrencian Villa, Drayton Green.* (*figs*. 42. to 62.). – This villa, of which *fig*. 42. is the ground plan, is unquestionably the most remarkable of its size in the neighbourhood of London, on account of the great variety and beauty which have been created in it, under the direction of Mrs. Lawrence, F.H.S., the lady of the celebrated surgeon of that name. The straggling little village of Drayton Green lies about seven miles to the west of London, at a short distance from the road to Oxford. The surrounding country is flat, or nearly so; and it is principally under pasture, in very irregular enclosures, with hedges, also very irregular in regard to height and breadth, and abounding, in most cases, with English elms, oaks, and limes. The soil is a thin loam on gravel, and the country generally is considered remarkably healthy, and agreeably picturesque.

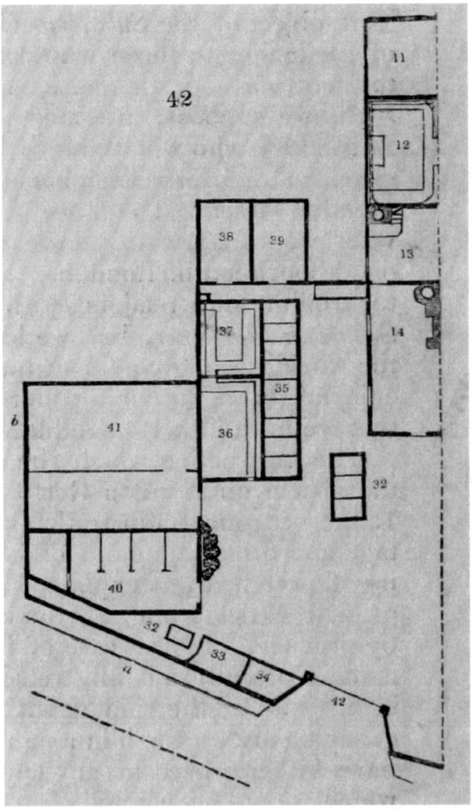

DOI: 10.4324/9780429198809-31

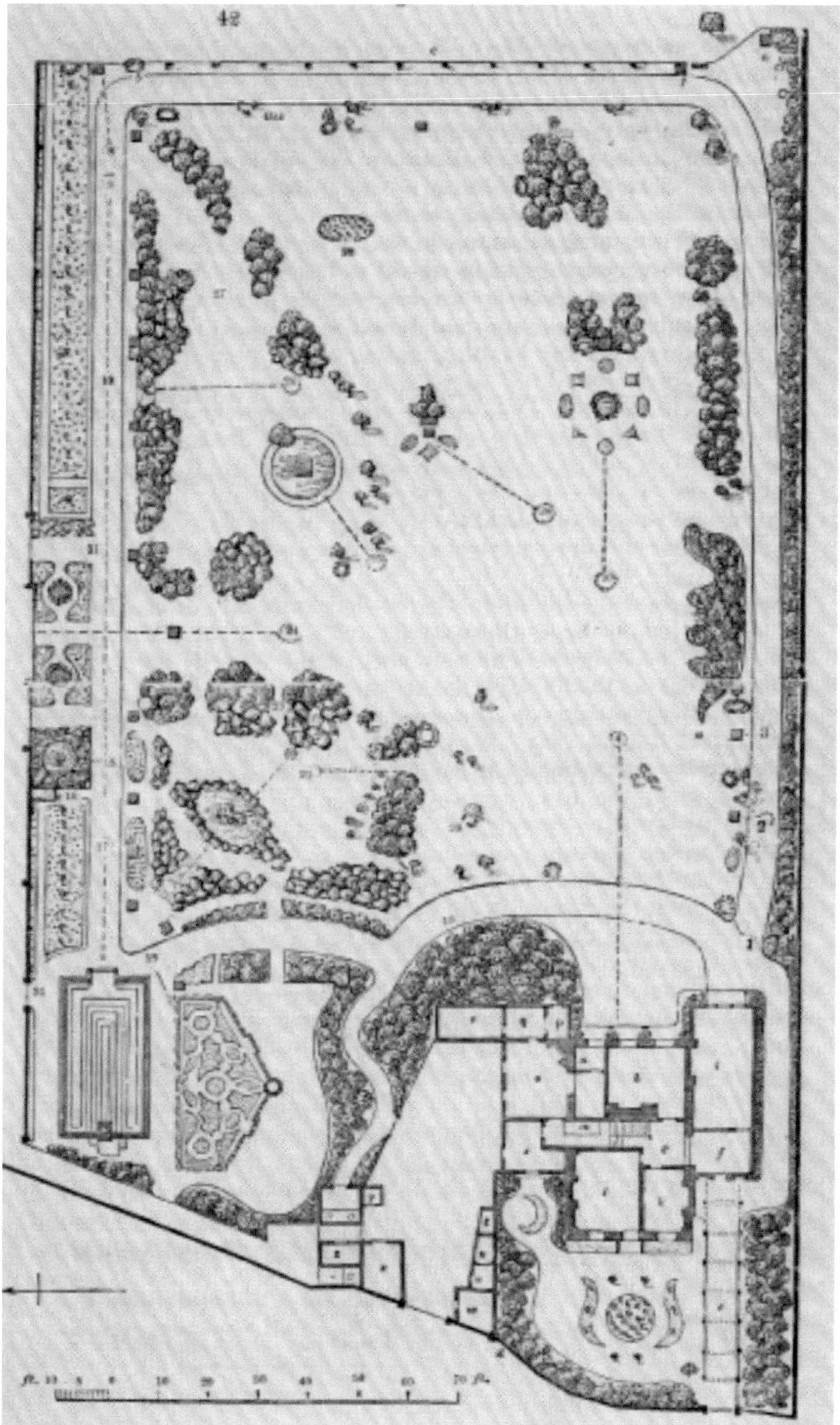
42
ft. 10 5 0 10 20 30 40 50 60 70 ft.

The Lawrencian Villa may be said to occupy in all about twenty-eight acres. The house, stable offices, and decorated grounds stand on about two acres; and at the distance of about two or three hundred yards, across the road, are the kitchen-garden, poultry-houses, and piggeries, occupying nearly two acres; and two pasture fields, containing twenty-four acres

. . .

Entering the lawn from the drawingroom (*h*), we find a gentle descent from the veranda to the walk. Turning to the right, at the angle at 1, we observe the foliated vase *fig*. 44., the base of which is concealed by a plant of tree ivy; proceeding onwards towards 2 and 4, we pass the pedestals and vases *figs*. 45. and 46. We are now at a sufficient distance from the garden front of the house, to see it to advantage by turning round; and, if we step on the lawn to the point 4, we shall find the view *fig*. 47., to the left of which will be observed the ivy vase, and a basket containing a pyramid of roses; and to the right an elevated rustic basket of pelargoniums. The large window on the left is that of the dining-room.

44

Looking from 1, 2, and 3, across the lawn, the eye observes an intricate maze of agreeable and beautiful objects, but sufficiently distant not to create the idea of being crowded or confused; the reason of which is, that there is always an ample surface of naked lawn in the foreground, or middle distance, to contrast with the ornamental groups, and to throw them sufficiently far from the eye.

The margin of plantation on the right is composed partly of evergreen trees and shrubs, and partly of deciduous flowering kinds. The groups on the left hand are, in part, of more rare sorts, and contain a great many fine hybrid rhododendrons and azaleas. All these plantations and groups are treated in the picturesque manner; there being scarcely anything in these grounds, except the single plants, such as the standard roses, and some rhododendrons and other shrubs, which can be considered as treated in the gardenesque style of culture.

At 5, there is a fine specimen of double-blossomed furze, and two splendid vases on elevated pedestals; both combining to form an interesting termination to the comparatively straight walk from the house to this point. One of these vases at 6, when the spectator is at a distance, appears to him to be the terminating point, while that at 5 comes into his view afterwards. The walk from 5 to 7 is several feet higher than the floor of the veranda in the front of the drawingroom; and hence the views towards the house, being along a descent, are less interesting than, from the number of objects on the lawn, they otherwise would be. The

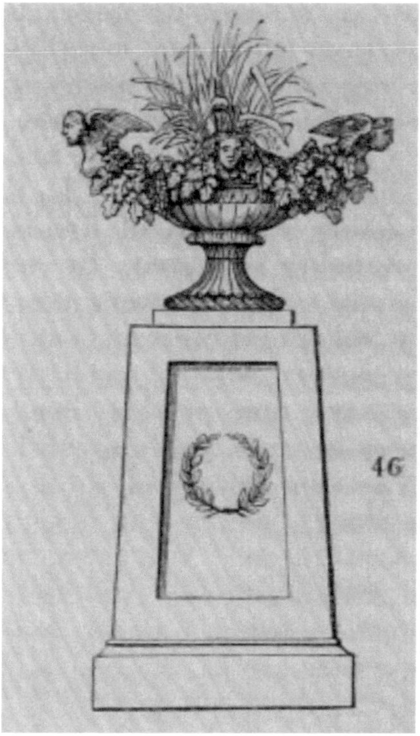

45 *46*

47

view into the paddock, to the right, affords an agreeable relief from the excess of beauty and variety on the lawn, as it consists of a plain grass field, grazed by some fine Alderney cows, and planted with two or three scattered elms, oaks, and aspens and other poplars.

The next scene of interest is the Italian walk, arrived at the point 8, in which, and looking back towards the paddock, we have, as a termination to one end of that walk, the rustic arch and vase *fig*. 48.

48

From the point 9, we have the view of the Italian walk *fig*. 49., with a span-roofed green-house as the termination at the farther end, and an elegant fountain on the right hand. The border on the left is planted with the most choice herbaceous flowers, interspersed with standard roses at regular distances; and the wall is devoted in part to the finer fruits, but principally to climbing roses, and other climbing or twining shrubs of fragrance or beauty. At the point 10, there is a rustic archway of rockwork on the right, from which an interesting view across the lawn is obtained. At the point 11, there is a walk across the border to the bath-house, adjoining which is a camellia-house (12); and beyond that two long sheds (13, 14), for tools, pots, &c., with potting benches, and other places for garden materials and operations, and for the boilers to heat the bath, the camellia-house,

49

&c. In the camellia-house there are sixty-seven of the finest species and varieties that can be procured.

At the point 15, we are immediately in front of the fountain *fig.* 50., supplied from a cistern which forms a small tower on the top of the tool-house; and beyond that is a walk to the stone cistern at 16, which supplies water for watering the garden. The water is raised to these cisterns by a forcing pump in the stableyard.

50

Farther on in the Italian walk, at 17, we have the front view of the span-roofed green-house *fig*. 51.; and a little to the left, at 18, the view of the French parterre *fig*. 52.

51

52

Proceeding towards the house, a view of a handsome weeping ash (20) is obtained from the point 19; and, at the farther extremity of the walk, the vases placed at 1, 2, 3 on the plan have an excellent effect, backed by the marginal plantation of evergreens. Leaving the walk at 19, and passing the weeping ash at 20, if we advance on the lawn to 21, and look towards the south, we have the pollard vista *fig*. 53.; and, changing the position to 22, we have the view of the rockwork, statue of Fame, &c., shown in *fig*. 54.

53

On the right and left of 23 are two groups of rockwork, with concealed springs, which drop from rock to rock, and from stone to stone, and form curious little moist places for aquatic plants. Advancing to 24, and looking northwards, we have the statue of Mercury in the foreground, and behind it the camellia-house, the wall on each side of which is heightened with trelliswork for creepers, as shown in *fig*. 55.

At 25, we have the view of the fountain and arch behind, shown in *fig*. 56. In the basin are nymphæas and other aquatics; and on one side is a Napoleon willow.

At 26, we have the view of the rustic arch and Cupid, shown in *fig*. 57.; and, at 27, the tent seen in *fig*. 56. is frequently pitched in the summer time, which gives the idea of the warm season, and of the enjoyment of coolness and refreshing breezes in the midst of intense sunshine.

54

55

At 28, there is a bed of *Rò*sa índica, in the centre of which is a large plant of *Yúcca* gloriòsa; and, proceeding across the lawn to 29, we have the view *fig*. 59.; and, at 30, we have the fountain, surrounded by baskets of flowers, with the two garden nymphs *fig*. 62.

56

57

We shall now suppose that the spectator walks across the lawn, and, passing the span-roofed green-house, enters the court of offices by the door at 30.

. . .

59

Remarks. This villa may be considered as a model of its particular kind; and, though it may not be in the power of many to imitate it in every thing, yet the humblest and most economical possessor of a villa residence of two acres may take a lesson from Mrs. Lawrence's taste, as displayed in the manner in which the trees and shrubs are grouped on the lawn. Every one cannot have so many fountains, or form rockwork of spars, fossil organic remains, and other geological specimens brought from distant parts of the country; but every one may sink in the ground a few small wooden cisterns lined with lead, and supply them with water by hand, as it evaporates in the summer season. Some of these may serve as brilliant spots to attract the eye, and others as habitats for aquatic plants

The margins of basins of this sort can be effectually disguised with rockwork, and this can be procured from the nearest brick-field, stone-quarry, or, perhaps,

62

from old houses, which are taking down, chalk-pits, ferruginous gravel-pits, &c. If suitable vitrified bricks cannot be procured, common bricks may be joined together, in masses of any size and shape, by cement; and there is no reason why blocks so formed, or any other materials to be substituted for rockwork, should not receive weather stains artificially, no less than the walls of a house, where the object is to imitate an ancient building. As to the wooden cisterns, they will last long enough: and we know, from experience, that it is cheaper, in the end, to form such cisterns of wood, lined with lead, than to build cisterns of brickwork and cement; for, unless these are of considerable size, the cost is as great as where lead is employed; and they are much more apt to leak, and receive injury from frost.

It is worthy of remark, that a good deal of the interest attached to the groups on the lawn of the Lawrencian Villa depends on the plants which are planted in the rockwork. Now, though every one cannot procure American ferns, and other plants of such rarity and beauty as are there displayed, yet there are hundreds of alpines, and many British ferns, which may be easily procured from botanic gardens, or by one botanist from another; and, even if no perennials could be obtained suitable for rockwork, there are the Californian annuals, which alone are sufficient to clothe erections of this kind with great beauty and variety of colouring.

With regard to the statues, vases, &c., though some of these, at Drayton Green, are of bronze, marble, or stone, and have cost considerable sums, yet others of composition, equal in point of taste, though far inferior in pecuniary value, may readily be procured, at a moderate cost, of Austin's artificial stone, or of earthenware.

We are aware that there are many persons, of a simple and severe taste, who will think that the Lawrencian Villa is too highly ornamented with statues and sculptures; but allowance must be made for individual taste, for devotion to the subject, and for the limited extent of the place. Were Mrs. Lawrence in possession of a villa of 100 acres, there can be no doubt that she would display on her lawn a taste as appropriate to a residence of that extent, as the taste she has displayed at Drayton Green is suitable for that place.

27

'EMBELLISHMENTS'; 'ARCHITECTURAL, RUSTIC AND FLORAL'

Andrew Jackson Downing

Source: Andrew Jackson Downing, 'Embellishments; Architectural, Rustic and Floral', *A Treatise on the Theory and Practice of Landscape Gardening* (1841), sect. 10 [348–403], excerpts pp. 351–357; 383–388

The highly decorated terrace, as we have here supposed it, would, it is evident, be in unison with villas of a somewhat superior style; or, in other words, the amount of enrichment bestowed on exterior decoration near the house, should correspond to the style of art evinced in the exterior of the mansion itself. An humble cottage with sculptured vases on its terrace and parapet, would be in bad taste; but any Grecian, Roman, or Italian villa, where a moderate degree of exterior ornament is visible, or a Gothic villa of the better class, will allow the additional enrichment of the architectural terrrace and its ornaments. Indeed the terrace itself, in so far as it denotes a raised dry platform around the house, is a suitable and appropriate appendage to every dwelling, of whatever class.

The width of a terrace around a house, may vary from five to twenty feet, or more, in proportion as the building is of greater or less importance. The surrounding wall which supports its level, may also vary from one to eight feet. The terrace, in the better class of English residences, is paved with smooth flag stones, or in place of this, a surface of firm well-rolled gravel is substituted. In residences where a parapet or balustrade would be thought too expensive, a square stone or plinth is placed at the angles or four corners of the terrace, which serves as the pedestal for a vase or urn. When a more elegant and finished appearance is desirable, the parapet formed of open work of stone, or wood painted in imitation of stone, rises above the level of the terrace two or three feet with a suitably bold coping. On this, vases may be placed, not only at the corners, but at regular intervals of ten, twenty, or more feet. We have alluded to the good effect of climbers, here and there planted, and suffered to intermingle their rich foliage with the open work of the parapet and its crowning ornaments. In the climate of Philadelphia, the Giant Ivy, with its thick sculpturesque looking masses of foliage, would be admirably suited to this purpose. Or the Virginia Creeper, (the Ivy of America,)

 DOI: 10.4324/9780429198809-32

may take its place in any other portion of the Union. To these we may add, the Chinese twining Honeysuckle, (Lonicera flexuosa,) and the Sweet-scented Clematis, both deliciously fragrant in their blossoms, – with many other fine climbers which will readily recur to the amateur.

There can be no reason why the smallest cottage, if its occupant be a person of taste, should not have a terrace decorated in a suitable manner. This is easily and cheaply effected by placing neat flower-pots on the parapet, or border and angles of the terrace, with suitable plants growing in them. For this purpose, the American or Century *Aloe*, a formal architectural-looking plant, is exceedingly well adapted, as it always preserves nearly the same appearance. Or in place of this, the *Yuccas*, or "*Adam's needle*, and *thread*," which have something of the same character, while they also produce beautiful heads of flowers, may be chosen. *Yucca flaccida* is a fine hardy species, which would look well in such a situation.

Where there is a terrace ornamented with urns or vases, and the proprietor wishes to give a corresponding air of elegance to his grounds, vases, sundials, etc., may be placed in various appropriate situations, not only in the architectural flower-garden, but on the lawn, and through the pleasure-grounds in various different points, *near the house*. We say near the house, because we think so highly artificial and architectural an object as a sculptured vase, is never correctly introduced unless it appear in some way connected with buildings, or objects of a like architectural character. To place a beautiful vase in a distant part of the grounds, where there is no direct allusion to art, and where it is accompanied only by natural objects, as the overhanging trees and the sloping turf, is in a measure doing violence to our reason, or taste, by bringing two objects so strongly contrasted, in direct union. But when we see a statue or a vase placed in any part of the grounds where a near view is obtained of the house, (and its accompanying statues or vases,) the whole is accounted for, and we feel the distant vase, to be only a part of, or rather a repetition of the same idea, – in other words, that it forms part of a whole, harmonious and consistent.

Vases of real stone, as marble or granite, are decorations of too costly a kind ever to come into general use among us. Vases, however, of equally beautiful forms, are manufactured of artificial stone, of fine pottery, or of cast iron, which have the same effect, and are of nearly equal durability, as garden decorations.

A vase should never, in the open air, be set down upon the ground or grass, without being placed upon a firm base of some description, either a *plinth* or a *pedestal*. Without a base of this kind, it has a temporary look, as if it had been left there by mere accident, and without any intention of permanence. Placing it upon a pedestal, or square plinth, (block of stone,) gives it a character of art, at once more dignified and expressive of stability. Besides this, the pedestal in reality serves to preserve the vase in a perpendicular position, as well as to expose it fairly to the eye, which could not be the case were it put down, without any preparation, on the bare turf or gravel.

. . .

Sundials are among the oldest decorations for the garden and grounds, and there are scarcely any which we think more suitable. They are not merely decorative, but have also a useful character, and may therefore be occasionally placed in distant parts of the grounds, should a favourite walk terminate there. When we meet daily in our walks for a number of years, with one of these silent monitors of the flight of time, we become in a degree attached to it, and almost look upon it as gifted with a species of intelligence, which looks out when the sunbeams smile upon its dial-plate.

. . .

Open and covered seats, of various descriptions, are among the most convenient and useful decorations for the pleasure-grounds of a country residence. Situated in portions of the lawn or park, somewhat distant from the house, they offer an agreeable place for rest or repose. If there are certain points, from which are obtained agreeable prospects or extensive views of the surrounding country, a seat, by designating those points, and by affording us a convenient mode of enjoying them, has a double recommendation to our minds.

Open and covered seats are of two distinct kinds; one *architectural*, or formed after artist-like designs, of stone or wood, in Grecian, Gothic, or other forms; which may, if they are intended to produce an elegant effect, have vases on pedestals as accompaniments; the other, *rustic*, as they are called, which are formed out of trunks and branches of trees, roots, etc., in their natural forms.

There are particular sites, where each of these kinds of seats, or structures, is, in good taste, alone admissible. In the proximity of elegant and decorated buildings where all around has a polished air, it would evidently be doing violence to our feelings and sense of propriety to admit many rustic seats and structures of any kind; but architectural decorations and architectural seats are there correctly introduced. For the same reason also, as we have already suggested, that the sculptured forms of vases, etc., would be out of keeping in scenes where nature is predominant, (as the distant wooded parts, or walks of a residence,) architectural, or in other words, highly artificial seats, would not be in character: but rustic seats and structures, which, from the nature of the materials employed and the simple manner of their construction, appear but one remove from natural forms, are felt at once to be in unison with the surrounding objects.

The simplest variety of covered architectural seat, is the latticed arbour for vines of various description, with the seat underneath the canopy of foliage: this may with more propriety be introduced in various parts of the grounds than any other of its class, as the luxuriance and natural gracefulness of the foliage which covers the arbour, in a great measure destroys, or overpowers the expression of its original form. Lattice arbours, however, neatly formed of rough poles and posts, are much more picturesque and suitable for wilder portions of the scenery.

The temple and the pavilion, are highly finished forms of covered seats, which are occasionally introduced in splendid places, where classic architecture prevails. There is a circular pavilion of this kind at the termination of one of the walks at Dr. Hosack's[1] residence, Hyde Park. Fig. 57.

We consider rustic seats and structures as likely to be much preferred in the villa and cottage residences of the country. They have the merit of being tasteful and picturesque in their appearance, and are easily constructed by the amateur, at comparatively little or no expense. There is scarcely a prettier or more pleasant object for the termination of a long walk in the pleasure-grounds or park, than a neatly thatched structure of rustic work, with its seat for repose, and a view of the landscape beyond. On finding such an object, we are never tempted to think that there has been a lavish expenditure to serve a trifling purpose, but are gratified to see the exercise of taste and ingenuity, which completely answers the end in view.

Figure 58, is an example of a simple rustic seat formed of the crooked and curved branches of the oak, elm, or any other of our forest trees. Fig. 59, is a seat of the same character, made at the foot of a tree, whose overhanging branches afford a fine shade.

[Fig. 57.]

[Fig. 58.]

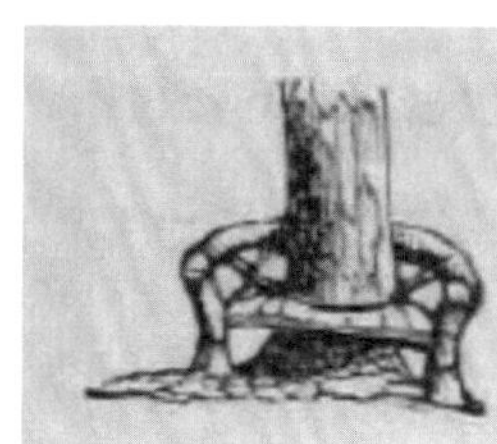

[Fig. 59.]

Figure 60, is a covered seat or rustic arbour, with a thatched roof of straw. Twelve posts are set securely in the ground, which make the frame of this structure, the openings between being filled in, with branches (about three inches in diameter) of different trees – the more irregular the better, so that the perpendicular surface of the exterior and interior is kept nearly equal. In lieu of thatch, the roof may be first tightly boarded, and then a covering of bark or the slabs of trees with the bark on, overlaid and nailed on. The figure represents the structure as formed round a tree. For the sake of variety this might be omitted, the roof formed of an open lattice work of branches like the sides, and the whole covered by a grape, bignonia, or some other vine or creeper of luxuriant growth. The seats are in the interior.

[Fig. 60.]

Figure 61, represents a covered seat of another kind. The central structure, which is circular, is intended for a collection of minerals, shells, or any other curious objects for which an amateur might have a *penchant*. Geological or mineralogical specimens of the adjacent neighbourhood, would be very proper for such a cabinet. The seat surrounds it on the outside, over which is a thatched roof or veranda, supported on rustic pillars formed of the trunks of saplings, with the bark attached.

[Fig. 61.]

Many of the English country places abound with admirable specimens of rustic work in their parks and pleasure-grounds. White Knight's,[2] in particular, a residence of the Duke of Marlborough has a number of beautiful structures of this kind. Figure 62, is a view of a round seat with thatched roof, in that demesne. Three or four rustic pillars support the architrave, and the whole of the exterior and interior, (being first formed of frame-work,) is covered with straight branches of the maple and larch. The seat on the interior, looks upon a fine prospect; and the seat on the back of the exterior, fronts the park.

[Fig. 62. Rustic covered Seat.]

There is no limit to the variety of forms and patterns in which these rustic seats, arbors, summer-houses, etc., can be constructed by an artist of some fancy and ingenuity. After the frame-work of the structure is formed of posts and rough boards, if small straight rods about an inch in diameter, of hazel, white birch, maple, etc., are selected in sufficient quantity, they may be nailed on in squares, diamonds, medallions, or other patterns, and have the effect of a *mosaic* of wood.

Among the curious results of this fancy for rustic work, we may mention the *moss-house* – erected in several places abroad. The skeleton or frame-work of the arbor or house, is formed as we have just stated; over this small rods half an inch in diameter are nailed, about an inch from centre to centre; after the whole surface is covered with this sort of rustic lathing, a quantity of the softer wood-moss of different colours is collected; and taking small parcels in the hand at a time, the tops being evenly arranged, the bottoms or roots are crowded closely between the rods with a small wooden wedge. When this is done with some little skill, the tufted ends spread out and cover the rods entirely, showing a smooth surface of mosses of different colours, which has an effect not unlike that of a thick Brussels carpet.

The mosses retain their colour for a great length of time, and when properly rammed in with the wedge, they cannot be pulled out again without breaking their tops. The only example which we have seen of a handsome moss-house in this country, is at Winship's[3] Garden, Brighton, near Boston.

A prospect tower is a most desirable and pleasant structure in certain residences. Where the view is comparatively limited from the grounds, on account of their surface being level or nearly so, it often happens that the spectator, by being raised some twenty-five or thirty feet above the surface, finds himself in a totally different position, whence a charming *coup d'œil*, or bird's-eye view of the surrounding country is obtained.

Those of our readers who may have visited the delightful garden and grounds of M. Parmentier,[4] near Brooklyn, some half a dozen years since, during the lifetime of that amiable and zealous amateur of horticulture, will readily remember the rustic prospect-arbor, or tower, *figure* 63, which was situated at the extremity of his place.

[Fig. 63. Prospect Arbor.]

It was one of the first pieces of rustic work of any size, and displaying any ingenuity that we remember to have seen here; and from its summit, though the garden walks afforded no prospect, a beautiful reach of the neighbourhood for many miles was enjoyed.

Figure 64 is a design for a rustic prospect tower of three stories in height, with a double thatched roof. It is formed of rustic pillars or columns, which are well fixed in the ground, and which are filled in with a fanciful lattice of rustic branches. A spiral stair-case winds round the interior to the platform of the second and upper stories, where there are seats under the open thatched roof.

[Fig. 64. Prospect Tower.]

Editorial Endnotes

1 Hossack [sic] – Dr. David Hosack (1769–1835) was a medical man of Scottish descent, whose extensive private garden at Hyde Park, New York, became one of the more well-known early 19th-century American gardens.

2 Whiteknights – the 80-acre estate at Earley, Berkshire, in England, owned by the Duke of Marlborough, George Spencer-Churchill (1766–1840), that was sold upon his death. An account of the estate in the form of a poem by Barbara Hofland with engravings originating from on-site paintings by her husband, Thomas Hofland, was published in 1819.

3 Winship's Garden – A 40-acre plot established by Captain Jonathan Winship (1780–1843/1847?) and his younger brother Francis in 1820 at North Brighton near Boston, it served as the base for their horticultural business. Jonathan Winship was one of the principal founders of the Massachusetts Horticultural Society.

4 M. Parmentier – Andrew Parmentier (1780–1830) was a landscape designer and an immigrant from Europe. He designed and planted a garden of around 30 acres near Brooklyn from which he sold a variety of plants, shrubs, and trees. His *Periodical catalogue of fruit & ornamental trees and shrubs, green-house plants, etc. Cultivated and for sale at The Horticultural and Botanic Garden of Brooklyn . . .*, was published in 1828.

28

'GARDEN SCENERY AND ORNAMENTS'

Shirley Hibberd

Source: Shirley Hibberd, 'Garden scenery and ornaments', *Rustic Adornments for Homes of Taste* (1856) [327–353] excerpts, pp. 327–329; 335–353

WHERE shady trees invite the wanderer to a seat, how pleasant is it to find the means of rest and shelter in a garden. What more delightful when the sun burns in Leo, or when the Virgin holds empire in the meridian, than to lounge in a cool shady recess, with a favourite volume and a canister of that seductive, sedative weed, which wafts us on its thin blue wreaths of smoke to the highest region of the most dreamy Elysium. The bees hum about full of business, reproaching us for the listless mood in which we watch their earnestness – everything but the bee is sleepy; the flowers nod as if napping, the air hums itself to sleep, and lo! when we thought we were fast anchored to that favourite book, we, too, have drifted, like a weed upon the wave, into that tropical region where sleep, the "comfortable bird," broods "over the troubled sea of mind, till it is hushed and smooth."[1] There may be a variety of opinions on such a subject, but to us there is no better vindication of a summer-house than the opportunity it affords for the quiet enjoyment of a book, or an afternoon nap. I have always had a bowery reading-room of this sort, and would give up flower-growing before I could relinquish this most exquisite of pleasures. There are some books which seem to have been written for such reading. I could not read Herrick with such pleasure in the presence of a lamp and curtains, nor Earl Surrey, nor Jean Paul, no, nor Tennyson. Did Alexander Smith write his "Life Drama" in such a spot; I think he could hardly have produced it with his feet on a carpet hassock, cramping his chest to a vile mahogany table. Yet there are proper books for the fireside. No one would attempt to read "Faust," or "Zriny," or the Latter Day Pamphlets, in a summer-house; we must adapt things to each other, and ourselves to them, to ensure happy results, and herein lies the art of ornamenting gardens.[2]

A garden is an artificial contrivance for purposes of beauty and pleasure. As an ornament it is a work of art, and into whatever vagaries we may be led by our enthusiasm, we must remember the first principles of the enterprise if we would avoid incongruities and absurdities. The ornaments of a garden should be

 DOI: 10.4324/9780429198809-33

appropriate to the style in which the ground is laid out, and the kinds of plants with which it is stocked. Tangled shrubs, rock-work, and tree stumps filled with ferns, would be as absurd in the immediate neighbourhood of a dwelling, as a marble statue of classic outline would be if stuck in the cleft of a tree in the wilderness.

There are many who like brakes, and glens, and molehills beneath the drawing-room window; but if such a style be in good taste, we may as well camp out in thickets, where mole-hills are more natural, and the brambles finer. The straight lines and angles of the house suggest similar lines and angles in that part of the garden which approaches it most nearly, and the transition from gay and formal parterres, and bright clean gravel paths, to shrubbery and wilderness, should be gradual, each advance preparing the eye for a change from the highly artificial to the semi-natural.

Unfortunately but few of us can develop the true theory of gardening in the neighbourhood of towns, for the terrace, the lawn, the fountains, and shrubberies, occupy more space than can be obtained, except by the lords of manorial acres. Yet this is not to deter us from the adoption of a correct taste in planning and ornamenting our circumscribed grounds; and we need not have a stagnant pool beneath the window of the breakfast-room, or a mass of gimcrack piled under distant elm trees, because we have not space wherein to do things on a scale of magnitude. If a few leading principles are observed at starting, everything will in time fall into its proper place, and there will be no eye-sores, no incongruities.

. . .

It is not our intention to write an essay on style in gardening, and we have been led into this discursive gossiping by the reflection that gardens are seldom so well planned, or so much ornamented as they might be; and when ornaments are introduced, it is but too often that they are inappropriate to the scene, and, if not inappropriate, are frequently rendered so by being injudiciously placed. We would see the house surrounded by elegance of form and colour, the truly rustic should have its place elsewhere, we should go to it, not have it come to us; for nothing can be more absurd than to imitate eastern luxury in carpets and furniture, and by one step pass from mirrors and gilt cornices, and rich tapestry, to honeysuckle hedges, hermits' huts, and bogs covered with toad-stools. A really rustic cottage may have rustic ornaments, even at the threshold, but a villa of formal design must have its garden in keeping at least in proximity to its walls. We only insist on propriety, unity, congruity. Secure fitness of style, and the individual taste may roam at pleasure through all the regions of statuary and fancy-work, till it finds pleasure in the smallest and humblest ornament that ever decked the threshold of an English home.

Where a rustic tone prevails, as it does in the good old English gardens, where the digitalis and the holyhock still assert their claims to admiration, there are no better ornaments than rustic baskets and tree-stumps. In such a garden the summer-house should be of simple design, with an elliptic arch, or domed roof of bark or thatch, the front opening to a rill of water on a green slope, beyond which the eye catches avenues of stately flowers, while on the other hand, the garden loses itself in a maze of verdurous branches. In such scenes, rock banks and ferneries, aquaria shadowed with willows, and stocked with water-lilies and British fish, are appropriate, while the fancy too may sport with ivy mounds, mossy tree-trunks covered with polypodies, bark baskets filled with British orchises, or shrubby exotics. The hum of the bee sounds natural and homely here, and the bee-house should be of the simplest rustic structure, and the hives of straw.

. . .

There are many makers of rustic furniture, but unfortunately, not being florists, they do not avail themselves of all the opportunities which present themselves for a display of rustic taste in such things. Mr. Curry, of Brook Street, Upper Clapton, has gone far in advance of his craft, in all the details of design and adaptation, through the help of his skill as a florist, and perhaps still more from his intense love of ferns and mosses. Mr. Curry's rustic chairs and baskets are as superior to the majority of such productions as might be expected from one whose hobby and passion is fern-growing, and who devotes as much enthusiasm to the stocking of stumps and baskets with rare and beautiful plants, as he does to the mere carpentry of their formation. His rustic arbours and bark-houses are of admirable designs, and he fits up bark-baskets and tree-stumps in a manner most tasteful and bewitching.

I bear this testimony to Mr. Curry out of gratitude for his obliging kindness, as much as approval of his taste in planting rustic ornaments with appropriate and well-selected plants. Those who love to see noble clumps of ferns thrusting their woolly croziers through the ancient bark of a decayed oak-bole, or who want a supply of ferns in pots, or chairs and baskets, thoroughly rustic in character, cannot do better than visit this industrious purveyor to the rustic appetite.

Where it is possible to make an ornament useful, or at least appear useful, it is always much more appropriate. Thus the summer-house or arbour offers shade and a place of rest, and hence should be placed in a spot where rest will be most agreeable: a good view is often appropriately chosen as the site for an invitation to halt.

Rustic arbours may be shaded with fruit trees, vines, currant, cherry, and any other bowery fruit trees, and shrubs. Ivy, clematis, everlasting pea, yellow and white jasmine stauntonia, virginian creeper, etc., are also useful, and make elegant screens. Italian and Chinese summer-houses should be more trimly decorated, and none but the most elegantly arranged climbers led over their finished arches and latticework. The object here is rather to produce an elegant display of colour; a thick rustic bower would be too English to be appropriate. Folding chairs are sometimes made for gardens, so arranged that the back folds down on the seat and keeps it always dry; the visitor has to throw up the back, when the seat is found dry and well protected from the effects of the sun.

As to summer-houses and arbours generally, there never was so much ugliness scattered about the world as of late years has been sprinkled over suburban gardens under the vain pretence of ornament and use. The majority are like toll-houses, or beer-shops, or sweet-meat stalls, destitute of elegance, use, and propriety. A summer-house need not be utterly hidden, but it ought not to stare straight out upon us from a back wall, its ugly lattice-work without one creeping tendril to cool and comfort it, and its interior visible to every gazer, as if it were anything but a place of shade and rest. Though you never use it, it must appear fit for use or it is no ornament. It should be well shrouded with greenery, be easy of access, sufficiently inviting to attract a stranger, yet quiet in tone, and of a chaste pleasing outline. Some of these suburban retreats look very much like the little cabins of the fly-boats on the junction canal; and one always expects to see a head pop up above them to take hold of the rudder and steer them into port. Arbours indeed! what a perversion of language, they are *h*arbours for earwigs, cats, and other vermin, not tree bowers and trellis shelters, as their philology imports.

For garden seats and arbours, where rustic wood-work would be inappropriate, those of metal-work will be found of the greatest value. Mr. Fox, of Snow Hill, produces some fine examples of artistic iron-work – chairs in gothic, vine-leaf, and oak patterns, at very low prices, and of incomparable beauty. The obdurate nature of the metal is entirely overcome in the elegant traceries, waving lines, and light flowing scrolls, which adorn these admirable pieces of furniture. Perhaps the most useful of all his metal adornments for gardens are the arches and canopies of wire-work for spanning paths, or as affording proper sites for the growth of elegant climbers. There can be nothing cheaper or more durable, while the elegance of these ornaments, when covered with fine climbing masses of stauntonia, or the verdurous tendrils of *Astrolobium niphon*, or even common ivy, is unmatched. The canopies are suitable for plants of less rapid growth, and are in themselves elegant structures, though of course still more so when skilfully covered with the tendrils and blossoms of suitable plants. We strongly commend these useful structures to all lovers of elegance in gardening.

Summer-houses in the Chinese style have a fantastic but pleasing effect if well placed. The construction is thoroughly artificial – rural, but certainly not rustic; and if such buildings are associated with moss mounds, rustic chairs, tree stumps, and rock-work, the effect will be anything but pleasant to a cultivated eye. The approach to a pagoda should be by a clean winding walk, with a well-cut box-edging, and a bright display of the finest evergreens. One or two fine aucubas, some rhododendrons, tree box, and laurel, may be grouped to support the scene, and give a welcome shade; but deciduous trees and shrubs, with the exception of lilacs, acacias, and jasmines, are scarcely appropriate in the immediate neighbourhood of such an edifice. The construction itself cannot be too elaborately decorated; quaint carving, geometrical ornaments, and filigree-work, the whole gaily painted and kept fresh, all give it an appropriate character; for if Chinese at all, it should be so all over in its design and adjuncts.

A well-kept patch of turf, with some gaily-disposed exotics, will complete the scene. Rose, jasminum, or clematis arches may lead off to the less artificial walks, so as to form a transition from the highly artificial tone of the house and its surrounding ornaments.

. . .

Where, in the divergence of paths, or in the passage from one distinct portion of the ground, to another equally distinct in character, it appears inadvisable to place a gate – a rustic arch of wood-work, or a natural arch may be formed. A natural arch of interlacing branches is appropriate to almost any scene; it recalls Arcadia, and we breathe freely under its shade. Nor is there any difficulty if the trees be first established at proper distances, in bringing them into proper union by judicious pruning and leading of the young shoots. Or, a more definite arch may be produced by a skilful topping of pairs of trees well-selected as to size, and then planting them so that the branches left for interlacing, shall reach across to each other at a suitable distance and angle. Treated in this way, they form thick umbrageous bowers, and may serve the double purpose of arches over walks and shady retreats; for, having lost their leading branches, they will grow freely into a compact head. Limes are perhaps the best for such treatment, but maples, elms, and most free-growing trees will serve the purpose.

It is a pity that arbours are not more frequently constructed in this way. Who does not prefer a seat under a tree,

> "Shaking its million leaflets in the sun,"[3]

to one in the confined air of a close timber box?

There is abundant room for design in the structure of ordinary timber arches. Whalebones were much the fashion years ago for spanning walks, but we have grown wiser, and prefer appropriateness.

Trellises and wire-work designs of light character suit for such climbers as *Passiflora rhodochiton, Cobea scandens*, the tuberous *Tropæolum, Maurandyas*, and fuchsias of pendant habit.

Trellises fitted to pots and covered with climbers, make noble window and terrace ornaments. The choice kinds of *Tropæolum* are among the best of plants for the purpose. To carry them over the entire surface it is necessary to watch the young shoots as they make progress, and lead them regularly. Fine lead wire will be found useful in all cases where creepers have to be led over trellises. The thread-like fingers of the *Tropæolum* make the best effect if trained over the interior of a sphere; but for plants of stouter growth, such as the pyramidal Bell-flower, a flat and somewhat fan-shaped design will be most suitable.

Vases for gardens are useless unless furnished with fine plants, and when so furnished, there are few objects more chaste and classical, or suitable as ornaments to those portions of a garden which are near the house. If filled with a variety of plants, they should all be growing, and in high condition, the blooms mingling and clustering thickly in well-contrasted colours. The verbenas, mixed in colours, trail over and produce a charming effect. Petunias, heliotropes, and other low-growing, half-creeping exotics are best suited. The dark nasturtium also looks well when partially covering the base with pendant stems and rich crimson blossoms. Geranium, cineraria, calceolaria, Noisette roses, orange and lemon trees, collinsia, nemophila, schizanthus, balsam, cockscomb, anemone, and all kinds of bulbs, especially hyacinths, are suitable. The plants should all be grown in pots, none of them on any account planted in the vases. The pots can be lifted in or out as occasion requires, and a succession kept up during all seasons.

Vases of very elegant designs are made by the Messrs. Seely, New Road, who use a durable kind of artificial stone in their construction, which is far preferable to the perishable material they are usually made of. The Messrs. Seely have at all times a vast variety of most beautiful patterns in stock, besides fountains, stone tanks, and garden statuary, all in sterling designs adapted to the several uses of the ornaments, and the positions in which they are to be placed.

Some very tasteful designs for garden vases are now executed in terra cotta, but we do not know who are the manufacturers. They have an elegant appearance, and admit of colouring. Mr. Fox, of Snow Hill, deserves the thanks of all lovers of rural elegance for the improvements made by him in metal vases for terraces, balconies, balustrades, etc. The fluted vase, and the Medici vase, manufactured in cast iron, are models of elegance, and admirable examples of the adaptability of metal to horticultural purposes.

Perhaps the best form for an ordinary flower-stand is that which consists of a series of shelves on an elegant receding frame, each shelf having perforations where the pots are to be placed, and from these perforations the drainage is carried away by concealed pipes to a reservoir in the foot of the frame. Such a stand may be rendered very ornamental, but the cheapness, lightness, and elegance of wire-work will always give it precedence over wood, to say nothing of its durability; for the latter soon splits in the sun, or rots in the wet, and sometimes long before it has repaid its cost is weather worn, and *hors de combat*.

Stands for the open air should be of cast iron, painted in imitation of bronze, or of real bronze or brass. Wire is too fragile for the larger plants which are expected to adorn a stand in the open air, and too light for the weather to which out-door objects are occasionally exposed.

Rustic stumps and baskets are highly ornamental, but must always be kept in gay trim. Nothing looks better in front of a cottage than a basket on a twisted pedestal of unbarked timber; the basket itself woven of stout osiers, or made of well-selected pieces from tree loppings. In November such a basket might be filled with bulbs, such as crocus, hyacinth, and snowdrop. When these come out it might be furnished with showy annuals, or a mixture of half-hardy shrubby exotics. The plants best suited for summer blooming in rustic baskets, and on hollow tree stumps, are fuchsias, calceolarias, Tom Thumb geraniums, hydrangeas, petunias, cinerarias, China roses, verbenas, nemophilas, mignionette, pimpernel of all kinds, mimulus, and hawkweeds.

The stiff-growing plants should occupy the centre, and those of drooping habit the sides. If ferns are grown with flowers, the soil should be soft peat, with a mixture of leaf mould and sandy loam. If it is not intended to grow ferns in them, the soil should be composed of three parts good turfy yellow loam, and one of well-rotted dung. The compost should be well mixed, and a layer of moss put in at the bottom and round the sides to keep the soil moist. These ornaments require frequent attention, and a copious supply of water. They are usually pitched inside to protect them from the ravages of damp.

Very elegant wire borderings and baskets are now to be had at very moderate prices, their lightness and durability rendering them very acceptable in grounds where neatness and finish of outline are wanted *instanter*, or to preserve the grace of any portion but just planted, to be removed when the edgings have attained sufficient strength. Large wire baskets make fine ornaments if covered with well-trained ivy, or with half-hardy showy creepers, one of their great advantages being that they preserve some of the gaiety of the summer scene during the winter, when the elegant wire tracery affords cheerful outlines for the relief of the ground.

Of the little grottoes and arches of shells, which now and then strive to please us in small gardens, it is unnecessary to speak. Mr. Loudon says on this subject: – "Eye-traps, painted perspectives, on walls or boards, as terminations; mock hermits, soldiers, banditti, wooden lions, sheep in stucco, or any other figures of men or animals intended to pass for realities, though still used in Holland and France, may be pronounced as too peurile for the present age. If they are still admired by the city mob in a suburban tea-garden, so much the better; the mob must be pleased as well as their superiors, and the rich vulgar may join with them; but the object of all the arts, whether useful or agreeable, is to elevate our tastes and enjoyments; and therefore, as soon as men's minds are prepared for any refinement on former things, the particular art to which these belong should prepare the way for their removal, by presenting appropriate substitutes. A few reading tents and portable coffee-houses, scattered in umbrageous and picturesque situations over the public parks round London and Edinburgh, as at Paris and Vienna, would be as fitting resources for one class of pedestrians as those crowded yards called tea-gardens are for others."[4]

My Dear Reader, The household gods require no further vindication at my hands; you and I are agreed as to the homage they are entitled to, and the mission they are to perform in our behalf. May your hours of rustic recreation profit you in body and soul. May your flowers flourish, your bees prosper, your birds love you, and your pet fishes live for ever. May the blight never visit the tendrils that make your arbours and porches leafy, your borders gay, or your fern-banks verdurous; and may you find in every little thing that lives and grows a pleasure for the present hour, and a suggestion of things higher and brighter for contemplation in the future. I herein reach my hand towards you with an affectionate Farewell!

Editorial Endnotes

1 Where sleep, the "comfortable bird" . . . "til it is hushed and smooth" – John Keats, *Endymion*, Bk. 1, 453 (1818)

2 In an extended comparison for appropriateness in garden ornament, Hibberd suggests that harmonious literature is good for the garden while controversial topics are more suited to the fireside. His list of authors is wide-ranging, including poets from older lyrical traditions such as Robert Herrick (1591–1674) and Henry Howard, Earl of Surrey (1516/1517–1547), to contemporaries such as Alfred Tennyson (1809–1892; poet

laureate from 1850 until his death) and Alexander Smith (1829/1830–1867). Smith was a Scottish poet whose poem 'A Life-Drama' (1852) sold thousands of copies but was mocked by some critics. The German writer 'Jean Paul' (pseudonym for Johann Paul Friedrich Richter, 1763–1825) was one of the German romantics most influential in Britain. For fireside reading, Hibberd recommends Goethe's *Faust* (1808) and 'Zriny'. The latter refers to 'Zriny' (1812), a drama written by another German author (and war hero), Theodore Charles Körner (1791–1813). It tells the story of Nicolas Zriny, a 16th century Croatian-born Hungarian aristocrat and Hapsburg commander, whose defence (and death) for the city of Sziget against the Ottoman Turks prevented their incursion into Vienna. The drama held particular currency in the light of the Crimean War (1853–1856) when, once again, the Western powers (Turkey included) were pitted against the east (Russia). Before the success of *Rustic Adornments*, Hibberd had written *An Epitome of War* (1856), an account of the Crimean War. 'Latter-day pamphlets' refers to a series of tracts (published in 1850) by Thomas Carlyle (1795–1881), Scottish historian, philosopher, and essayist, attacking what he considered to be the failings of his time.

3 "Shaking its million leaflets in the sun", not identified.

4 J. C. Loudon, "Eye-traps . . . tea gardens for others." *Encyclopaedia of Gardening*, most editions, Chapter 3, Part 2, Book 3, 'Science of gardening', p. 361 (1827 edn.).

29

'ON ARBOURS OR BOWERS'

Frances Jane Hope

Source: Frances Jane Hope, 'On Arbours or Bowers', *Gardeners' Chronicle* (6 March 1875), pp. 305–306

ARBOURS are a necessity abroad, but even in this country many love such open-air retreats. At Interlaken, last autumn, I saw a very nice natural one, formed by a Plane (P. occidentalis); the trunk formed the centre and sole support, and on three sides the branches were trained down to the ground and formed a complete shelter from the sun: four or five persons could sit under it. On the same self-supporting plan, at Spa, I was startled by a huge crimson umbrella of Virginian Creeper. It occurred to me if there had been, in the case of the Vine, two or three plants set originally for the centre support, twined so so as to form one stem, a greater height of bower might have been attained and strength to bear the weight of the summer growth. As it was, two people could sit in comfort. One cannot expect deciduous alleys, arches, or arbours, to be strictly symmetrical, or on the perpendicular, at the end of autumn. The young wood requires arranging and curtailing, but, I bethought me, how superior such shelters were to the sentry-boxes of trellised woodwork one meets with, over which the creepers never do grow as expected, and which are so often either fresh painted and nailed out of the carpenter's hands, or in a state of dishevelled disrepair. With a real tree each season only strengthens and beautifies the arbour. A Laburnum would form in our country a beautiful golden umbrella, or arbour, as likewise would a Mountain Ash. Both are free flowerers and certain fruiters, easy to train and wattle, and around each bower there would be sure to spring a perfect sod of seedlings, pleasant to watch, and useful for wattling when the wall gets thin as the tree grows older. Those who love bowers sit under them, I presume, and had better do so early and late, when the rowans begin to colour, and so keep in check the voracious thrushes. Possibly a tame winged hawk might keep them off, or a well-fed tabby with kittens trained merely to blink at and watch the birds. For my own part I like to see the blackbirds gorging themselves, and struggling, with wings out-stretched to keep their balance on a weakly shoot, which they do not intend to leave until every berry is devoured.

Weeping trees of course are all ready and natural for sun shelters, and have always been used. Planted and grown for bowery purposes their position as

 DOI: 10.4324/9780429198809-34

to sun and wind should be particularly considered. Ash and Elm are the best I have ever seen; Thorns, Birch, Willow – good weepers all – would be, I fancy, too small-leaved, and Beech too close in growth. The dark impervious evergreen bower, of Yew for instance, I cannot like. There is always a stuffy, dusty, unventilated air in it, and one pants for the open. Possible eaves-droppers outside prevent freedom of conversation; and inside are certainly spiders – old, grey, dusty ones, that have woven their first and now know they they are in their last web, in that very bower – desperate spiders, made up of eyes and nipping jaws. In the fresh deciduous arbour, on the contrary, however closely grow the branches, you can see as well as be seen – can hear all the gentle summer sounds and feel "the air nimbly and sweetly" passing through the leaves (*Macbeth*, act i., s. 6). There the spiders are young and active "spinners in the sun." Heavy-bodied "white cross" Arachne (a beautiful insect) makes her web as you sit; and another garden spider, of small body and preternaturally long legs, will harmlessly stalk over you. True, caterpillars dropping down by their swinging threads, on cocoon "thoughts intent," may alight on your neck, an unwelcome Eve (*Paradise Lost*, book v.), but sitters in bowers must lay their account for such trivial annoyances as insects. Always busy when in the garden, I have never had time to "take mine ease in mine" arbour (*Henry IV.*, part i., act iii., s. 3); but nothing would tempt me into fog-houses or rustic Heather erections,* with benches of decaying Pine logs, out of which protrude rusty old nails; or, if of sound Fir, then resinous oozings greet you. Wooden houses, like enlarged bathing-machines, are likewise odious, scribbled and whittled over in all directions. When such are the garden sun-shelters, I think one is better off indoors at once, with wide-open windows, in a comfortable chair. To sum up these desultory remarks, the best bowers surely are –

1. A noble tree, not far from the house, in the style of the grand Beech at Newbattle Abbey. 2. A Weeping Ash or Elm. 3. The umbrella, or trained and wattled arbour of a growing tree. 4. The clipped evergreen bower. (For invalids who must have such close shelter, and no circulation of air, a conservatory or cool plant-house would surely be more pleasant.) 5. The purely artificial house of moss, wood, or plaster.

6. " 'A Temple to Friendship,'
  Cries Laura, enchanted,
'I'll build in this garden –
  The thought is divine.'
The temple was built;
  And it now only wanted
An image of Friendship
  To place on the shrine."

(*Author unknown.*)[1]

Source Endnote

* Such structures are very suitable for glens and woods or points of view where shelter from a storm or rain may be frequently required.

Editorial Endnote

1 'Author unknown' is Thomas Moore (1779–1852), an Irish poet and writer.

30

'LAMPORT'

'P.'

Source: 'P.', 'Lamport', *Gardeners' Chronicle* (25 September 1897), pp. 209–210

The Lamport rockery is not remarkable for an extensive variety of alpine plants. The intention has been to clothe it suitably, that the effect of the whole may be pleasing and picturesque, and that every plant when viewed individually should appear to be in just its right position. It is more correct to say that the plants have been selected to clothe the rockery, than that the latter was constructed for the growth of a collection of plants.

In the manner in which the plants have been cultivated there is much novelty. It may be said here that every stone of which the structure is composed has been placed in position by the owner himself, or by his direction, and in his presence. He has done the planting, and no other person has anything to do with it unless by his instruction. Close to our feet there is the common Veronica spicata. It happened to be in bloom, and it was noteworthy because growing luxuriously apparently upon the flat surface of a large stone. We should have been none the wiser had Sir Charles not explained that he had chiselled a small hole through the centre of the stone, and put soil into it, so that the roots of the plant could by that means reach the ground through the stone. This sort of thing occurs frequently, and in the case of different species, and affords an illustration of the amount of mechanical work that has been expended – some of the stones so treated being 14 inches thick. The hammer and chisel are always at hand, and they account for many of the plants succeeding in very strange situations. No plant that grows quickly is a favourite for this structure. Everything is in miniature, and if the plants are not so naturally, then their cultivation is directed to that end. It is full of plant curiosities. A stunted individual that refuses to make free growth is just the kind of plant that is sought. Dwarf Conifers form one of its features, and Sir Charles has been at some trouble to procure them. Some of them are known to be upwards of seventy years old, and have not made more than 3 feet natural growth. But besides being of this great age, many of them have also to support Ivy, which, planted at the foot, has run – well, just as much over its host as it is allowed to do, for the smallest plant is "cultivated," and limited as to size. These aged Conifers include

DOI: 10.4324/9780429198809-35

several Spruce Firs that came from the Royal Exotic Nursery of Knight & Perry at Chelsea, now owned by Messrs. J. Veitch & Sons; also several Retinosporas, silver and gold-coloured forms of Lawson's Cypress (Ivy-clad), and others. But the strangest looking object amongst the Conifers is a diminutive fir, probably seventy years old. It has five straggling roots, the ends of which only have penetrated between the stones, whilst a few inches of each, and the base of the plant are suspended and disclosed. Evidently when the plant had become established, the stones have been picked away from the roots so far as it was safe to do this. Thus to a large extent, the root system as well as the part that is usually above ground, may be seen.

The Utah Agave (A. utahensis) grows on a piece of rock a few feet from the ground, in as perfect a position as this Alpine species could be given. A few plants of the Japanese Maple (Acer polymorphum) in close proximity appear very suitable, and have made little growth, although planted as long ago as 1869. Here and there is the variegated Box; and occasionally over the rocks and caves, hangs a variegated Bramble. At the present time this Bramble is a very pretty sight, the foliage is beautiful, and the canes bear abundance of fruit. Yonder one sees the curled-leaved variety of the common Tansy (Tanacetum vulgare), and the alpine Strawberry in more than one spot clothes a projecting piece of rock with its pretty leaves. Over a good part of the rockery the Spider House-Leek (Sempervivum arachnoideum) creates a silvery effect, distinct and pleasing. We have not previously seen this species growing so freely and happily. A Thalictrum, or Meadow Rue is noticed, and a hardy species of Euphorbia, about 6 or 7 inches high, with dark green leaves, gives another aspect entirely to the part it covers. This species never flowers here, and contrary to the circumstance being considered a disadvantage, it is appreciated, for, said Sir Charles, "Species that do not flower on the rockery usually present a good appearance for the greater part of the year, but flowering species are apt to look 'weedy' directly they have bloomed." In the spring, however, almost the whole face of the rockery is clothed with Aubrietia deltoides, and it is a sight not to be forgotten. But it is not necessary to further remark upon the species of plants, for they have been chosen from different reasons entirely to those that would count with the average gardener. Two other features of this rockery, moreover, call for mention: the first is that of the Crystal Caves formed of quartz, &c., which are certainly uncommon, and produce a distinct and admirable effect. The second, is that of the pretty miniature figures or models a few inches high, that represent gnomes or fairy miners at work in the caves and crevices; some have caught the trade union spirit and are "on strike," as will be seen on reference to fig. 63. The demand they make (a familiar one) is prominently displayed on a board hoisted at the entrance to a crevice. The use of miniature figures was advocated by Loudon in his *Encyclopædia of Gardening* as an appropriate set-off to pigmy Fir-trees, and his idea is explained by woodcuts on pp. 28 and 29 of the edition of that work published in 1850. They certainly increase the weirdness and novelty of the scene; whilst the positions some of them have been placed in at Lamport are suggestive of reality. One of these is lying

Fig. 62. – Section of Rockery at Lamport Hall.

Fig. 63. – Section of Rockery at Lamport, Showing the Fairy Miners "On Strike." (The figures are about 3 inches high.)

at full length upon a rock gazing over the ledge; at others, apparently walking beneath. But the first object that fixes the attention, and even startles most visitors, is a life-size figure, in terra-cotta, of a young lady sitting upon a rock reading a book, which is held in her hands. She has evidently been engaged in tending the plants, for a watering-can stands close to her side. This figure was seen some years ago at an exhibition at Brussels, and later Sir Charles Isham purchased it from a dealer in such things in Oxford Street, London; since being stationed upon the rockery, two presents have been made her; a hat and a diamond ring. The figure is so good that it is invariably mistaken for an actual person.

We must hasten to notice features other than the rockery, however, yet not before mentioning an old plant of Ampelopsis Veitchi that covers the wall opposite. We said "cover," but that is not strictly correct, for Sir Charles, as he pulled away a few leaves and young shoots, remarked, "It would grow all over the wall if I would let it." That is not desired; the wall must be made visible in some places, and the thick stems of the plant are objects of admiration that require no clothing, at Lamport, at least.

Before leaving the scene we noticed a flag displayed, and in letters upon the wall may be read, "Fifty years old! This is the Jubilee of the rockery, commenced in 1847."

. . .

Part 2

INDOOR GARDENS

a)

PLANTS UNDER GLASS (AND WATER)

Editorial Headnote

31. Daniel Ellis, 'Description of a Plant-case or Portable Conservatory, for growing Plants without fresh Supplies of Water and Air, according to the Method of N. B. Ward, Esq.; with Physiological Remarks' *Gardener's Magazine*, vol. 15 (September 1839), [481–505] excerpt, pp. 481–486
32. Nathaniel Bagshaw Ward, 'On the Application of the Closed Plan in Improving the Condition of the Poor', *On the Growth of Plants in Closely-Glazed Cases* (1842), [57–64] excerpt, pp. 57–61
33. Rev. Thomas James, 'The Flower Garden' . . .', *Quarterly Review*, vol. 70 (1842), [196–243] excerpt, pp. 214–216
34. 'An Admirer of Flowers', On the Cultivation of Plants in the Windows of Living Rooms, showing their Tendency to Promote Health, with their Poisonous Effects when introduced into Sleeping Apartments, *Floricultural Magazine and Miscellany of Gardening*, vol. 1 (March 1837), pp. 217–219
35. Donald Beaton, 'Plants in a Bed-room', *The Cottage Gardener* (6 March 1855), pp. 428–429
36. E. A. Maling, ch. 1 'Introductory' and ch. 2 'Filling a Drawing-room Case . . .'; ch. 7, 'In-door Plant Case . . .', *Flowers and Foliage for In-door Plant Cases* (1861), pp. 1–14, [43–57] excerpt, 43–4837.
37. John Lindley, 'The Belgian Window Garden', *Horticulturalist and Journal of Rural Art and Rural Taste* (March 1849), pp. 427–430
38. Dr. Regel, 'Culture of Plants in Rooms', *The Garden* (9 March 1872), pp. 359–362
39. Shirley Hibberd, 'Preface', 'The Aquarium', 'The Freshwater Aquarium', *Rustic Adornments for Homes of Taste* (1856), pp. iii–iv, 3–5, 15–20, 87–91
40. C. Schickler, 'Flower Stand with Fountain Operated by Air Pressure' ('Blumentisch mit Springbrunnen durch Luftdruck'), *Illustrirte Garten-Zeitung*, vol. 1 (1856–1857), pp. 234–237
41. 'Automatic Room Fountains' ('Selbsthättige Zimmerfontäne'), *Dr Neubert's Deutsches Garten-Magazin*, vol. 1 (1882), p. 186

Conditions indoors in the 19th century were not ideal for the cultivation of plants. One remedy was to put them under glass. The practice was long established by artisan weavers who deployed relatively cheap bell jars inside (and outside) to force and protect plants. The Wardian plant case, in contrast, was an especially constructed device which prevented dehydration by allowing condensation to form on the interior surface of the glass of the plant case. The design featured in the illustrations of 'Description of a plant-case', *Gardener's Magazine* (1839), is by James McNab (1810–1878), based on Ward's plan. Daniel Ellis (1771–1841), a Scottish botanist with medical training, introduces the Case and adds extensive 'physiological remarks' not included here. Variations on the model were developed before and after Ward patented his design. The ornamental glazed case designed by Scottish lawyer and amateur botanist Allan Alexander Maconochie (1806–1885) preceded Ward's invention by ten years, but Maconochie did not publish his findings. The Farquharson Plant Case (see Volume 3), E. A. Maling's Heated Case (see below), and the Waltonian Case, which was heated by oil lamp (1857), were other variations. Ward's patent endured, however, partly because it was key to the import and export of plants from around the globe (see Volume 3). Several factors outlined in the introduction contributed to the development of the Wardian case in the home; the impact of imported hardwoods (which were more resistant to damage by water) as a result of increased colonial trade in the Tropics is one example.

At the time of the initial publicity about the Wardian case, glass was heavily taxed and the ownership of it a luxury. However, its inventor, Nathaniel Bagshaw Ward, suggested that it might be developed within a limited budget. In a chapter from *On the Growth of Plants in Closely-Glazed Cases* (1842), Ward (1791–1868), a medical man, positions the case within the context of Enlightenment science and the benefits of natural light to all living creatures. He argues that the poor could be beneficiaries as long as they conform to middle-class norms, exemplified by his critique of the culture of florists' flowers.

There were concerns about the effects of living with indoor plants. In an excerpt from 'The Flower-garden', *Quarterly Review* (1842) (an unsigned review of thirteen different gardening books), the Rev. Thomas James refers to the Wardian Case and recommends plants under glass as a cure for invalids as does the author (unidentified pseudonym 'An admirer of flowers'). In 'On the Cultivation of Plants in the Windows of Living Rooms', *Floricultural Magazine* (1837), James is keen to promote the health-giving properties of indoor plants generally; he warns, however, of the dangers of indoor plants at night based on the idea that plants without sunlight produce carbon. Donald Beaton, 'Plants in a Bed-Room', *The Cottage Gardener* (1855), suggests otherwise.

With the removal of the glass tax in 1845 and developments in ironwork, the spread of plants under glass expanded (see Volumes 3 and 4). The use of glass provided plants with winter protection, extending the bloom of exotics and reducing their damage from indoor pollution. There was, however, confusion among amateur gardeners between the bringing in of plants to provide temporary display throughout the year and the bringing in of specialist plants during the winter for protection. Elizabeth Anne Maling (1829–1866) clarified these issues In *Flowers and Foliage for In-door Plant Cases* (1861). With a self-deprecating voice, she establishes the rigour of her experimental methods to find the best way to keep plants healthy under glass. There is a suggestion that she, too, is an invalid and that indoor planting has given her agency. The final excerpt details her patent for an indoor plant case. Note her recommendation of the Veitch nurseries (referred to elsewhere in the collection). Maling was the author of a number of compact guides on the management of indoor plants (and songbirds), as well as a novel, *Cragstone Cottage*, between 1860 and 1863 before her marriage in 1865. See also IIc.

John Lindley's two articles on 'A Belgian window garden,' *Gardeners' Chronicle* (1846) were published soon after the abolition of the glass tax in 1845. Three years later, they appeared in the *Horticulturalist and Journal of Rural Art and Rural Taste* with superior image reproduction (the source used here). Lindley credited his report, in part, to Victor Paquet, *Almanach Horticole* (1846). The movement of text and image indicates the extent of the dissemination of horticultural ideas through the press across the globe. The illustrated contraption originated in the apartment blocks of Europe among the 'ladies' of France and Belgium and, Lindley argues, could benefit less wealthy urban dwellers in Britain. The idea of a window screen (also noted by Ward) had much to recommend it as the building of terraced houses continued apace. Lindley's critique of 'inexperienced amateurs' in relation to plant care contrasts with the inclusiveness of Maling and other garden commentators writing for an amateur audience.

Lindley's critique was surprising given that through his publications and other numerous activities he disseminated a scientific approach to gardening (most notably in his support for the Natural system of classification of plants) to a broad audience that included women (see Volume 3). The son of a nurseryman, Norfolk-born Lindley (1799–1865) developed the capacity to closely observe plants and the ability to illustrate them, talents which secured him the patronage of William Jackson Hooker and subsequently Sir Joseph Banks in London. He became assistant secretary at the Horticultural Society of London from 1822 and organised its flowers shows. By 1849 (when this article was republished), he had been in post as professor of botany at the secular University College, London, for 20 years, curator at the Chelsea Physic Garden for 10 years, and editor of the *Gardeners' Chronicle* (est. 1841). He went on being the assistant secretary, vice-secretary, or deputy secretary all that time, eventually becoming full secretary in the late 1850s.

In one of a series of articles on the 'Culture of plants in rooms', *The Garden* (1872), Dr. Eduard August von Regel (1815–1892) chronicles subsequent, more elaborate and costly developments of the 'double window' and of plant cases. Regel was a botanist at the Imperial Botanic Garden of St. Petersburg and its director from 1875.

The last three sources focus on water: from Shirley Hibberd's aquarium (*Rustic Adornments for Homes of Taste*, 1856) to the development of the representation of indoor fountains (mentioned by Hibberd, but not illustrated) in the German press. The first edition of *Rustic Adornments* opens with the section on aquaria and exemplifies the delivery of scientific discourse into the domestic realm. In later editions the section is positioned towards the end (and the numerous references to prices and suppliers removed). C. Schickler (n/d), a nurseryman operating in Stuttgart during the 1850s and 1860s, provided details of 'Flower stand with fountain operated by air pressure', *Illustrirte Garten-Zeitung* (1856–1857). The editor of *Dr. Neubert's Deutsches Garten-Magazin* (1882), over 25 years later, indicated how automation had entered into the domestic environments with his recommendation for 'Automatic room fountains'. The reference to 'universal refrigerators' suggests that engineered domestic refrigeration was not completely unknown in Germany at this date.

31

'DESCRIPTION OF A PLANT-CASE . . .'

Daniel Ellis

Source: Daniel Ellis, 'Description of a Plant-case, or Portable Conservatory, for growing Plants without fresh Supplies of Water and Air, according to the Method of N. B. Ward, Esq.; with Physiological Remarks.' *Gardener's Magazine*, vol. 15 (September 1839), [481–505] excerpt, pp. 481–486

TO THE EDITOR OF THE "GARDENER'S MAGAZINE."

SIR,

IN compliance with your request, I now send you an account of the experiment made here of the mode of growing plants in close glass cases, as practised by Mr. Ward. The delay in answering your letter has arisen from the desire of continuing the experiment to the end of the year. The paper, as you will see, was read to the Edinburgh Botanical Society a few weeks ago, nearly in the state it is now communicated to you. It is accompanied by a copy of the design of Mr. M'Nab for a glass case, which, as forming a piece of ornamental furniture, has been much approved. To this account I have appended some physiological remarks, being an attempt to explain "the appearances on scientific principles," which, in your opinion, has not yet been satisfactorily done. I am, Sir, very truly yours,

Edinburgh, June 29. 1839.
DANIEL ELLIS.

---

1. *Construction of the Plant Case.*

THE plant case, or portable conservatory, which I am about to describe, was constructed somewhat more than a year ago, under the immediate direction of Mr. James M'Nab, superintendant of the Caledonian Horticultural Society's garden. Mr. M'Nab had repeatedly visited the collection of Mr. Ward, at his house in Wellclose Square, London, and expressed so much admiration of the success of his plan, that I became desirous of getting a case made, more especially as Mr. M'Nab kindly offered to give the design for its construction, and superintend its execution. From the first, it was my wish to consider it as a sort of experiment,

DOI: 10.4324/9780429198809-38

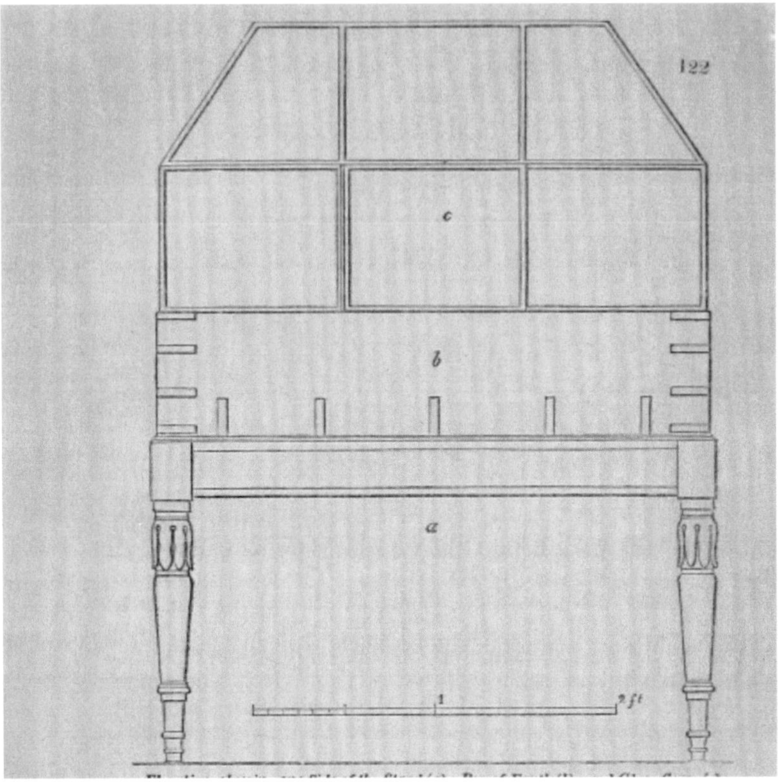

122 Elevation, showing one Side of the Stand (a), Box of Earth (b), and Glass Cover (c).

open to the inspection of such persons as might take an interest in the subject; and, as its success involves questions interesting alike to the botanist, the physiologist, and the horticulturist, I have thought that a concise account of the results of one year's experience, accompanied with some physiological remarks on the growth and condition of plants placed in such novel circumstances, might not be unacceptable to the Botanical Society, which has so recently done me the honour of admitting me as one of its members.

Some gentlemen having expressed an intention of fitting up cases for the same purpose, I have requested Mr. M'Nab to furnish me, not only with a copy of his design, but also with a statement of the leading particulars which require to be attended to in the construction. As the apparatus was to be placed in the window of a drawingroom, its form and dimensions were determined in accordance with that idea. It is composed of three parts: the stand (*fig*. 122. *a*), the box (*b*), and the glass roof, or cover (*c*). The stand is 1 ft. 10 in. in height; the box, 8½ in.; and the cover, 1 ft. 7½ in.; making the total

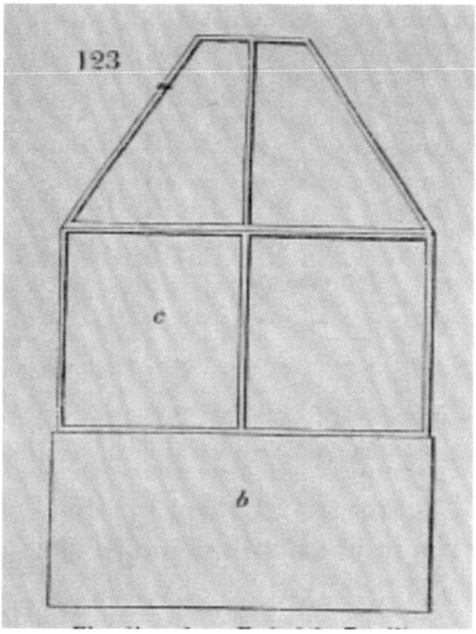

123 Elevation of one End of the Box (b) and Cover (c).

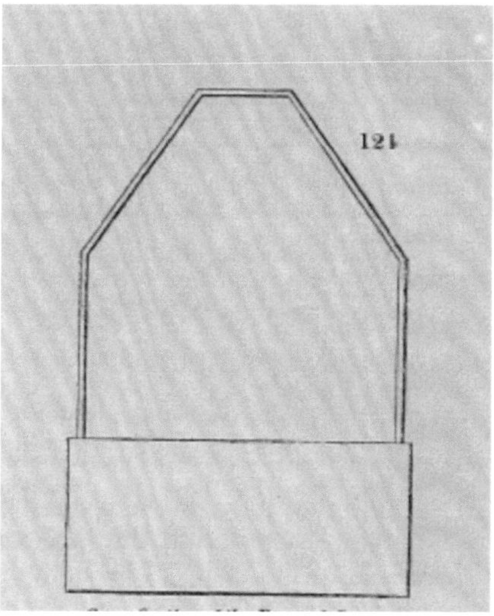

124 Cross Section of the Box and Cover.

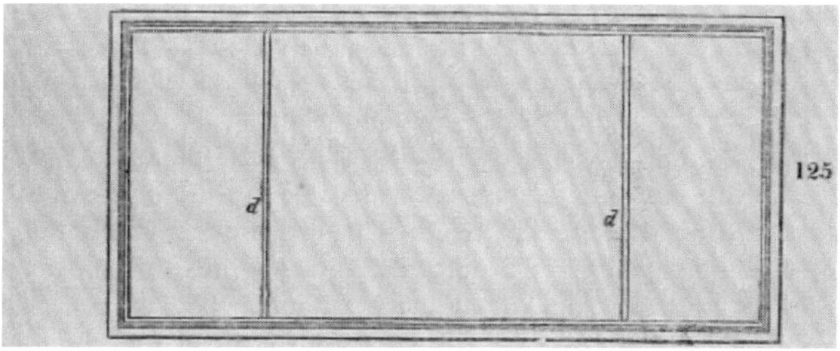

125 Plan of the Box, showing the two Partitions (d d).

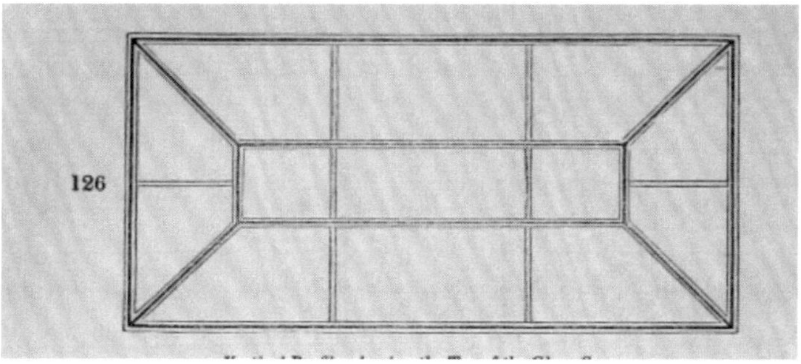

126 Vertical Profile, showing the Top of the Glass Cover.

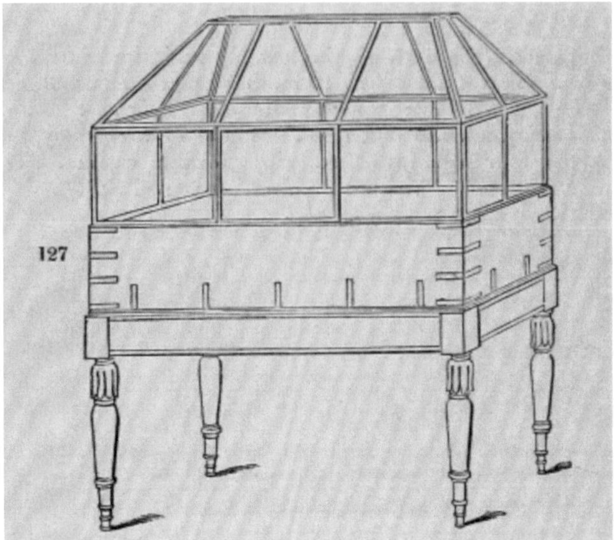

127 Perspective View of the Stand, Box, and Cover, complete.*

height 4 ft. 2 in. The stand, on which the box rests, is made of mahogany, and supported on four legs, furnished with movable casters. The box contains the soil; and is made of well-seasoned St. Domingo mahogany, previously steeped for a fortnight in Kyan's preservative composition. Its sides are 1¼ in. thick, mitred and dovetailed together at the corners. The bottom of the box is of Honduras mahogany, 1 in. thick, and is formed of numerous small pieces, framed and flush-paneled, and so arranged as best to resist the yielding of the wood, in consequence of the mass of moist earth which it has to bear. To give it greater strength, two cross, or tie, pieces stretch from side to side, and are dovetailed into the sides: they are placed at equal distances from the two ends, and thus divide the box into three compartments; but, as they have large open spaces at the bottom, and through their centres, they permit the moisture to percolate freely through the whole of the soil. The bottom, being properly fitted, was fixed to the sides by brass screws; and the brass bands, at the corners and bottom, are fixed on with brass nails; no iron being used in any part. Along the upper edge of the box a groove is sunk, to receive the lower edge of the glass roof, which rests securely in it. This groove is lined with lead; its inner lip is $^1/_{16}$ of an inch lower than the outer; and at each end is a notch, ⅛ of an inch only above the bottom of the groove, to allow the condensed moisture, which trickles down the inside of the glass, to flow back

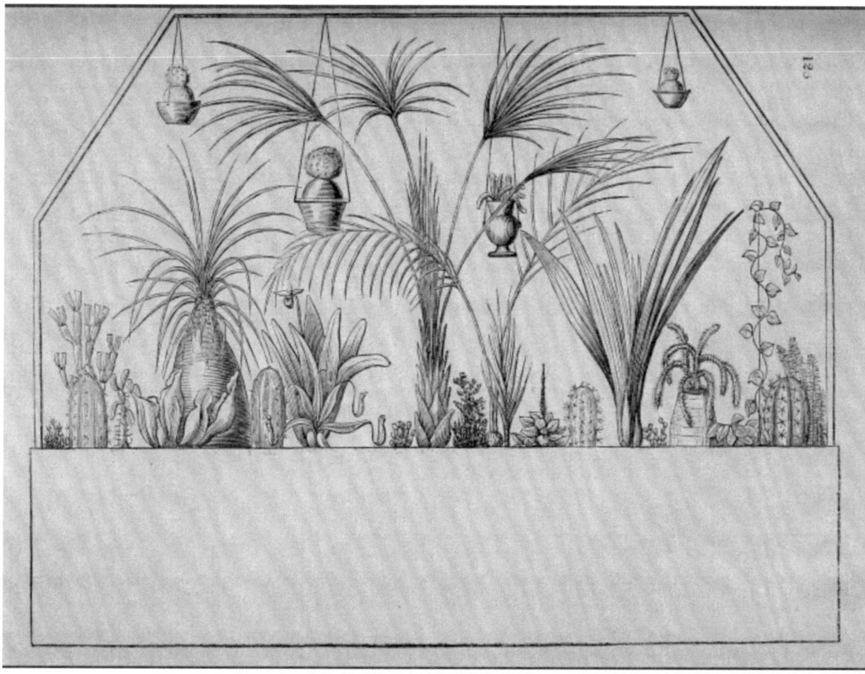

128 Longitudinal Section, showing the Plants growing in the Soil, and suspended from the Brass Rod under the Top of the Glass Case.

to the soil. Instead of lead, the lining of this groove should be of brass, which would prevent the galvanic action which arises from the contact of the two different metals.

It only remains to speak of the glass cover: its frame-work is made of brass, with a door on one side, made to fit close, but which can easily be opened when needed. The glass used for the cover is flattened crown-glass, except for the door, which is plate-glass. The glass panes were fitted into the frame with great care, and with a putty specially made for the purpose; this putty received afterwards three coats of paint. Along the top of the roof two brass rods extend, from which small pots containing plants may be suspended. The whole of the framework is well fitted, and nicely put together, so as to preclude, as far as could well be done, all interchange between the air in the case and that in the room. It was before stated, that the total height of the case is 4 ft. 2 in. Its length is 3 ft.; and its breadth, 1½ ft. Its form and dimensions are accurately represented in the beautiful drawings (*figs*. 122–128.) of Mr. M'Nab, which accompany this paper.

Source Endnote

* The reader will bear in mind that these plant stands are not intended to be used for packing plants for sending to a distance. Boxes for this purpose are differently constructed, as may be seen by consulting the *Encyclopædia of Gardening*, edit. 1835, pp. 539, 540.

32

'ON THE APPLICATION OF THE CLOSED PLAN IN IMPROVING THE CONDITION OF THE POOR'

Nathaniel Bagshaw Ward

Source: Nathaniel Bagshaw Ward, ch. 5, 'On the Application of the Closed Plan in Improving the Condition of the Poor', *On the Growth of Plants in Closely-Glazed Cases* (1842), [57–64] excerpt, pp. 57–61

AMONG the numerous useful applications of the glazed cases, there is one which I believe to be of paramount importance, and well deserving the attention of every philanthropist: I mean its application to the relief of the physical and moral wants of densely crowded populations in large cities. Among the members of this population there are numbers, who, either from early associations, or from that love of Nature which exists to a greater or less degree in the bosom of all, are passionately fond of flowers, and endeavour to gratify their taste at no small toil. But before I proceed to show how this may be effected, there is one point of so much importance to the well-being of every individual, that I cannot pass it by without notice. This is the free admission of light* into their dwellings. I have already alluded to the effects of light upon vegetation; and its influence upon the animal economy, although not so immediately obvious, is not the less striking. This has been proved by the experiments of Dr. Edwards.[1,2]

"He has shown that if tadpoles be nourished with proper food, and are exposed to the constantly renewed action of water (so that their branchial respiration may be maintained), but are entirely deprived of light, their growth continues, but their metamorphosis into the condition of air-breathing animals is arrested, and they remain in the form of large tadpoles. Dr. Edwards also observes, that persons who live in caves and cellars, or in very dark and narrow streets, are apt to produce deformed children; and that men who work in mines are liable to disease and deformity beyond what the simple closeness of the atmosphere would be likely to produce. It has recently been stated, on the authority of Sir A. Wylie,[3] that the cases of disease in the dark side of an extensive barrack at St. Petersburgh, have been uniformly, for many years, in the proportion of three to one to those on the side exposed to strong light. On the contrary, the more the body is exposed to the influence of light the more freedom do we find, *cæteris paribus*, from irregular

DOI: 10.4324/9780429198809-39

action and conformation. Humboldt has remarked that among several nations of South America, who wear very little clothing, he never saw a single individual with a natural deformity; and Linnæus, in his account of his tour through Lapland, enumerates constant exposure to solar light as one of the causes which render a summer's journey through high northern latitudes, so peculiarly healthful and invigorating."[†]

I will now endeavour to show how the glazed cases may be made subservient to the benefit of the poor, and to point out how cheaply and easily this may be effected. A box lined with zinc, and having three or four openings in the bottom, will be required for the reception of the plants; and glazed frames can be procured anywhere, well painted and puttied, at about one shilling the square foot. The plants to furnish it can be procured abundantly in the woods in the neighbourhood of London. Of these I will mention a few. The common Ivy grows most beautifully, and can be trained over any part of the case, agreeably to the pleasure of the owner. The Primroses,[‡] in early spring, will abundantly repay the labour of fetching them, continuing for seven or eight weeks in succession to flower as sweetly as in their native woods. So likewise does the Wood-sorrel, the Anemone, the Honeysuckle, and a host of other plants, independently of numerous species of mosses and of ferns. Some of these latter are more valuable than others, in consequence of the longer duration of their fronds, such as *Lastræa dilatata* and its numerous varieties. There are likewise many cultivated plants procurable at little or no cost, which grow without the slightest trouble, such as the *Lycopodium denticulatum*, the common Musk-plant, Myrtles, Jasmines, &c. All the vacant spaces in the case may be employed in raising small salads, radishes, &c.; and I think that a man would be a bad manager who could not, in the course of a twelvemonth, pay for his case out of its proceeds. These remarks apply chiefly to situations where there is but little solar light. Where there is more sun, a greater number and variety of flowering plants will be found to thrive, such as several kinds of Roses, Passion-flowers, Geraniums, &c., with numerous beautiful annuals, viz. – *Ipomœa coccinea*, the species of *Nemophila, Convolvulus*, and a host of others: the vegetation in fact can be diversified in an endless degree, not only in proportion to the differing degrees of light and heat, but likewise by varying the quantity of moisture; thus, with precisely the same aspect, ferns and bog plants might be grown in one case, and Aloes, Cactuses, Mesembryanthemums, and other succulent plants in another.

These cases form the most beautiful blinds that can be imagined, as there is not a window in London which cannot command throughout the year the most luxuriant verdure. The condensation of the moisture upon the colder surface of the glass effectually obscures the view from without, and at the same time admits far more light than is allowed to enter by ordinary blinds. Nothing can be conceived more cheerful than the appearance of rooms thus furnished. As these cases become more general among the higher and middle classes, a new field of healthful industry will thus be opened to the poor, who might not only be employed in procuring plants for these cases from the country, but whose ingenuity might be called into play in executing various models of old towers, ruins, &c., in sand-stone, chalk, or other suitable material, which, at the same time that it served to ornament the case, would afford a suitable place for the growth of little Sedums and any plants that require less moisture than those which are planted in the mould. I need however say no more upon this point, as these various applications must be sufficiently obvious to every one.

But I must here caution the poor against indulging a taste for what are called fancy flowers, – things which this year are rewarded with gold medals, and the next are thrown upon the dunghill. Believing that all human pursuits ought to be estimated in exact proportion as they tend to promote the glory of God, or the good of man, let us for a moment compare the empty chase after fancy flowers with the legitimate pursuits of horticulture and floriculture. So far from the love of God, and the good of his fellow creatures, being the end and aim of the fancy florist, he values everything in proportion as it is removed from Nature, and unattainable by the rest of mankind. "A long time must elapse ere the world can hope to see a perfect Pansy"!!![4] says one of these fancy writers. How the world is to benefit by this Phœnix when it does arrive he will of course inform us in his next publication.[§] Let me entreat the poor to remember that their single talent should be well employed;[¶] let them learn to estimate things according to their true value, and devote their time and attention to the legitimate pursuits of horticulture and floriculture.

Source Endnotes

* "Let in the sun and you shut out the doctor," says an old Italian proverb.

† Carpenter's Physiology.

‡ There is, perhaps, no plant which offers so striking an illustration of the protection afforded by the glass as the common Primrose. Place, side by side, in a tub outside any smoky window with an eastern aspect, and where there is no artificial heat, two roots of Primroses, supplying them with water if needed. Cover one of these with a glass; and the difference in flowering is so great, that I cannot illustrate it better than by comparing it to the difference which takes place in the burning of charcoal, or any other combustible substance, in oxygen gas and in the open air.

§ That I have not expressed myself too strongly upon this point any one may convince himself, by consulting the history of *Tulipomania*.

¶ "His virtues walked their narrow round,
Nor made a pause, nor left a void;
And sure th' eternal Master found
The single talent well employed."
– Johnson. 'On the Death of Dr. Robert Levet' (1792).

Editorial Endnotes

1 Dr William Frédéric Edwards (1777–1842) was born in Jamaica and received a Unitarian education at William Hazlitt's (Unitarian) New College in Hackney before he moved to Europe and became known as a French physiologist and a pioneer in anthropology.

2 His experiments were described in this quotation from *Principles of Human Physiology* (1842) by William Benjamin Carpenter (1813–1885), also a Unitarian; the precise edition has not been identified.

3 Sir A [sic] Wylie – Sir James Wylie, 1st Baronet (1768–1854), a Scottish physician who served in Russia from 1790 until his death. He is reputed to be one of the organizers of military medicine in Russia.

4 Fancy writers – The advice and competition to produce the best or perfect pansy (Viola) were similar to those for the Auricular which are referred to elsewhere in the collection.

33

'THE FLOWER GARDEN' (A SERIES OF REVIEWS)

Thomas James

Source: Rev. Thomas James, 'The Flower Garden', *Quarterly Review*, vol. 70 (1842), [196–243] excerpt, pp. 214–216

Our notice of ferns might seem rather foreign to the subject of ornamental gardening (though we shall have something to say of a fernery by and bye), were it not for the opportunity it affords us of introducing, probably for the first time to many of our readers, a botanical experiment, which, though for some years past partially successful, has but lately been brought to very great perfection for the purposes both of use and ornament. We allude to the mode of conveying and growing plants in glass-cases hermetically sealed from all communication with the outer air. There are few ships that now arrive from the East Indies without carrying on deck several cases of this description, belonging to one or other of our chief nurserymen, filled with orchideous plants and other new and tender varieties from the East, which formerly baffled the utmost care to land them here in a healthy state. These cases, frequently furnished by the extreme liberality of Dr. Wallich,[1] the enterprising and scientific director of the Hon. Company's gardens in the neighbourhood of Calcutta, form on shipboard a source of great interest to the passengers of a four-months' voyage, and, after having deposited their precious contents on our shores, return again by the same ship filled with the common flowers of England,

> 'That dwell beside our paths and homes,'

which our brethren in the East affectionately value by association above all the brilliant garlands of their sunny sky.

This interchange of sweets was a few years ago almost unattainable, the sea-air and spray, as is well known, being most injurious to every kind of plant; but their evil effects are now completely avoided by these air-tight cases, which admit no exterior influence but that of light. Without entering into any deep physiological explanation, it may be enough to say that vegetable, unlike animal life, does not exhaust the nutritive properties of air by repeated inhaling and exhaustion; so that

DOI: 10.4324/9780429198809-40

these plants, aided perhaps by the perfect stillness of the confined atmosphere, so favourable to all vegetation, continue to exist, breathing, if we may so say, the same air, so long as there is moisture enough to allow them to deposit every night a slight dew on the glass, which they imbibe again during the day. The soil is moistened in the first instance, but on no account is any further water or air admitted. The strangers which we have seen thus transmitted, being chiefly very small portions of succulents and epiphytes, though healthy, have shown no inclination to flourish or blossom in their confinement; but it must be remembered that the temperature on the deck of a ship must be very much lower than what this tribe requires, and the quantity of wood-work which the cases require to stand the roughnesses of the voyage, greatly impedes the transmission of light. As soon as the slips are placed in the genial temperature of the orchideous house, they speedily shoot out into health and beauty.

But while this mode of conveyance answers the purposes of science, a much more beautiful adaptation of the same principle is contrived for the bed-room garden of the invalid. Who is there that has not some friend or other confined by chronic disease or lingering decline to a single chamber? – one, we will suppose, who a short while ago was among the gayest and the most admired of a large and happy circle, now through sickness dependent, after her One staff and stay, for her minor comforts and amusements on the angel visits of a few kind friends, a little worsted-work, or a new Quarterly, and, in the absence or dulness of these, happy in the possession of some fresh-gathered flower, and in watering and tending a few pots of favourite plants, which are to her as friends, and whose flourishing progress under her tender care offers a melancholy but instructive contrast to her own decaying strength. Some mild autumn-evening her physician makes a later visit than usual – the room is faint from the exhalations of the flowers – the patient is not so well to-day – he wonders that he never noticed that mignionette and those geraniums before, or he never should have allowed them to remain so long – some weighty words on oxygen and hydrogen are spoken – her poor pets are banished for ever at the word of the man of science, and the most innocent and unfailing of her little interests is at an end. By the next morning the flowers are gone, but the patient is no better; there is less cheerfulness than usual; there is a listless wandering of the eyes after something that is not there;[2] and the good man is too much of a philosopher not to know how the working of the mind will act upon the body, and too much of a Christian not to prevent the rising evil if he can; he hears with a smile her expression of regret for her long-cherished favourites, but he says not a word. In the evening a largish box arrives directed to the fair patient, and superscribed, 'Keep this side upwards – with care.' There is more than the common interest of box-opening in the sick chamber. After a little tender hammering and tiresome knot-loosening, Thompson has removed the lid; – and there lies a large oval bell-glass fixed down to a stand of ebony, some moist sand at the bottom, and here and there over the whole surface some tiny ferns are just pushing their curious little fronds into life, and already promise, from their fresh and healthy appearance, to supply in their growth and increase all the beauty and interest of the discarded flowers, without their injurious effects. It is so. These delicate exotics, for such they are, closely sealed down in an air-tight world of their own, flourish with amazing rapidity, and in time produce seeds which provide a generation to succeed them. Every day witnessing some change keeps the mind continually interested in their progress, and their very restriction from the open air, while it renders the chamber wholesome to the invalid, provides at the same time an undisturbed atmosphere more suited to the development of their own tender frames. We need scarcely add, that the doctor the next morning finds the wonted cheerful smile restored, and though recovery may be beyond the skill, as it is beyond the ken, of man, he at least has the satisfaction of knowing that he has lightened a heart in affliction, and gained the gratitude of a humble spirit, in restoring, without the poison, a pleasure that was lost.

Editorial Endnotes

1 Dr Wallic – Dr Nathaniel Wallich (1786–1854)

2 ὀμμάτων δ' ἐν ἀχηνίαις
ἔῤῥει πᾶσ' ἀφροδίτα. – ÆSCH. *Agam*. 408.

'a listless wandering of the eyes ...' * - Aeschylus, ancient Greek playwright, possibly from the play *Agamemnon* (text unclear). James provides the original Greek which suggests his erudition.

34

'ON THE CULTIVATION OF PLANTS IN THE WINDOWS OF LIVING ROOMS'

'An Admirer of Flowers'

Source: 'An Admirer of Flowers', 'On the Cultivation of Plants in the Windows of Living Rooms, showing their Tendency to Promote Health, with their Poisonous Effects when introduced into Sleeping Apartments', *Floricultural Magazine and Miscellany of Gardening*, vol. 1 (March 1837), pp. 217–221

Very much has been said about plants in windows, some asserting their tendency to injure health, and others the contrary. There is one point, however, in which I think all will agree, and that is, their beautiful appearance. Whether in the splendid halls and drawing-rooms of the wealthy, or in the humble cottage of the poor, there can be but one opinion respecting their appearance. What artificial splendour can compete with a number of handsome flowering plants in the windows of our rooms? Certainly none. Their beautiful green leaves, contrasted with their blossoms of various forms and colours, present an appearance altogether beyond the reach of art; and during the winter months, when the ground is covered with snow, and the gardens present an appearance more of desolation than beauty, the trees all leafless, and the flowers cut down by the frost, *then* what a feast is presented by plants blooming away in your house, (heedless of the chilling blasts without) enlivening the dreariness of winter, spreading an odoriferous perfume through your apartments, and rivalling in beauty many of those tender kinds, who only delight us with their fragrant flowers during the genial summer weather, not able to bear up against the chilling and frosty air of winter. With regard to the choice of flowers for windows, of course, that is a matter of taste, but for the cottager, I think, he will find the Fuchsia, Hydrangea, Chinese Primrose, Cactus speciosa, and Monthly Roses, to be not only cheap but elegant acquisitions to his apartments.* We will now proceed to consider the principal question on the subject, viz: – are they beneficial or injurious to health? and I shall endeavour to show clearly and upon philosophical principles, that in moderate quantities they are decidedly healthy. Our atmosphere is simply a mixture of oxygen or vital air, (so called because no animal can live without it) and nitrogen (called also azotic gas, because it would immediately deprive any one of life who was to breathe it)

 DOI: 10.4324/9780429198809-41

with an exceedingly small quantity (about 1 part in 1000) of carbonic acid gas, which is also poisonous, and, of course, several adventitious substances, such as watery vapour, &c. Now in breathing this atmospheric air, man, and all other animals retain the oxygen which enters into the blood, and return the nitrogen, which being lighter than the air ascends and waits for fresh combinations; thus man deprives the air of its constituent which supports life and returns the poisonous part; on the contrary plants, (not flowers only, but plants generally,) give out during the day a large quantity of oxygen, which combining with the nitrogen which man exhales, preserves the equilibrium, and reforms atmospheric air. Here, then, we see that "the plant purifies what the animal had poisoned." The loss of the vast quantity of oxygen which is absorbed by the breathing of animals, would soon render the air totally unfit for our use, if the Almighty Framer of the universe had not in his infinite wisdom appointed an antidote in the vegetable creation. This is, I think, a sufficient evidence that plants in rooms are decidedly healthy, but when I state this, I must also assert, that in bed-rooms they are exceedingly deleterious, as during the *night* many plants give out carbon instead of oxygen, and by that means instead of purifying the air help to poison it. The fact, that man exhales air unfit for being re-breathed, is too clear and generally admitted to need an experiment; but perhaps it may be as well to mention one, as some readers of this article may be sceptical on this point, to shew that during the day plants give out oxygen; I therefore select the following from Parke's Rudiments of Chemistry: – "Invert a glass bell full of water in a flat dish of water, and introduce leaves under it. Expose the apparatus to the sun's rays and very pure oxygen gas will be disengaged, which will displace the water in the glass and occupy its place. In like manner, a sprig of mint corked in a small portion of carbonic acid gas, will render it capable of supporting life." Thinking, as one of our present authors writes, that "Floriculture is amongst the most innocent and humanising of all pleasures," and that "everything which tends to diffuse such pursuits amongst those who have too few amusements, is a point gained for happiness and for virtue," I consider it a duty which I, as a Floriculturist, owe to my favourite recreation, to endeavour to remove that prejudice which so generally exists against flowers in windows, by showing, that (except in sleeping apartments) it is destitute of foundation; and further, by proving on chemical and philosophical principles, that they are not only interesting and beautiful, but actually conducive to the salubrity of the atmosphere. Should these remarks, in any measure, dispel that prejudice, it will afford pleasure to

AN ADMIRER OF FLOWERS.

Source Endnote

* Agreeably with a suggestion of our much-valued Correspondent, we might add many others to his list; but, in the successful cultivation of plants in windows, far more depends on care and attention than on the kind of plants. Those which an experienced cultivator would regard as tender, are yet, when placed under the care of fair and attentive cultivators, often seen to display a degree of rich luxuriance, and a delicacy of tint both in flower and foliage, that may be sought in vain in the greenhouses of the wealthy. – EDITOR.

35

'PLANTS IN A BED-ROOM'

Donald Beaton

Source: Donald Beaton, 'Plants in a Bed-room', *The Cottage Gardener* (6 March 1855), pp. 428–429

THE first person who told me that plants were poisonous in a bed-room was Lord Lovat's housekeeper, at Beaufort Castle, in 1821. Her name was Amelia McPhail. I had a few plants at home at that time which I was learning to strike from cuttings, and I did not much like the idea of having poison in-doors, although none of us knew of it but myself; still, I was loth to turn them out. Besides, I knew my father kept *nux vomica* in the house to poison dogs with, such as might stray about, and kill lambs, or frighten sheep, and no harm came to any of us from that kind of poison being in the same house. But the following summer one of my sisters was taken very ill, and the doctor ordered all my plants out of the house, but he could not save her; and I had a lurking suspicion, for a long time, that my plants, after all, were the cause of her death. Hence it was, that for many years I had a great horror of plants being kept in a living room.

When I was in Edinburgh, I lodged, part of the time, with two medical students, who attended the University, and the botanical lectures of Dr. Graham. Before we parted, I was almost a doctor myself. I learned how plants were supposed to poison the air of a bed-room; that their leaves sucked in the life-sustaining portion of the air (oxygen gas), and gave out "foul air" (carbonic acid); hence the reason why unthinking people took up the notion that plants vitiated the air of the room. If the quantity of either life-sustaining or foul air which 75 pot plants could thus destroy, or produce, in twenty-four hours, was computed, and found to be so much, the self-same quantity of air could be proved to be vitiated in ten minutes by an infant three months old, and by a full-grown man in two minutes! If this be so, and happily there is not the slightest question on the subject in these days, the most dangerous of all the poisons for the air of a bed-room is a full-grown man.

Although it is quite true that plants do vitiate the air of a room to, comparatively, a fractional degree, it is equally well ascertained that they consume and destroy a very great deal of foul air; and that without foul air, such as would kill a man, plants could not be kept alive at all. We gardeners know this fact from our

 DOI: 10.4324/9780429198809-42

every-day experience; we cannot grow plants so well, or so quickly, in the sweetest air as in a stinking hotbed. All the animal creation vitiate the common air every time each one breathes the breath of life, or life-sustaining air; and were it not that all the vegetable kingdom depend on this vitiated air for part of their subsistence, and a great part, too, this world would have been at an end as soon as animals covered the face of the earth. Therefore, and without the shadow of a doubt, plants are the best purifiers of all the agents that have yet been known to cleanse the air of a bed-room, or any other room in a house, provided always that such plants are not in bloom, or, at least, do not bear bloom with a strong scent.

The scent of flowers is now almost the only thing which defies the art of man to compute by measure or weight. Every one must compute the scent by the sense of smell, the most varying of our senses; so varying, indeed, that one could sleep on a bed of Roses, another among Syringa, or Magnolia, or Lilac-blossoms, and so forth, while a third party could not sit ten minutes in a room with a Hyacinth in bloom. I have known such an one; and another, who is compelled to leave England altogether during the Hay season. But, in all this, there is no poison in the smell of these flowers; nothing but that some constitutions cannot bear them.

And now for the practical value of this kind of doctoring. I saved more than two hundred good plants in my own bed-room during the late frost; there they were, day and night, during five weeks, and I have never been so healthy these thirty years. I seldom went through so much without some slight cold, or a little sore throat; but the air of the rooms was kept so fresh and so pure by the requirements of so many plants crowded together, that an invalid would certainly have gained strength in it during the time. A medical gentleman, who likes a chat about flowers, called on me in the midst of the frost, and when I heard the announcement I began to tremble. He always asks to see my flowers, and I thought to myself, I shall "catch it" now, when he finds them all over the house. I never asked him if he studied in Edinburgh, but I think half the doctors in the world go there; but that is not "here nor there." When I asked him what he thought of my plan, "Not much;" he replied, "for if all the world knew the value of plants so well as you seem to do, our profession might take to the plough." "Well," said I, "all the world shall know just what you say, and judge between us."

But I am not quite sure that I would have said any thing about it, were it not that Mrs. Robert Moody told me, to-day, that Mrs. Brown had lost her Geraniums with the frost. Mrs. Brown is fond of plants, and Mr. Brown is very fond of her; but they have not been married long, only getting towards the ninth month; and Mr. Brown vowed that he would not risk the chance, or the life – I forget which they told me – of an heir, for all the flowers in England; and so the plants could not be admitted into his bed-room on that score. And judging there might be more of his views respecting an heir or heiress, I judged it right and proper to give a practical illustration against one of the most ridiculous of all the popular prejudices of the present day. If I was compelled to sleep in a close room, without a fireplace, to keep the air of the room in motion, I would not only keep it well stocked with strong, healthy plants; but if I could afford it, I would change them twice a-week, for more security against the evil effects of close, vitiated air. And if there is one medical gentleman alive, who still holds a different view of the subject, it is dangerous to let him be at large during a hard winter, or, indeed, at any time. To be permitted to misconstrue the truths of science, to the prejudice of the invalid, is as dangerous as to trust in the haphazards of empiricism; but to be allowed to attend to the wants of some favourite flowers would be, to some invalids, as useful and valuable a medicine as any within the range of pharmacy.

36

FLOWERS AND FOLIAGE FOR IN-DOOR PLANT CASES

(Excerpts)

E. A. Maling

Source: E. A. Maling, ch. 1, 'Introductory' and ch. 2, 'Filling a Drawing-room Case . . .'; ch. 7 'In-door Plant Case . . .', *Flowers and Foliage for In-door Plant Cases* (1861), pp. 1–14; [43–57] excerpt, 43–48

INTRODUCTORY

THE In-door Plant Cases appear before the public with more than one plea for their kind reception; not only on account of their economy and their picturesqueness, but also because of their simplicity and their use.

These Cases are thoroughly practical in their plan, having been built up piece by piece, and year by year, with most watchful interest; not for the public, but simply for my own use in pursuing a favourite occupation: indeed, my chief claim to be heard in speaking of their management lies in the fact of my having had more time than other people to try experiments as to the plants they suit.

The original idea was of course taken from the Wardian Case, to which, for many purposes, I thought that the addition of heat would be a great improvement; as many persons wish to adorn their room with the bright colours and graceful forms of tropical flowers and leaves, for which both heat and moisture are at times required.

The cold closed Cases are effectual in preserving many plants from injury from without, during many months; although I have not found those Cases answer well for flowers in general, when quite without heat in winter. Under the present system the plants may either be cool, if they are of a nature to bear such temperature; or at ten minutes' notice the heat can be raised to any degree up to 90°.

The remembrance of the beautiful Japanese plants exhibited at Kensington Gore last summer within a week after they arrived in England, will be enough to prove the value of shelter and of still quiet air alone, for preserving plants in beauty for a very long time, as these were brought over in the Wardian cases.

For my own part I always thought that those Cases might be improved on to an extent that would quickly render them universal favourites for home use; and having been for some years confined to a single room for all my gardening

DOI: 10.4324/9780429198809-43

pleasures, and forbidden also to keep plants there at night unless shut in by glass, I had both time and opportunity to experimentalise freely. I should be really afraid to say how many plans, from first to last, I *have* tried: glasses admitting no air, and glasses always open; heated Cases and cold Cases; Cases warmed from beneath, and Cases where the object was rather to heat the air; square Cases and round Cases; large Cases and small Cases; flat Cases and sloping Cases. It would be almost impossible to enumerate here all the variations that my plan has seen: I can only hope that the result of these experiments will be found successful.

Many people besides myself have tried, of course, to keep drawing-room plants in blossom; for even where one fresh flower can succeed another without an instant's trouble, one naturally dislikes to see the bright things fade. Those who have tried, however, to prolong their beauty, know best with what success their efforts have been rewarded.

The object of the In-door Plant Cases, of which I am now writing, to begin with the negatives, is *not* to exclude the air except under special circumstances; when, for instance, it is desirable for the occupants of the room, or when great heat, or smoke, or fog, or dryness (as in a warm room on a frosty day), has made the air injurious to the plants themselves.

In these cases too, I am told by a learned friend, the air itself is actually not excluded; though the vapour is kept in and the plants are saved from the exhausting influence of a drying atmosphere.

Secondly, the object of these Cases is not to keep plants wet: there is a certain degree of moisture needed, both by the roots and foliage, which is generally desirable for all healthily growing plants; but this degree is far removed from wetness: those who have tried moisture, as I have done repeatedly, know how very quickly all their plants "damp off."

The least degree of moisture which will keep the plants unflagging, is the proper amount for these Cases to receive, and I may safely say that weeks have elapsed between the times when my own get watered. Standing in sand, which is always moister the lower we get down in it, the porous flower-pots absorb quite sufficient water; and were it not for the heat which gradually dries the sand at one end of the Case, even the sand itself would be often long without the moisture needing to be renewed.

This is the general rule. When plants are being rapidly forced on, or when they are of a nature to drink up very quickly all the water near them, some modification must be made to suit them. Strange to say also, plants require water at no time so much as when they are in blossom, and the first result of neglect is seen by the vexatious token of the flowers drooping, sometimes falling, frequently turning yellow, and often being even past recovery, if their distress has been long unnoticed. Wet sand, however, generally keeps all safe; and it is very easy to pour a little water upon the sand outside the pot which one wishes to preserve.

The actual objects of the In-door Plant Cases are easily to be shown. Every one knows how, in a town, the leaves of trees become all clogged and blackened; the combined influence of smoke and dust soon rendering them far from sightly objects.

In a room we expect that much of this is escaped; but alas! the housemaid's brush is very little less injurious, as the fine dust rapidly clings to and spoils the plant: so much of its beauty lies in the glossy leaves and that look of bloom and freshness which is so soon destroyed.

The glass Case then effectually preserves the plants from this, as well as from another danger – more vital in its harm – of being placed outside a window, or being left in a draught from one, while the room is put in order.

Gas is very injurious, and so is a heated, fire-dried atmosphere, as they absorb from the leaves far more moisture than the roots can replace. From all these dangers, the mere glass enclosure is an effectual shelter.

Many foreign plants also, Begonias and Ferns, and many brilliant flowers, are used to live in the moist, warm atmosphere of shaded woods, where a steam of vapour often hangs, and where they draw their moisture chiefly from the air. For plants like these a warm, moist air, is *home;* and it is really delightful to watch how they thrive in it: they look so glossy, so luxuriantly green; and even if we keep the front side of the Case open to let the sweet air escape, on closing it for awhile we have back again all the mist-bathed foliage.

And it is so economical! When we think of the fleeting flowers that grace our open drawing-room stands all the summer long, we begin to see that some reform is wanted.

I have till quite lately been interesting myself in these Cases, contriving to work them up to something like completeness, without much noting the time the flowers lasted; still the lamentations of friends who found, time after time, this flower and that flower, and another flower lasting with me, while theirs meanwhile had gone through many a change, – all this made it quite evident that two months were long for Primroses, and five weeks for Hyacinths; while for Geraniums and Gloxinias, Fuchsias and Begonias, I venture not to speak.

I may say most truly that I know no other means of keeping flowers anything like so long as two months, neither in a drawing-room with its dry air, nor in a greenhouse with its passing currents, and certainly not in a garden-bed, in the finest season.

Yet, after all, the secret is a very simple one. It is merely the stillness of the air around them which preserves the flowers so very long unchanged; while the peculiar arrangement by which the heat is supplied provides that soft dewy atmosphere in which so many of the prettiest plants delight.

FILLING A DRAWING-ROOM CASE, OR A NICHE OR RECESS, WITH FLOWERS.

THE first point in beginning to prepare a stand of flowers is to lay a foundation such as may last for some time. The box, whether for a glass Case or one fitted to a recess, being all ready to receive the plants, a layer of charcoal should be laid on the floor, all over it, carefully avoiding raising that charcoal higher than the level of the sized pots used: if thus the box is made about eight inches deep, the probable size of the usual flower-pots being five inches high, a three-inch layer would be quite allowable. If small pieces are used, they can be easily taken out should it become desirable to lower a larger pot. But charcoal being light, and also very purifying, it is the best thing possible to place it where the water given may be often supposed to collect. Charcoal is moreover one of the very best of actual manures for plants.

Having arranged the charcoal, packing it closely in and preserving a level surface, the next operation is to fill up with the silver sand, which should be poured in by degrees till every space is filled.

Having raised the sand about an inch above the charcoal, my own invariable plan is to water thoroughly; giving always as much water as the sand absorbs without getting very wet. The charcoal is useful, it must be remembered, not only in occupying space lightly, but also in preventing too much water from being thus admitted.

The next process should be to arrange in the tray a series of empty flower-pots. I think it is best to put a large one at each corner, where this can be done, keeping the top rim always just *below* the woodwork, and then a number of pots, from three to five inches in diameter, should be disposed about; not exactly to touch one another, but yet so arranged as to leave no large vacant space. Around these empty pots, the sand again has to be shovelled in: a lady's trowel does as well as anything, or a jug or tea-cup may be used for pouring this in quickly between the pots.

The sand being filled in level with the top, or very nearly so, many people advise a layer of cocoa-nut dust to be spread over its surface, and even, perhaps, gently stirred in amongst the sand itself. This cocoa-nut dust is obtained by beating and pulling the rough outer husks; and, for my own part, I know nothing like it for bringing on the plants. The Ferns I have had in it have really thriven wonderfully, as well as other things described in *In-door Plants*.

The surface, then, being prepared with moist sand and cocoa-nut fibre (which latter must be first thoroughly well steeped in water), the next consideration is to plant the mossy bed; and for this the beautiful Lycopodiums are the best (*L. denticulatum, L. atroviridis, L. stoloniferum*). All of these are delightful in their different ways; they grow and spread very rapidly in a warm, moist atmosphere, forming a living carpet far superior to anything we could fashion. On the whole, perhaps, *L. denticulatum* is one of the very best, as it forms so close a mass of perennial green. *L. apoda* is pretty, but so dense a moss, that it is, perhaps, more suited for merely filling pots of myrtles, for instance; in which case it is useful.

The ground-work being arranged, the next thing to provide for is the supply of evergreens and fine-foliaged plants, whether for a Case, or for a recess or stand. Nothing gives a more tropical and beautiful tone to these kind of glass Cases especially, than a good selection of bold and telling foliage.

I am myself fond of mixing with the red-veined, spreading leaves of the tuberous-rooted Begonias, the tall cold leaf of the Arum, or water Calla, and the dark green massive shadow of the Lignum vitæ, or of little Cedars. These little evergreens may be met with continually at a most trifling cost; and really I think it is seldom that these drawing-room Cases, or, yet more, recesses, look their best without them.

The best rule, as far as I know yet, for disposing these groups is to arrange for high trees, and an undergrowth of low flowers; or to form a background of deep foliage for the gay flowers in front, broken here and there by the foliage and the flowers verging mutually on each other.

The sides should in either case be tolerably clothed with green, and when the Cases can be so placed as to have a box of dark, close evergreens standing on either side, or a group of these trees filling the window or back of the recess before which it stands, the good effect is very much increased.

When a glass Case is required to look equally well on both sides, or when it stands in a window facing a sitting-room as its permanent place, it is of course essential to adopt the former plan. One of the best plants, then, for the centre, is a beautiful arched Begonia, rising like a fountain, and drooping all round gracefully.

When this cannot be obtained, or where it would be thought unsuitable, an evergreen should replace it, in shape of a Deodara: and here I must remark that nothing is more striking than such a plant for this place, when surrounded in the spring or winter with the delicate bulbs which come most naturally amidst such sombre foliage.

Perhaps it may not have been remarked how very much there is in suiting plants to seasons, and to each other. The Snowdrop and Scilla at the foot of the dark Pine are in most perfect harmony; and when again the red-veined foliage of the grand Begonias, the exotic Lilies, and the Trumpet-flowers, blend with scarlet Cactuses for their undergrowth, we feel at once that they are all in keeping: that it is really a miniature nook borrowed from some Indian forest.

These latter are the flowers we rejoice in for the summer months; then autumn comes with its fading leaves and its keen remembrance of beauties scarcely past: I always find it very difficult to admire any autumnal plant stand.

The best plan is to look forward. Heaths are natural – at the least in name – Ferns and Moss too; and, where they can be obtained, the beautiful little Harebells and many red-berried plants, as well as blue and white ones. All these are suitable amidst the mingled foliage and the wreathing climbers; but the best thing, I think, is perhaps for once to forget the season, and having kept on late the gorgeous summer flowers, – Achimenes and Gloxinias, Geraniums and Roses, lingering with us long, – to go at once to our winter stores, and trust to Camellias and Azaleas, Poinsettias and Begonias, Cyclamens and gay bulbs, for the November stock.

But when spring bursts on us, the plants are at their loveliest: Violets and Primroses, Lilies of the Valley and the sweet pink Roses, Hyacinths and scented Tulips, Sweet-briars and Carnations, – all come rushing on us till we know not which to choose, and heartily long for all.

Spring is certainly my favourite time for these delightful stands. For one good reason, it is just the season when it is most difficult to keep plants at all in rooms in any other way. We may see them in shops, and buy them, and they may be most exquisite; but the Lily bells soon hang limply, the Rose leaves are scattered, and each succeeding Primrose grows smaller and paler.

Of course it must be so: at what time of year are our rooms so dry, and so full of currents of fresh air by day as windows fly wide open, while a heated and not fresh atmosphere comes later in the evening?

I believe I am not exaggerating when I say that a week would be a long time for such plants to retain their beauty, if kept on exposed stands. I am not speaking now of the plants which "go out to parties;" though most people may sometimes have pointed in some nursery to a dismal, shrivelled thing, and will have been duly told on asking of its history, how "it was at a party last night, or the one before." A lamentable example, truly, of the course which their own plants at home are more or less slowly following.

The glass Cases, however, while they exclude exhausting currents of air, and include the soft, warm vapour in which the plants luxuriate, preserve in beauty for very long the flowers they shelter. No dust or blight falls on their shining petals, and the verdant freshness of the leaves remains for long unfaded.

Every decaying flower or leaf should instantly be removed, a little fresh air should be daily given, and water, sometimes at least, to keep moist the sand that surrounds the roots. Shading from the sun is almost needless, so thick a dew rises up at once when it shines out warmly; and if you desire to inhale a delightful perfume you have but to unclose the case, and the warm scented breath of the air within will rush out into the room.

The general rules for the care of these Cases are therefore not many: they have to be aired just when they look rather steamy; and if any plant shows signs of spotting, or of decaying, or of mildewy leaves, those leaves must be removed and more air given; though in some instances, if all the other plants seem thriving, it may be supposed that the plant affected is itself in fault, and it should then be exchanged.

One thing, too, is particularly to be noticed: the various bulbs and annuals, plants that die down when their blossoms are over, would naturally then be placed in a dryish air and get little if any water, from above at least, while ripening their seeds or tubers; remaining therefore in a damp, moist atmosphere, they damp and decay directly, and not only become themselves unsightly objects, but injure other plants which may happen to be near them. Partly on account of this liability of leaves to injure each other, and partly because plants thrive best the more air they have and the less they are crowded and shut out from the light, it should be a universal rule never to add a plant that can be well dispensed with. If a Case once looks well without crowding, the addition of more plants only risks spoiling both its looks and them.

I have myself very seldom more than from eighteen to two dozen plants of all kinds in a four foot stand or Case, and even of these many are very small.

IN-DOOR PLANT CASE, AND LISTS OF PLANTS.

I WILL begin this chapter by an exact description of the In-door Plant Case. These Cases are so made as to be either a most useful "forcing house" for supplying stands and conservatories, bringing on at a time more than sixty pots, in a temperature of from 65° to 90°; or else when used for ornament containing a most charming group of from eighteen to two dozen flowers and plants, such as would hardly be preserved for many hours without some such assistance in a dry or heated atmosphere.

These Cases are of two sizes, the larger ones being about four feet long, by two feet wide and high, while the boxes on which the glazed frames rest are eight inches deep.

These boxes are lined with zinc, and fitted with a hot-water apparatus most accurately balanced. No lamp or anything of the kind is required, as hot water is poured in through a concealed opening from without, being also let off in the same way when cold.

This water maintains the temperature at a height sufficient for most plants for four and twenty hours without refilling; but when greater heat is required it can be raised to a degree varying from 90° down to 65° by adding more hot water after the first twelve hours.

A very simple contrivance for raising the upper glass guards against undue condensation and consequent dropping of moisture on the plants within, and this pin, attached to the upper panel, also affords means of giving air to the plants inside. Although the Case is so strong that it may readily be moved about from place to place, it is put together by means of four small screws and in less than three minutes' time; and the whole affair being perfectly self-contained, with the exception of a common funnel, there is nothing wanted extra. Arriving one day at its destination, it may look next morning as if it had been filled for months.

As these In-door Plant Cases seem likely to form a tolerably important manufacture, steps have been taken to have them patented, with all the recent improvements; and they are to be had solely at the manufactory, 4, All Saints Place, Caledonian Road, London. The price is four guineas for a plain useful Case of the size described, which is made in the strongest manner, to take to pieces, and pack in a shallow box, for perfectly safe conveyance to any distance, either by rail or water.

The Cases can be made, I believe, in any style of ornament. A good sort for general use is formed of plain mahogany, and looks very well. When used the floor requires to be covered with at least an inch of sand, to prevent the pots from resting too closely upon the hot-water Case, and also to preserve the required amount of moisture. After an inch or so, sand is not *necessary*, though very often *useful*. Flower-pots of the same size as the pots which sink into them are no bad contrivance for securing to the plants which need it a strong degree of heat. But in buying plants of an unknown kind, it is always best to ask the heat to which they have been accustomed. A small thermometer will be very useful for suspending in the case.

Silver sand may be got at sixteen shillings a ton at Kennard's, Swan's Place, Old Kent Road, and peat, sphagnum moss, loam, &c. can be also got there. If the sand in use has been long saturated with moisture or mixed by any chance with decaying leaves, it can be thrown into a heap in the sun and dried, when, having been aired and sifted, it is as good as ever. And this is really convenient, as it is an awkward and tiresome thing to be often changing.

My own Cases are always filled with sand nearly level with the tops of the pots themselves; but then I often grow both moss and bulbs in the sand in mine, either with or without the addition of a little cocoa-nut dust.

If moss is *not* grown in them, the top has to be covered with some pretty wild moss, carefully freed from insects. This freedom from insects is certainly most essential, for one single slug in a plant case rapidly accomplishes a world of damage; every bitten leaf being quite spoiled for ornament.

I will now proceed to name a few of the most favourite flowers; mentioning such kinds only as are really excellent. I feel that in some respects I am a very bad person to give lists of plants, because it is years since I have had any opportunity of visiting first-rate gardens, or of seeing new plants displayed; at the same time the same reasons which lay me under this disqualification will serve to prove that if I can get nice flowers from the mere help of catalogues and books, others surely with more advantages may do far better; and there is one advantage here at any rate to my readers, that the plants I name, except where otherwise stated, have been tried and proved; not by seeing them in hothouses and conservatories, but as fairly kept entirely in a room, some few merely for a prolonged time of flowering, others for many months, or entirely grown, as in the case of bulbs and others, in my windows and in my little "stoves."

As far as it is possible to resist the temptation of getting other plants, I always now try myself to keep to Evergreens and Begonias, Ferns and Bulbs, or tubers of some description. Many plants are so hard to keep and to stow away when they are out of blossom, while tubers lie dormant in their own dry pots, and are ready at the proper season to take into rapid growth; the summer and winter stocks of tubers or of bulbs thus changing places regularly in the plant case and in the cupboard; for a good dry cupboard is as good a place as any for a store of dry bulbs and tubers.

Roses and Geraniums, tempting as they are to grow, cannot be so easily put aside in their less interesting stages; and the same objection applies to Fuchsias and to some other plants.

Evergreens, Ferns, and Begonias, are, however, always beautiful, the evergreens at their worst time not being really ugly, while happily that worst time is just in the summer, when Roses and Geraniums may be so easily bought in blossom. It seems to me to be a particularly great object in fitting up these cases to have at the same time foliage of so much beauty that if there were no flowers, the case would still be beautiful, and yet to guard against such a display of *colour*, even in the leaf, as would weaken the effect when brilliant flowers were used.

Any one fitting up a case with foliage, should therefore have a notion of the plants which they want to use; and I cannot too strongly recommend the Catalogue published this year by Messrs. Veitch, of Chelsea and Exeter, for giving exact descriptions of almost every kind of stove and greenhouse plant, "foliage plants" being included with the most splendid Camellias and every brilliant flower that can be imagined.

Stove plants sound grand for a little glass case to grow; but it is, in fact, much easier for an amateur to grow a plant of this kind than to get one to blossom of the hardier kinds, which crave so much air and light, and are always ready to suffer from confinement.

37

'THE BELGIAN WINDOW GARDEN'

John Lindley

Source: John Lindley, 'The Belgian Window Garden', *Horticulturalist and Journal of Rural Art and Rural Taste* (March 1849), pp. 427–430

THOSE who are debarred from the enjoyment of a GARDEN by sickness, residence, or fortune, should take a leaf out of the book of the French and Belgian ladies, who succeed, by means of double-glazed windows and other contrivances, in providing themselves with an ample supply of fresh flowers at all seasons of the year. With us, the first object of the dwellers in towns is to buy plants, the next is to provide for them. Elsewhere it is thought more advantageous in the first instance to secure the means of keeping a plant in health, and that being accomplished, to obtain it. We will not be so uncivil as to reproach our fair countrywomen with herein indulging in that sort of caprice which is vulgarly called putting a cart before a horse; but we shall confine ourselves to an explanation of the manner in which other persons proceed, leaving all who are concerned to form their own judgment in the matter.

"In Belgium," says M. VICTOR PAQUET, "wherever you go, you see spaces between double-sashed windows filled in the winter time with the most charming flowers. Elsewhere the balconies are turned into green-houses, and you may find on the fifth or sixth floor a miniature green-house gay with the brightest flowers and the greenest foliage. In Paris there are many such contrivances, especially two on the fourth floor of a house in the Boulevard de la Madeleine, at the corner of the Rue Caumartin. Here are to be found the rarest plants. Camellias grow in the open ground. Passionflowers cling to the columns; the creeping fig forms a carpet upon the walls, and water-plants start up from tiny basins curiously contrived in the solid brickwork. By turning a screw a stream of limpid water flows down a rock, from whose crevices start up Ferns and Lycopods and such things. And what is it that adjoins this little paradise but a bed-room! The first beams of the morning sun throw upon the bed of the owner the shadows of Palm-leaves and Bananas, or of garlands of Passionflowers."

This sort of garden, though on the fourth floor, is, however, rather too ambitious for everybody's taste, and therefore we agree with M. PAQUET that the little

 DOI: 10.4324/9780429198809-44

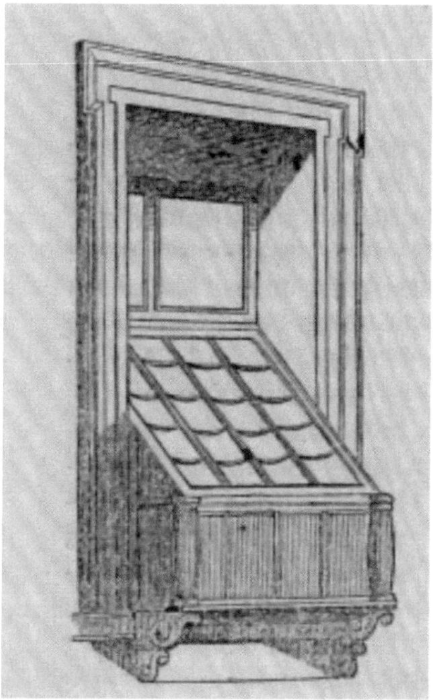

Fig. 46.

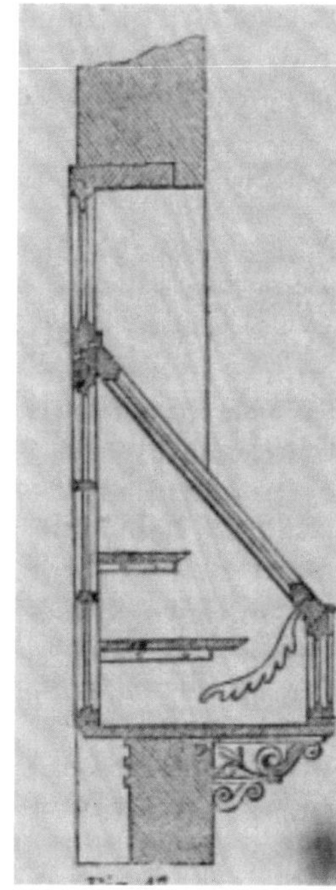

Fig. 47.

Belgian window-gardens are upon the whole more likely to meet the means, if not the fancy of the mass of mankind. We therefore borrow from him the following figure and description of one of them, which has now become extremely common. Let figure 46 represent the outside of this window-garden, and 47, a section of it, together with the window to which it belongs. A sloping roof of glass is carried outwards from the middle cross bars of the sash in such a way that the upper sash allows light to enter the room freely. The lower sashes open as usual by a pair of folding leaves as wide as the window frame. A pair of brackets carries the floor which projects beyond the walls of the house. Shelves are fixed to uprights next rhe window-leaves, and the sloping roof is raised or depressed by means of a rack, which is easily reached from the inside."

It is obvious that this contrivance possesses all the advantages of a Ward's case, without its inconveniences. Being placed on the outside of a house, it occupies no space that is otherwise required. As the glass next the room will always be warm, there will be no condensation upon it to conceal the plants which the window-garden contains. The folding leaves render it easy to get at the interior, and in towns water could be readily laid on for the purpose of moistening the plants when it is required. Such little contrivances would suggest themselves to any ingenious mechanic. A place of this kind might even be heated in severe weather by a pan of boiling water, or protected by a woollen covering thrown over it at night. In short, it seems to be of all things the best adapted to the purposes, as well as means, of those to whom window-gardens would be acceptable. In such a place it would be perfectly practicable to have all sorts of forced flowers in spring and winter; Roses in summer, Pelargoniums in autumn. A thousand pretty plants might be selected for variety, and the whole would be a most agreeable amusement for those who are neat, careful, and skilful.

A model of this Belgian window-garden was most obligingly forwarded to us from Boulogne, by our good correspondent, Mr. CRUCKSHANKS, and shown to the Fellows of the Horticultural Society. It teaches us several things. A rolling blind is made to work in the inside, so as to screen the plants from too much sun, when placed in a southern aspect. A trellis-work of wires is carried up the ends, and adds much to the appearance of the little garden. Here thrives Maurandya Barclayana, together with the little Tropæolums of Chili. The bottom should be guarded by a raised edge next the sitting-room, so as to form a shallow box at least 6 inches deep; this is to be filled up with damp moss, in which the pots are plunged, and by which they may also be covered. Attention to these matters has been found by experience to be essential. We would add moveable ventilators at the front and sides, so that air might be given when it would not be prudent to raise the sash.

Such little details are what constitute the secret of success; and now that the plans of construction are intelligible, and that many are, as we believe, preparing to adopt these gardens, we must offer some hints in a general way.

The advantage of such an apparatus consists in its separating plants from a sitting-room when they are not wanted, and introducing them there the instant they are wished for. Open the doors, and the garden forms part of the room; close them, and the apartment is relieved from the presence of the plants. But, what is of much more importance, in the latter case the plants are themselves completely relieved from the fatal atmosphere of the apartment, – not fatal, indeed, if inhaled for a few hours at a time, but certainly destructive if endured for long.

What, it may be asked, is there in the air of a sitting-room which plants are thus unable to endure? Can anything be purer than the atmosphere of an English drawing-room? Perhaps not; but it is this purity which in part inflicts the injury. Plants would thrive better if it were otherwise, but it is more especially its dryness. Let any one measure the moisture of a sitting-room and the open air, and he will see how great a difference prevails. We have this moment tested it by Simmons's hygrometer;[1] in the open air this instrument indicates 40°, in a sitiing-room 60°.

When plants are kept in a dry atmosphere, they rapidly lose their water of vegetation; the sides of their pots are robbed at the same time; and it is impossible for plants to suck out of soil thus partially dried the moisture demanded for the sustenance of their exhausted foliage. Such a state of things is inseparable from a sitting-room. To render the latter congenial to plants it would be uninhabitable by ourselves. The extent to which plants are injured in a common sitting-room is strikingly illustrated by the condition of cut flowers. Let two clusters of fresh gathered flowers be introduced into a sitting-room; place the one in the mouth of a narrow necked jar of water, and arrange the other upon such a shallow pan of water as a deep dish will furnish. It will be found that the latter will be perfectly fresh days after the former are faded. The reason is that in the narrow necked jar the flowers have no access to water except through the ends of their shoots, and are surrounded by a very dry air; while in the flat dish they are able to absorb abundant water, because a large part of their surface is in contact with it, and are moreover surrounded by air incessantly moistened by the vapor that continually rises from the dish.

Of this we may be sure, that darkness, dust, heat, want of ventilation, and all the other calamities to which plants in sitting-rooms are subject, are as nothing compared with the inevitable dryness of the air; which indeed acts injuriously, not merely by exhausting plants of their water of vegetation, but by lowering the temperature of the pots in which they are grown, in consequence of the evaporation constantly taking place there.

What makes the evil greater is, that the plants which are purchased for sitting-rooms are invariably brought into high condition by being grown in a damp atmosphere. They are transferred from the hands of skilful gardeners, armed with the most perfectly constructed forcing-houses, into the care of inexperienced amateurs, whose means of maintaining a plant in health are something considerably less than nothing.

A case will illustrate this: A Rose bush is bought in the market, fresh and trim, with one or two flowers open, others in bud, more still younger, and many but just peeping out. From such a specimen nothing, it would seem, can result but a long succession of beauty. But this charming thing, so fresh and promising, was, perhaps, a few hours before, the inhabitant of a damp green-house or pit, where its leaves were formed in shade, and their surface softened by a daily bath of artificial dew. It is suddenly conveyed to a sitting-room; its leaves shrivel up under the withering influence of its new habitation; the fountains of life become dried; the young flowers, starved by want of their accustomed food, drop off, the leaves follow them; the green-fly or red spider attacks the suffering remains, and a week or two are sufficient to witness the destruction of all the buyer's hopes.

We appeal to everybody's experience for our proof that this is an ordinary case. But a Belgian window-garden removes the difficulty; in such a place a plant is kept in precisely the circumstances most conducive to its health; light and moisture foster the young shoots, and the softened air provides a due supply of all that is indispensable to vigor.

To those who propose to engage in this kind of amusement we would add a very few words of empirical advice. 1. Always use rain-water; 2. Always let it be milk-warm; 3. To every quart of rain-water add half a grain of nitrate of ammonia, or sulphate of ammonia; 4. Invariably keep up a continual current of *warm* fresh air through the garden whenever the circumsrances of the season will permit; the easiest method of accomplishing this, which is equally important in large as in small houses, and yet is almost universally neglected, we shall consider on a future occasion. – *Gardener's Chronicle.*

Editorial Endnote

1 The hygrometer was an instrument for measuring humidity designed by E. Simmons that had recently been tested in the gardens of the Horticultural Society and reported in the *Journal of the Horticultural Society*, vol. 1 (1846), pp. 127–130.

38

'CULTURE OF PLANTS IN ROOMS'

Eduard Regel

Source: Dr. Regel, 'Culture of Plants in Rooms', *The Garden,* vol. 1 (9 March 1872), pp. 359–362

DOUBLE WINDOWS

THE following general rules may be found useful in the case of plants grown in double windows: –

1. When the mercury stands below the freezing point, the ventilators should not be opened.
2. During hoar frost, if the wind strikes the window, the ventilators should be closed even if the mercury stands above 32°; as otherwise the immoderate influx of the wind cannot be prevented.
3. In mild, fine weather greenhouse plants may receive as much air as possible, but hothouse plants only so when the temperature outside is higher than that within, and when the air is calm and not too dry. In spring, before the greenhouse plants are placed out for good in the open air, the double window should be previously ventilated by day and by night.
4. In opening the ventilators from without, care should be taken not to open them too much, as too great an influx of dry summer air would destroy all the advantages which the double window affords. In summer, during stormy or very dry warm weather when ventilation from within is impossible, fresh air may be freely admitted through all the ventilators which have been made to open inwards.
5. The windows which open inwards should not be opened during dusty summer weather, nor when, in cold weather, the other windows of the room are opened.

SHADING.

In intimate connection with ventilation and the regulation of the temperature stands the subject of shading. In bright, strong sunshine, when there is no ventilation, the temperature inside the double window is raised not only to a considerable

DOI: 10.4324/9780429198809-45

degree, but it will also dry the air very much. If, at the same time, there is a flow of air, either from within or without, the greater number of greenhouse plants in the double window require no shading in the course of the autumn, winter, and spring. But when, during constantly cold nights and clear, open sunny weather, greenhouse plants must be kept longer in their winter quarters, they may have a slight shading in spring by giving the outer window a thin coat of lime-wash. As soon as the weather becomes permanently warmer, greenhouse plants are put out for good in the positions intended for them in the open air, having been previously hardened off by the admission of air.

As the double window must always occupy a sunny position, it must always be artificially shaded when it is used in the end of spring and during the summer for the culture of tender hothouse plants. This shading may be effected either by means of lime-wash, as mentioned, or by suspending a covering of thin canvas, or some suitable material. The first method is the simplest, but it has this disadvantage, that it remains during gloomy weather, and then does not look well either from within or without. This disadvantage may, to some extent, be diminished by laying on as thin a coating as possible, so as to break the force of the direct rays of the sun, and yet admit sufficient light.

The plants in a common double window are best shaded on hot and bright summer days by hanging a shade of sailcloth outside which can be raised and lowered by a cord. The same mode of shading may be used for the double window, only here the raising and lowering of the shade must be from the outside, or the cord may be passed over rollers, and brought in through the frame of the window, so that the shade may be raised and lowered from within at any time when necessary. This arrangement is, of course, the best.

As the erection and maintenance of such a shade is always costly, we may suggest two other methods. The first of these resembles the limewash plan in being a permanent shade formed of some shading material fastened over the outside of the window. For this purpose wide-meshed muslin, or a shade woven of reeds, or twigs with interstices at least as wide as the width of the materials, is used. This shade, which is to remain fixed all the summer, and which merely breaks the force of the sun's rays, and yet looks better than the lime-wash, is first applied in spring, when there is reason to dread that the increasing heat of the sun may injure the plants.

The other mode consists in hanging on the inside of the glass a shade of thin linen or cotton material. This plan is easily adopted, but the shade can only be properly managed by raising and lowering it with cords close to the glass, as the ordinary method of doing so by means of rollers would be injurious to many of the plants in the window, or could not be used at all. If this inner shade be raised and lowered by some contrivance similar to that which is employed for the windows of railway carriages we may consider it the best kind of shade, as it can be used or removed at all times, is not affected by the weather, is easily managed, and, lastly, is not expensive.

With reference to shading in general, it is to be particularly observed that it promotes a luxuriant and strong growth of the plants by inducing a moist atmosphere;

but if it is carried to too great an extent, it is unfavourable to the production of flowers. Therefore, wherever a permanent shade is employed, it should not be used sooner than the season of the year renders it absolutely necessary, and should be removed early in autumn; at the time when no shade can be given, ventilation both from within and without should be employed according to the weather. But where a movable shade is used, the plants should have the benefit of the morning and evening sun, and should be shaded only from eight or nine o'clock in the morning until three or four o'clock in the afternoon. In winter, the sun may be allowed full access during the entire day.

INTERIOR ARRANGEMENTS.

Double windows, which always possess a greater depth than ordinary room windows, should bring safely through the winter as large a number of plants as possible, or supply flowers during the winter months for the other windows or for flower stands. For this purpose, in addition to the lowest shelf, or that which rests immediately on the window-sill, one or more shelves should be added. In order not to interrupt the circulation of the air, instead of boards, the upper shelves should be made of laths placed at the distance of an inch from each other. These laths should, moreover, be moveable, so that they may be all or partially removed at any time when the growth of the plants below them renders this necessary.

The uppermost of these shelves will occupy the warmest and the driest position, and on that account are the most suited for the forcing of flowers. The plants placed on them are reached by means of a movable ladder with steps on both sides.

Finally, in "inserted" double-windows, at the bottom of which there is no reservoir for heating purposes, the taller plants are placed on a board which lies on the ground, and only single small plants on the window-sill, which is not more than 1½ foot broad, and immediately in front of the lowest panes. The height of the lowest shelf from the ground depends on the height of the plants, the tops of which should receive as much light as possible. By the arrangement of the plants on separate shelves, the smallest and weakest specimens will be next to the window, and the larger ones placed farther back, so that each may have as much light as possible.

Further, the plants in the double window should not be placed so closely together that their branches will touch each other, or even that some of them may lie against others, otherwise only thin bad specimens will be grown, as occurs in plant houses under the same circumstances. We may here repeat what has been already said, that it will afford the cultivator more pleasure and satisfaction to raise a few good specimens than many bad ones.

PLANT CASES.

A contrivance much adopted for the culture of plants in rooms, especially for such plants as do not thrive in the dry air of a room, and which has been frequently employed with the best results, is a large glass case arched at the top; or it may be of small size and constructed of glass and iron, and placed on a table or stand near the window. When it is of small size, the whole of the top may be taken off in order to attend to the plants, and when of large dimensions, windows or openings are made in the sides for the same purpose. The particular form of the case and the table or stand may be varied in a manifold degree.

The figure [...] is an illustration of one of these plant cases of moderate size. The stand bears, firstly, like most flower-stands, a box or case, made to fit it exactly. Inside this case is another made to lift out, and formed of zinc or of tinned iron, with a pierced bottom, and a tray underneath to receive the water which drains down, and which is carried off through a cock underneath. In order to expedite the drainage from the upper case into the tray, the upper case should stand on hollow supports. We have also given a vertical section of this arrangement, on which the permanent thriving of the plants cultivated in the plant case depends, for unless

Plant Case.

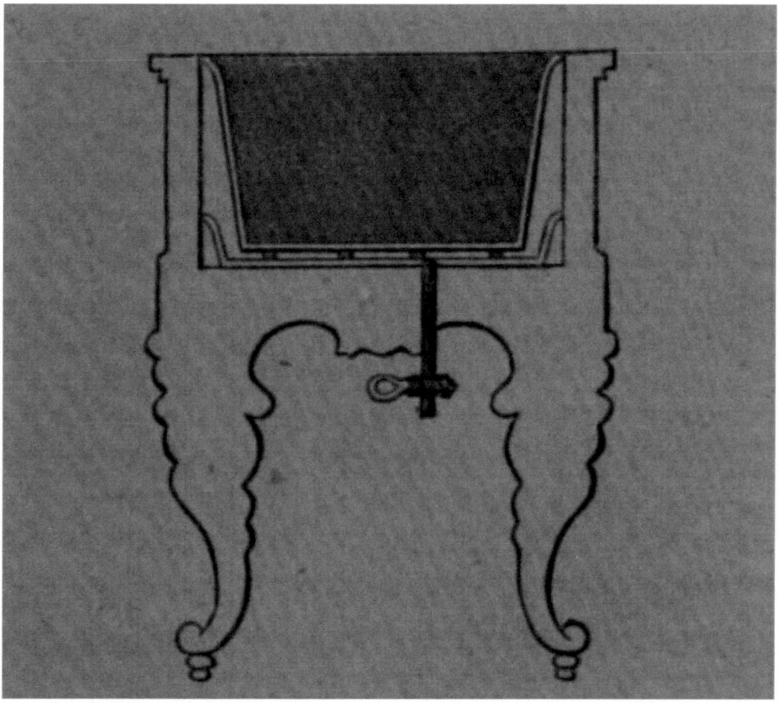

Section of a Plant Case.

the drainage is perfect, the earth ceases to be percolated by air, and becomes sour and unfit to furnish proper nutriment to the plants.

In the culture of stove plants, this most important point must be attended to, not only in reference to the plant case, but also wherever plants are grown in iron boxes which stand in wooden cases (as, for instance, for forming arbours in windows, &c.), similar precautions must be taken with regard to the drainage.

The iron case is now filled with earth, in which are planted Ferns, Aroids, Marantas, Bromelias, Selaginellas, Palms, &c., or pieces of stone may be placed in it, between which earth is to be filled in and planted with a suitable selection of plants which can be grown between the stones.

In plant cases of larger dimensions, little landscapes with fountains, cascades, and tiny mountains may be represented. Dead branches of trees, of not too great a size, and knotty roots, may be placed here and there, and planted with Bromelias, Orchids, and Aroids; handsome baskets and other contrivances containing Orchids, Æschynanthus, &c., may be suspended at intervals, while dwarf Selaginellas and neat Ferns cover the turf between the plants.

On the [...] page is an illustration of the gable end of one of the largest plant cases that we have seen, and which Herr Graveur Zimmerman has constructed and set up in a room of his dwelling-house in St. Petersburgh. It is 9½ feet long,

Large Plant Case

5¼ feet wide; the windows are 5¼ feet high, and the height of the gable end is 7½ feet. The stand supports a reservoir of zinc filled with water, the surface of which is for the greater part covered, with the exception of an irregular space in the middle, where numerous fountains play and gold fish swim about, while water drops continually from the cliffs in the foreground. This water passes into the under basin. Where water is not used, the basin is covered over with a perforated plate, on which a layer of soil is placed, and in this many kinds of handsome indoor plants, such as Ferns, Palms, Aroids, Scitamineæ, Lilies, &c., are planted, the surface under them being covered with a green carpet of Selaginella, Ficus, and Tradescantia, while here and there handsome small specimens of variegated Caladiums, Eranthemums, and Gloxinias in flower add greatly to the effect. Between these plants arise several higher cliffs, of which the one represented in the foreground contains a zinc reservoir at the height of four feet from the bottom, which feeds the fountains and the little cascades which fall from the rocks. The water is forced into this reservoir from the larger one by means of a pump placed under the stand. On the cliffs and in their crevices grow all sorts of ferns, Dasylirions, Aroids, &c., and here and there knotty branches are furnished with climbers, Orchids, Bromelias, and similar plants, which attain a luxuriance equal to that which ones sees in a regular plant house.

We know that the Japanese delight to give a representation of a whole landscape on the smallest scale in their little gardens; but in the plant cases we may have before our eyes, the whole year through, the luxuriant vegetation of the tropics on a small scale; and, confined to a town life, may cultivate for our enjoyment a pleasing collection of the finest and most tender plants from all countries.

CONSTRUCTION OF A PLANT CASE.

The roof of the plant case may be of various shapes – angular, round, vaulted, &c. – but it should always be made as light as possible. The frame may be of cast iron, wrought iron, or sheet iron, and should be painted and varnished to preserve it from rust. In plant cases of small size the entire glass cover can be taken off, or a large bell-glass may be used; in those of moderate size single panes are used as windows to open and shut. In those of larger size, a cast iron or wrought iron frame is used, covered with lightly constructed iron framed windows, which can be raised at pleasure to put in and take out plants. In Herr Zimmerman's, the framework is made of wrought iron, and all the lower windows open on hinges. A similar structure of wood would be too clumsy, and would, from the effects of the summer air and the moisture inside, soon warp and require to be repaired.

POSITION OF PLANT CASES.

In order to secure good results in the plant case, it should be placed near the window in a position where the sun may shine on it during the greater part of the day. Where the room is not exposed to the sun, the plant case should be placed near the window, and a selection made of such plants as thrive in the shade, such as Ferns, Aroids, &c.

Plant cases of large dimensions, like that of Herr Zimmerman's already described, should be placed at a distance of about three feet from the window, as if placed nearer, the shade of the window piers would be injurious to the plants, and yet the window should be near enough to allow of the beneficial influence of the sun. Where the locality permits, a corner room exposed on both sides to the sun should be selected for plant cases of large size. While small plant cases on stands as high as the window-sill are best placed in the window recess, those of larger size, which must be placed some feet from the window, should be on stands not more than from one foot to one and a half foot high, this being the best height for light, management of the plants, and also for the decoration of the room.

WARMING PLANT CASES.

As the plant case stands in the apartment, special heating is usually unnecessary. Only when variegated tropical plants like Anæctochilus, Marantas, Caladiums, &c., are grown in cases of smaller size, or when East Indian orchids, such as Vandas, Saccolabiums, &c., are grown in those of greater dimensions, can heating be employed with advantage. For this purpose, as in the case of the double window, a hot-water apparatus only can be employed, and that may be either such a one as was there described, or a small covered reservoir may be placed underneath, at a few inches distance from the bottom of the plant case, and kept heated by a lamp. To conduct the heat, pipes should be laid under the soil of the plant case and along the surface of the stand. If the reservoir or the pipes are placed in immediate contact with the soil in which the plants grow, so much vapour will be developed from the soil, which will be condensed on the inside of the glass, as to obscure or totally hide the plants from view, and so all the effect of the arrangement will be lost. Finally, wherever large cases are placed in rooms with a temperature of from 60° to 65° Fahr., instead of employing artificial heat, it will be sufficient to water and sprinkle the plants daily with tepid water and to keep closed vessels of hot water near the cases.

ADVANTAGES OF PLANT CASES.

The glass covering of plant cases not only shelters the plants from the effects of dry air, from great changes in the temperature produced by air and heat, and lastly from the hurtful dust, which in all dwelling-rooms covers the plants more or less, but it also prevents an undue exhalation of moisture from the soil and the plants, so that in the inside a moderate, equable, moist, and warm temperature is maintained, which, for a large number of the handsomer tropical plants, is an essential condition of successful culture and strong healthy growth. By the double glass covering the quantity of light which reaches the plants is somewhat diminished. But as the greater number of plants which love a moisture-laden atmosphere do not naturally grow in localities exposed to the full sun, but rather in the shade of shrubs and forests, such as ferns, the greater number of the monocotyledonous plants of warm countries, and many of the most admired fine foliage plants (Aralias, Rhopalas, &c.) of our hothouses, they will not at all suffer when the case is placed in an otherwise favourable sunny position near the window, if a suitable selection of plants has been made. Thirty years ago bell glasses and water glasses were very much used in rooms, and placed over cuttings to induce them to strike root. The plant case is merely an expansion of this idea, which however, was first practised on a large scale by Dr. Ward, of Clapham Rise, about thirty years ago.

MANAGEMENT OF PLANT CASES.

With respect to special treatment, we add some instructions which, however, are particularly applicable in the case of plants from the warmer latitudes. Every day the plant case should be visited once, and every withered or fallen leaf and shoot should be removed. Moreover, should any mouldy or disagreeable smell be perceived, the cause must be sought for and removed, and all the windows should be opened for some hours, or the glass covering removed, so as to give free access to the fresh air. The windows must always be kept clean, and be regularly wiped inside and outside. Small cases should be washed on the inside once a week. A mouldy smell, foul air, and a diminution of light are the consequences of neglecting to cleanse the cases. Plants in a plant case do not require to be watered so often as those in the open room. In dull weather, and during the short days of winter, care should be observed in this respect, and no more water given than is absolutely necessary. In fine, sunny spring and summer weather, it will be proper to sprinkle and water carefully with tepid water. This will also remove any sourness that may exist at the bottom of the case. As it is chiefly plants from the warm latitudes which should be grown in plant cases (Ferns, Orchids, Marantas, &c.), a watering and sprinkling of tepid water, will prove very beneficial even in winter. When the room windows are opened in mild, warm, dull weather, the windows of the case should be opened, and the bell glasses raised in order to give the plants some air. In dry, hot summer weather, the plants should be slightly sprinkled in the morning and evening, and air should be admitted at night as well, to renew the air in the case as well as to lower the temperature.

When the growth of individual plants is so luxuriant as to interfere with that of other plants, or to produce crowding and choking, they must either be trimmed or removed altogether. If in bright, hot summer weather the case comes under the direct rays of the sun, it should be shaded during mid-day by suspending a curtain in front. If no shade is used, then air must be admitted, and the plants should be sprinkled.

ARRANGEMENTS WITH REGARD TO PLANTING.

When, for purposes of planting, earth is to be placed in the case, the perforated bottom which serves for carrying off the water should first be covered with potsherds, in such a manner that a concave piece may be placed over each of the holes. In this way the bottom will be covered with a layer of potsherds half an inch deep. On these should be placed a layer of clean sphagnum, and on this the soil, the most suitable mixture for which is meadow loam with some loose peat and sand. A stiffer mixture may be used between the stones of the rockwork, consisting of heath soil or turf or well-rotted leaf mould and loam.

In consequence of too strong a growth, deterioration of the soil, and the decay of individual specimens, certain parts, or even the whole, of the plant case must be planted afresh from time to time; in which case, fresh soil must be added under the directions already given. The most suitable time for this operation is in February and the beginning of March, when the new growth commences. The Selaginellas, which form the green turf, must be removed as often as they grow too thick and too long, as this spoils the appearance of the turf. With these (*i.e.*, the kinds generally used for making a turf) there is very little trouble. They are simply pulled up, and in their place fresh roots or growing cuttings are put in in tufts, and from these new and similar turf will soon be produced. This however, should not be done from the middle of October to the begining of January. – *From the German of Dr. Regel.*

39

RUSTIC ADORNMENTS FOR HOMES OF TASTE

(Excerpts)

Shirley Hibberd

Source: Shirley Hibberd, 'Preface', ch. 1, 'The Aquarium', ch. 2 'Construction and furnishing of aquaria, 'The Freshwater Aquarium', *Rustic Adornments for Homes of Taste* (1856), [3–14] excerpt pp. 3–5; [15–28] excerpt pp. 15–20; [83–192], excerpt, pp. 87–91

PREFACE.

WHATEVER serves to heighten the enjoyments of home, and add fresh graces to the domestic hearth, must be worthy of encouragement and culture. The last twenty years have been marked by such a progress in domestic æsthetics as to be worthy of designation as a new era in social life. The improvements in the structure of houses and the sanitary government of towns, have been accompanied by the quick march of the popular mind in the appreciation of beauty, and the promotion of every means of intellectual and moral refinement. If we have arrived at the culminating point in our civilization, we may yet feel some assurance that as long as luxury takes such innocent shapes as it now begins to delight in, our descent towards the horizon is certainly far off. But we cannot hope that the culminating point has yet been attained – we are surely still in the ascent towards it, and our course is becoming steadier and our light purer as we rise.

An eminent author has called this the Age of Veneer,[1] and another dignifies it with the title of the Age of Fustian.[2] I shall take the privilege afforded me in this Preface to call it the Age of Toys. We have nearly exhausted the means of morbid excitement, and are growing simpler, because purer in our tastes. Our rooms sparkle with the products of art, and our gardens with the curiosities of nature. Our conversation shapes itself to ennobling themes, and our pleasures take a tone from our improving moral sentiments, and acquire a poetic grace that reflects again upon both head and heart. The mark of our progress is seen in our love for toys, plant-cases, bird and bee-houses, fish-tanks, and garden ornaments, – they are the beads in our Rosary of homage to the Spirit of Beauty.

Happily in this country the Home of Taste is not a merely ideal creation; our domestic life is a guarantee of our national greatness, and as long as we shall

DOI: 10.4324/9780429198809-46

continue to surround that life with emblems and suggestions of higher things, so long will the highest teachings of knowledge, elegance, and virtue be attainable at the fireside. Our pleasures and pursuits have as powerful an influence on our national character as the precepts of sages at our seats of learning; and the simple toys that afford recreation for our hours of leisure may prove worthy advocates of morality and religion. In the contemplation of the wonders of nature, and the cultivation of domestic elegances, the intellect and moral nature must advance towards perfection, just as Ingomar and all his bandit savages[2] are humanized by the presence among them of the chaste Parthenia. Such is the ministry of the Toys that embellish the Home of Taste, and to aid that ministry will, I hope, prove to be the tendency of the present work.

Though many of the recreations treated of in the following pages are those which pertain more exclusively to country life, I have endeavoured to help the townsman in his participation of them, and trust my humble labours may aid in what has already been accomplished for the promotion of rustic life in towns – the realization, as far as possible, of the much-sought *Rus in urbe*

S. H.

The Nursery, Tottenham, Feb. 1st., 1856.

THE AQUARIUM.

"A sounding grotto, vaulted, vast,
O'erstudded with a thousand, thousand pearls,
And crimson-mouthed shells with stubborn curls
Of every shape and size, even to the bulk
In which whales harbour close, to brood and sulk
Against an endless storm. Moreover, too,
Fish – semblances of green and azure hue,
Ready to snort their streams."

Endymion.[3]

CHAPTER I.

The Aquarium is one of the latest and choicest of inventions for the Rustic Adornment of a home, and affords, at a comparatively small expenditure, an immense return in the way of instruction and amusement. Its very difficulties enhance its power of interesting, and add a zest to the enjoyment of success. Considered as a domestic ornament it is insurpassable, and, while in its humblest form it presents a constant succession of beautiful and novel objects, so to all the accessories of artistic decoration, it adds the charm of life in some of its most beautiful and strange developments. The merest glimpse of water is always refreshing to the eye; its clear, cool aspect, the mingling of many colours and forms; the peculiar growth of aquatic plants, and the still more curious forms and movements of aquatic animals, combine to form an assemblage of delightful and ever-changing pictures.

The Naiads need no longer dwell in forests lone, dipping their white feet in streams haunted only by the robin and the humble bee, but may sport in gay drawing-rooms, in homely parlours, in the study of the recluse, or the chamber of the valetudinarian. No longer need they fear winter storms and March hurricanes, but shall henceforth have homes within sheltered walls, impervious to frost, and shadowed by curtains, where love whispers, and young children play.

To the naturalist the Aquarium opens up new studies of the choicest wonders of the deep sea. Those departments of zoology which have for their regard the creatures of mid-ocean, or even of the pebbled shore, have hitherto made the slowest and least satisfactory progress of any; now they are to experience a "sea-change," for the dredge brings up the

> "Pale glistening pearls, and rainbow-coloured shells."[4]

and by the preservation of the creatures in their own element, and under circumstances almost as natural as that in which they were produced, we may study their habits and economy even to the minutest particulars.

We used to study the tenants of the sea by means of wretched specimens, shrivelled up in spirits, or crushed flat between the pages of books. Occasional festoons of sea-weed suspended from the ceiling, or set out upon the mantel, so as to absorb every stray wisp of smoke; a few corals and madrepores, and occasionally a queer-looking stuffed fish, the shape distorted and the colour gone, have long marked the extent of the means for the domestic study of "deep sea wonders;" while the ill-arranged specimens in the British Museum have, for years, been little better than such collections as superannuated sailors delight in, for awakening or directing public attention to a class of creatures having so few analogies to those of terrestrial origin.

The Aquarium exemplifies, in an instructive manner, the great system of compensation which, in nature, preserves the balance of equilibrium in animal and vegetable life. Indeed the recent adoption of this plan of studying the characteristics of creatures hitherto placed beyond our reach, except when dead and mutilated, has arisen out of the experiments of philosophers as to the nature of that duality of forces which render the two great departments of organized existences essential to each other, not merely for ordinary sustenance, but for the continuance of refined chemical operations essential to their respective organisms. Researches into the chemistry of animal and vegetable bodies, and especially of the effects they severally produce, by respiration, on the medium surrounding them, have resulted in the conclusion that animals and vegetables supply each other with the gases most essential to existence; what the one exhales as effete and obnoxious, the other absorbs for the highest uses of vitality. Animals take oxygen from the medium in which they live, and in return exhale carbonic acid. Vegetables also absorb oxygen gas, and give out carbon; but they also absorb the latter in greater quantity than they exhale it, and during their season of greatest activity throw off more oxygen than they take up at other times. Herein is the first element in the management of an Aquarium, which, to be successful, must contain a sufficient number of plants to supply the animals with atmospheric air for respiration.

CHAPTER II.

THE size and form of the tank must depend on the place for which it is intended, and the peculiar taste or fancy of the proprietor. If for ornament only, it must combine elegance with amplitude, for a very small tank has too much the appearance of a toy to improve the appearance of a room. A cylindrical form is the cheapest and most easily obtained. Mr. Gosse[5] describes one made for him by Messrs. Pellatt. It was the largest size they would undertake to blow, namely, twelve inches diameter and ten inches high. They may, however, be obtained of any height; the risk increases with the increase of diameter: the cost of this was 10s. 4d.

Dr. Badham,[6] in his clever paper, "Periwinkles in Pound," describes a still better form of a blow tank: – "A Marie Louise pear, cut horizontally across somewhere below the middle, will convey, in its broader segment, a tolerably correct notion of the contour. The dimensions are as follows: – Diameter at the base twenty-four inches, depth of sides fifteen inches, diameter at top fifteen inches. They hold about nine gallons of water, are fabricated of clear common glass, with a slightly green tinge, and placed on a raised rail and spare shelf about four feet from the ground, in a glass passage bearing a south-east aspect, opening into the garden. The total expense of these glasses, including packing and a long land journey, is only fifteen shillings apiece." These were furnished to Dr. Badham by Mr. Bowers, Corn Market, Ipswich; he having failed to obtain them on application to houses in London.

Globes and cylindrical vessels afford the cheapest forms, but apart from this commendation, they have few qualities to recommend them. They are liable to break, and there is a limit to their size. They distort the objects they contain by unequal refraction; and, being blown, their sides are never regular in form, never perfectly globular or truly cylindrical. Hence our little favourites change their shape and magnitude at every movement. Now they pass us in front, and appear simple lively creatures; but as they recede they assume monstrous proportions, and surpass all that we picture of the ugliness of gorgons and "chimeras dire."[7]

The best form is undoubtedly that of an oblong or octagonal box, formed of plate or sheet glass, and slate, the joints bound with zinc, and cemented with Scott's cement or common putty. Mr. Gosse describes his largest tank as measuring twenty-four inches by eighteen, by eighteen. The sides and ends are of three-sixteenth plate glass, the bottom of slate, the corners of birch wood turned into pillars, each surmounted by a knob, and united by a frame-top going all round. The glass is set in grooves in the slate and wood, and fastened with white-lead putty. The cost of this was £3 10s. A smaller one, fifteen inches by twelve, cost 21s.

A convenient size for an ordinary room is from two feet six to two feet eight in length, by sixteen or twenty inches wide. The depth should be the same as the width. Among painters and sculptors the rule for rendering an angular object graceful, is to give it the form called a double cube, in which the length is exactly double that of the width; the width and depth being equal. Such an object, if cut into two equal parts, would form two perfect cubes, and presents the most graceful outline that straight lines and right angles are capable of assuming. Such a tank placed on a table or stand, forms a reservoir or miniature lake, and requires only a little skill and patience to render it an ornament of a most refined and pleasing character. A large one involves no more trouble than a small one, but of course, is only suited to rooms and windows of greater capacity.

One great impediment has hitherto operated against the adoption of ornaments of this description, namely, the difficulty of obtaining them properly constructed at a fair price. Even an ordinary Ward's case is not to be easily obtained in London. I have been repeatedly solicited by friends, to advise them how and where to obtain a Wardian case, or a tank for aquatic products, so few houses are there on whom we may depend for intelligently carrying out our instructions, or for meeting our wants in such matters at a fair price. The importance of this matter induces me to refer to the Messrs. Treggon and Co., of 57, Gracechurch Street, London, who have recently given their attention to the construction of marine tanks with considerable success.

I have been supplied by them with a tank, the size and shape of which I think would suit the majority of persons, who do not care about the expense of a fountain or other elegant fittings. It measures two feet and a half by one foot and a quarter in width and depth. The two ends and the bottom are of slate, the sides of plate glass. The cost of this was £3. They make a smaller size, measuring two feet by one foot by one foot, for 20s., which is far cheaper, considering the advantage of flat glass and large capacity, than Dr. Badham's globes at 15s. They undertake to construct and fit up the tank according to the elegant design of Mr. Voyez, including fountain and supply pipes, and all necessary fittings.

This tank measures thirty-two inches by sixteen by sixteen. The fountain is kept in play by means of a cistern fixed above the level of the tank, whence a supply pipe descends, and passing beneath the floor, enters the tank from the wall against which it is placed, or up one of the legs of the stand, hollowed for the purpose. The portion of the tube which passes through the water must be of gutta-percha; and this, perhaps, would be the best material for the whole of the tubing, as there is then no fear of corrosion. The nozzle of the fountain is of glass; the dolphin is formed of artificial stone, and is firmly placed to the rocks on which it rests, but so arranged as to be easily moved away for alterations and repairs. Mr. Voyez has chosen the fresh-water tank for illustration, on account of the beauty of the ferns which flourish under the drip, and add vastly to the fresh brilliancy of the scene. The Marine Acquarium might be made after the same design, the ferns, of course, being unfit for its embellishment.

For a conservatory, Mr. Gosse recommends an Aquarium of a vase-like or octagonal form, either with or without a fountain. They may also be formed wholly of slate, earthenware, or stone, though, with either of these materials, the view must be from above the surface of the water. Such tanks should be more shallow, in order that objects at the bottom may be plainly visible. If surrounded with rock-work planted with suitable plants, such tanks are very ornamental, and well suited to the conservatory or aviary.

Wherever there is any liability to dust or soot, the tank must be covered over the top. The best cover is a sheet of glass of the requisite dimension. This may be merely laid on, or fitted with a hinge on one side and a handle on the other, with a jointed support, so that it may be left open, if necessary, without requiring to be held up by the hand. Messrs. Treggon make framed covers to some of their tanks; these are formed so as to make a kind of sarcophagus top, the slanting sides being joined at the edges with zinc and cement. Such a lid should never fit very close, otherwise the lives of the animals would be endangered by the shutting out of fresh air; and even, if admitting air freely at the edges, they must occasionally be lifted off for a few moments, to allow a change of air at the surface of the water. When this is done, some water should be taken up in a vessel, and poured back from a considerable height, several times, so as to agitate the surface, and mix with the water some of the superincumbent air. Fine muslin may be used in the absence of a glass lid. A fountain will, of course, keep the water in a fresh and sparkling condition, so as to render the aeration by hand-labour unnecessary; much to the benefit of the specimens, which thrive with greater certainty, and exhibit more lively movements under the splashing of a small jet.

A window is unquestionably the proper situation for a tank. An abundance of light promotes the growth of the plants and the health of the animals, and secures that succession of tiny globules which the vegetation on the stones at the bottom sends up, and which rises rapidly to the surface when the sun shines. This is one of the prettiest features of a well-stocked Aquarium; every stone is seen studded with little diamond points, even if its surface appears destitute of vegetation; and from some of the plants the bubbles ascend in continuous streams, reflecting the most beautiful hues as the sunlight glitters through them. These are globules of oxygen, and are produced by the plants, which decompose the water into its elements, retaining the hydrogen for their own use, and liberating the oxygen for the benefit of the fishes. In a dark corner, or if not sufficiently exposed to the sun, none of this natural aeration takes place; but instead of it, the water gets exhausted of its oxygen, the plants get cankered, the animals perish, and the whole becomes rapidly a mass of obnoxious putridity.

. . .

Among the plants suitable for a Fresh-water tank, *Valisneria spiralis* and the new river-weed, *Anacharis alsinastrum*, are of great value, for supplying oxygen, which constantly refreshes the liquid, and maintains the health of the fishes.

In a moderate-sized tank, the only plants grown on the surface should be those which occupy a small space, and which will find sufficient room for their roots, without straggling over and choking more delicate productions. Among the floating weeds, there are two of great service. The first is the floating pond-weed, *(Potamogeton fluitans,)* of which a little patch may be thrown in, and moored by a piece of bass, or slender stick, to the place you have chosen for it. The other is the common duck-weed, *(Lemna,)* a graceful and cheerful ornament, and though very common, both beautiful and useful in an in-door lake. A small quantity of this may be lifted from a neighbouring pond, in a vessel, and placed in the tank to shift for itself; and it will usually give birth to some small creatures, that will prove useful to the fish. It consists entirely of small fleshy fronds or leaves, floating on the surface; each leaf sends down a few white rootlets, and during sunshine the fishes shelter themselves under its emerald green patches, and glitter vividly between its interstices.

> "Their scaly armour's tyrian hue,[8]
> Through richest purple to the view
> Betrays a golden gleam."

There is a Cape plant admirably suited for both large and small Aquaria; it is the *Aponogeton distachyon*. This may be procured of the nurserymen, and will thrive in-doors to perfection. It has oblong deeply-ribbed leaves, on long foot-stalks; the flowers are produced in two-cleft spikes, with snow-white bracts, which are very fragrant and ornamental. The root is tuberous, and reaches down to any earthy matter it can touch. It looks well if fixed at one corner of the tank by means of a thread, till it becomes established. It then throws its fine foliage over the edge, and breaks the angular outline, a service that will be appreciated by the lover of elegance in form.

Another plant suitable for a small space, for though it extends rapidly is readily thinned and kept in check, is our own native water ranunculus, *(Ranunculus aquatalis,)* a most elegant creature of the waters, which, unfortunately, has caused the death of many a swimmer by entanglement with its extensive masses of submerged roots and branches, and which, in small rivers, frequently impedes the rower by clinging to the oars. This forms little floating islands covered with flowers like newly-fallen snow-flackes, and is, after the white water-lily, the most elegant of our native aquaties. Those who have visited Stratford-on-Avon, will remember with delight as long as they live, the wonderful effect which, during the latter part of the summer, this plant produces in the picturesque bays and inlets of the Avon; and how it seems to hover over the quiet surface of that classic stream, like a fairy oblation to the memory of the great poet. It is to be met with on almost every pond and river in the suburbs of London. There is abundance of it in the pond at Clapham Common, in the Ravensbourne at Rushy Green, in all the branches of the Lea, and in the Thames, above Fulham.

It has some points of interest to the botanist, which make it all the more acceptable in a position which enables us to study its structure and growth, without wading or wetting the soles of our shoes. Its lower leaves are divided into thin thread-like segments, while those on the surface are entire and handsomely shaped. The use of these divided leaves is to enable it to present a greater surface to the water for the absorption of the gases which it cannot obtain from the air above; and by the activity of these leaves it elaborates from the water a large quantity of oxygen, which gives it additional acceptability. It will send down its fibres to the bottom mud, and bloom abundantly, if planted early in summer. The blossoms should be clipped off as fast as they lose their beauty, and the foliage occasionally trimmed and circumscribed.

Another native plant of great service is the water-violet, *(Hottonia palustris.)* It is easily domesticated, and is a rare beauty. It sends down a series of elegant white threadlike rootlets, the whole of its foliage being under water. If you keep minnows and gudgeons, these will be constantly gambolling amongst its leaves, and afford much amusement. The flowers rise above the surface of the water in elegant clusters of a pale lilac or flesh-colour. A few seeds thrown in will insure a sufficiency of plants, which, if too numerous, are easily thinned out. Though this is a British plant, it is somewhat rare, and I do not know any locality near London where it is to be found, but as it is cultivated by nurserymen there will be little difficulty in obtaining it.

To grow the white water-lily, the stately queen of our British lakes and streams, requires great depth and extent of surface – more than an ordinary tank will afford. It is a delicate plant too, and needs a very pure air, so as to fit it only for a large tank in a conservatory, or an open air pond; but the yellow lily, *(Nuphar lutea,)* will occupy little more than a square foot of surface, and one of these will flourish in-doors with ordinary care, and add a bold grace and reality to the mimic lake. To grow it provide four inches of rich loamy clay, where the plant is to be placed; in two inches of loam it can be grown, but not well. The root must be buried in the loam, and a stone or piece of broken flower-pot placed over it till it begins to put out its stems, when the covering may be removed. It will soon display its elegant leaves upon the surface, and should be kept as clear of weeds as possible.

The common water-cress may make a fit companion for a lily, and is easily grown on a bottom of rich sandy mould; a few seeds or cuttings being sufficient to start a growth during April or August. As the seed is not to be obtained of every seed-merchant it may be as well here to mention that Mr. Waite, seedsman, of High Holborn, keeps it always in stock. For myself, I find as much interest in the growth of a few cresses as in the culture of the most rare aquatics. There is something choice in the idea of growing such an old favorite within doors, where we may see it put out its humble but pretty white blossoms all the summer long, reminding us of the quiet rills and splashy pebbles that fertilize the green nooks so dear to the robin and thrush, full of sweet flag and "cresses that grow where no man may see them."

The pretty water-lobelia will perhaps find better favour than our ancient salad. It is showy and delicate, and as it does not root deep, this is one of the plants that must be inserted in a crevice half-a-dozen inches below the surface of the water.

But the choicest of all plants for a Fresh-water tank are ferns, and in choosing them we must select such as haunt water-falls, dripping caves, and morasses. If a few ferns are placed in the centre, under the light spray of a small jet, the effect is beautiful "beyond compare," but requires skill to produce it, and patient watching to insure permanent success. Under a drip-glass it is less easy to grow them, and where neither drip-glass nor fountain aid the tank it is still possible, but less pleasing. In any case, ferns should not be planted till the tank has been established a week or two.

For ferns, the rock-work must be built up above the surface some four or six inches; a compost of sharp sand, small chips of limestone, pounded charcoal, and pieces of tile or broken flower-pots, reduced almost to dust, must be prepared and laid in masses between the rocks above the water. Care should be taken that the root-stocks are prevented reaching the water, either by the upper surfaces of the stones themselves on which the compost is placed, or by some pieces of broken flower-pots laid horizontally, with the compost on them and concealed by the superstructure, but sloping from the centre, so that the water of the fountain will readily run off into the tank; should it form pools about the roots, it may sodden and destroy the plants: small clefts are sufficient for the escape of the young fronds.

Editorial Endnotes

1 Age of Veneer – refers to an article titled 'Age of Veneer', *Fraser's Magazine*, vol. 42 (September 1850), pp. 237–245 (author not identified)
Age of Fustian – not identified
2 Ingomar, the Barbarian (1845) was a play, translated and adapted from German by Maria Lovell.
3 '*Endymion*' (1818), bk. 2, John Keats (1795–1821).
4 'Pale glistening pearls and rainbow-coloured shells' – verse 1, 'The Treasures of the Deep' (1823), Felicia Hemans (1793–1835).
5 Mr. Gosse – Phillip Henry Gosse (1810–1888), English naturalist (Volume 4).
6 Dr. Badham – Dr Charles Badham (1805–1857), mycologist and entomologist. His article "Periwinkles in Pound" was published in *Frasers Magazine* 1 (January 1855), p. 531.
7 Ugliness of gorgons and "chimeras dire." – John Milton (1608–1674), *Paradise Lost*, Book II, line 628 (16367). Quoted by Johnson and other writers since.
8 "Their sealy amour" – Thomas Gray (1716–1771), 'Ode on the Death of a Favourite Cat Drowned in a Tub of Goldfishes' (1748).

40

'FLOWER STAND WITH FOUNTAIN OPERATED BY AIR PRESSURE'

C. Schickler

Source: C. Schickler, 'Blumentisch mit Springbrunnen durch Luftdruck' ('Flower Stand with Fountain Operated by Air Pressure'), *Illustrirte Garten-Zeitung*, vol. 1 (1856–1857), pp. 234–237

The desire for fresh greenery and floral decoration during the season which keeps us in our rooms has long since resulted in the creation of flower stands, and nowadays these are not infrequently augmented by the addition of fountains.

 DOI: 10.4324/9780429198809-47

These fountains are fed from above by water reservoirs, but, because of the necessary connection of the individual parts, they either cannot be moved at all, or only by inserting shorter or longer connecting tubes, which is always causes difficulties and which makes the value of such fountains very doubtful in apartments whose space is constricted. This problem has occupied me for a long time, and now, in the construction of such devices, in which the water is made to jet by air pressure, I have found the means of eliminating it thoroughly. The flower stands that I have constructed in this way can be moved from their place at any moment and even taken outside the house without needing any modification and with no interruption to the flow of water.

The accompanying illustration shows such a flower stand sufficiently for an understanding of the interior arrangement.

The frame itself is made of any kind of wood and can decorated, with or without carving, simple or luxurious. The water equipment, on the other hand, is made of strong zinc, which must be painted so that it does not rust; the basin is best made of sheet metal and unpainted porcelain.

The water arrangement consists of two containers of zinc (e) and (f) of equal size, one of which is used in the upper, the other in the lower part of the flower stand. Both are connected to one another by means of the two tubes (c) and (d), which run through the pillars of the stand, and, with the exception of the tubes, must be completely airtight and watertight. The container in the upper part of the stand also has a basin-shaped attachment, on the bottom of which there is a screw (a) with a small elevation, through the opening of which the water is flushed by means of a funnel belonging to the apparatus. When the container is completely full, the screw is closed again and then water is poured into the basin itself until it flows down through the opening tube into the lower container.

As soon as this flow has started, the fountain also begins to work. The water that collects in the lower container forces the air contained in it through the tube (d) into the upper container (e), where it presses with the same weight with which the water flowing from the inner container becomes massive on the air located there, on that in the water trapped in the upper container and forces it to jump into the sky through the mouth of the extension pipe in the middle of the basin.

When it falls, however, this water collects again in the basin, from which it also flows down into the lower container, and continues until all the water that was present in the upper container is flushed.

When making this device, it is important to ensure that the pipes enter and exit at the right place, otherwise the fountain will, in the best case, only jet imperfectly, if at all. The upper tube (c) must reach all the way to the bottom of the lower container, so that it can let the water flow out, but absolutely no air can go backwards. The other tube (d), through which the compressed air passes from the lower to the upper container, must be soldered into the ceiling of the lower container so that it opens into the ceiling of the upper container (e), so that all the air below can flow upwards, but it remains quite impossible that even a drop of water could find its way down through this tube. The short tube in the middle of the basin, from which the water jet rises, must be soldered to the bottom of the upper container, reach up through it and rise a little above the level of the basin itself. It hardly needs to be noticed that it has to end in a rather fine, well-worked nozzle.

The larger the container and the narrower the tip of the tube, the longer the stream of the fountain will last with uninterrupted strength. It only stops when all the water has flushed out of the upper container, whereupon the water that has now accumulated in the lower container is drained through the small taps (b) at the bottom into the scoop, which is also part of the apparatus, and the upper container is then refilled, so the cycle begins again.

The flower stand shown here has a suitable size and a pleasing shape. The water jet is driven to a height of 1 ½ feet and plays for 4–5 hours without interruption: the whole area remains clean and dry, so that I have no problem recommending this new device as a true ornament for salons, especially in high summer when room fountains are very beneficial to health.

The basic price for these new flower stands with air pressure fountains is from 45 fl. to 80 fl.

Incidentally, similar flower stands with the finest artistic carving and gilding are made to order, and those cost 100 to 150 florins.

C. Schickler in Stuttgart

41

'AUTOMATIC ROOM FOUNTAINS'

Anon.

Source: Selbstthättige Zimmerfontäne ('Automatic Room Fountains'), *Dr Neubert's Deutsches Garten-Magazin*, vol. 1 (1882), p. 186

We cannot fail to draw the attention of our readers to the automatic indoor fountain constructed by Paul Lochmann. This newly constructed indoor fountain removes all the inconveniences which hitherto stood in the way of the extensive use of such utensils useful for health.

We will skip the more detailed description of the item and only mention that the cost of operating such a fountain is around 1/2 pfennig per hour. The essential advantage of such a room fountain is the reduction of the dryness of the atmosphere, which, according to eminent doctors, has an extraordinarily beneficial effect on health; furthermore, the air of the room is purified to a high degree. As the illustration shows, these indoor fountains come with a flower table and an aquarium. Due to the constant supply of air, the water in the aquarium rarely needs to be renewed. Certainly many friends of aquatic animals would keep them if they weren't so difficult to care for. The automatic indoor fountain considerably reduces the known inconveniences. Of course, the water will have to be renewed regularly as soon as it is dirty. The aquarium owner Gassner in Munich speaks very favourably about these products, and so we can with a clear conscience recommend to all lovers of indoor plants, aquatic animals and the best possible indoor air, these indoor fountains and universal refrigerators, manufactured by Louis Heinrici[1] in Zwickau in Saxony.

DOI: 10.4324/9780429198809-48

Editorial Endnote

1 Louis Heinrici (1847–c. 1930) was a German industrialist, known for manufacturing hot air engines. In 1914, his Heinrici engines were powering decorative fountains in the landscaped gardens of his factory.

b)

DECORATION WITH PLANTS AND FLOWERS

42. J. C. Loudon, 'Management of Town Green-houses, and of Plants in Chambers, Cabinets, and at Routs', *The Green-House Companion* (1824), sect. 8 [241–251] excerpt, pp. 246–251
43. Jules Lachaume, 'Decoration de bals et soirées' ('Decoration for Balls and Soirées'), *Les Fleurs Naturelles* (1847), pp. 75–78
44. Louis Van Houtte, 'Treillis Mobile' ('Movable Trellis'), *Flore des Serres* (August 1848), p. 380
45. George M'Ewen, 'A Plea for Pyramidal Pelargoniums', *Gardener's Magazine of Botany, Horticulture, Floristry and Natural Science*, vol. 1 (January to June 1850), pp. 68–70
46. Annie Hassard, ch. 10, 'Screens for the Fire-place', ch. 17, 'Pot Plants in Rooms', *Floral Decorations for the Dwelling House* (1875), [60–66], excerpt, pp. 63–64, [109–114] excerpt,109–114
47. John Wills, 'Plants for House Decoration', *Journal of the Royal Horticultural Society*, vol. 15 (1893), pp. 84–88
48. Anon., 'Un drame en cinq actes' (A Drama in Five Acts'), *Le Jardin* (1891), p. 223
49. Josiah Conder, 'Arrangement of Flowers, Introduction', 'Lineal Distribution', *Flowers of Japan and the Art of Floral Arrangement* (1891), pp. 21–23, [43–50] excerpt, 45–50

In common with some other sources in this collection, the advice on decoration with plants and flowers in the home is associated with aspiration and the rise of consumerism and global trade. Overall, in early 19th-century Britain, the fashion for the middle and upper classes was for mixed flowers in ornamental pots, with the proportion of greenery or foliage increasing as the century unfolds. In elite households, lavish floral displays gained precedence, eliciting a variety of responses at the very end of the century (for artisan floristry, see Volumes 2 and 3).

The excerpt by J. C. Loudon, *The Green-house companion* (1824), includes recommendations for plant arrangements for parties. In this context the term 'art' refers to 'artifice', as in man-made, and is applied to both plants and perfumes.

The connections between the feminine and the floral world through the elision of 'young girls' with flowers is heightened in the excerpt 'Decoration de bals et soirées' from *Les Fleurs Naturelles: Traité sur l'Art de Composer les Couronnes, les Parures, les Bouquets etc. de Tous Genres pour Bals et Soirées, Suivis du Langage des Fleurs* (*Natural Flowers: A treatise on the Design of Wreaths, Ornament and Bouquet of all kinds for Balls and Soirées, followed by A language of Flowers*). The book by Jules Lachaume (n/d) was the first to focus almost entirely on flower arrangement and served to publicise Lachaume's Paris-based floristry business (see IId and Volume 3).

The following year, Louis Van Houtte (1810–1876) of Belgium reported on the 'Treillis Mobile', a floral portable room divider, a development of the 'berceaux', a series of enclosed shaded gardens formed by screens within a larger space, seen in 17th-century France and adapted to 19th-century urban interiors. The art of 'Treillage' or trelliswork, screens or garden structures made out of wood including flexible willow, remained part of the vernacular of French (and Belgian) gardens throughout the 18th and 19th centuries (mentioned by Repton, Id). The brief article was published in *Flore des Serres* (1845–1880) founded by Van Houtte and based at his nursery in Ghent. Initially the journal was a lavishly illustrated catalogue of plants; its scope broadened over the decades to include articles on the situation of plants in the home and the exotic environments from which they originated (including accounts of his own exploration) and technical and horticultural innovation (see Volume 4).

George M'Ewen (d. 1858) was one of the lead gardeners for the Horticultural Society, and he was employed by the Duke of Norfolk on his estate at Arundel Castle in Sussex. His article 'A plea for pyramidal pelargoniums' was originally published in the *Gardener's Magazine of Botany, Horticulture, Floristry and Natural Science*, 1 (1850); it was recirculated over a decade later in the *Journal of Horticulture, cottage gardener and country gentleman* (1862; formerly *Cottage Gardener*) which had a more inclusive readership (see Volume 3). M'Ewen's 'Plea' includes illustrations and 'notes' by Henry Noel Humphreys (1810–1879), draughtsman, author, and naturalist.

A screen of living plants features in Annie Hassard's *Floral Decorations for the dwelling house* (1875). Industrialisation and imperial trade impact this household item which includes the use of zinc, wood, tiles, wire mesh, and cocoa-nut fibre; likewise for the materials of 'Pot plants in Rooms' featuring gold-coloured 'Ormolu', an alloy of tin, zinc, and copper. The expense of some of the items indicates the aspirations of the intended readership of 'ladies and amateurs' living in relatively modest accommodation. Hassard herself lived in Upper Norwood which was one of a number of areas of cheaper suburban housing in London of recent construction. Her compact book incorporates articles from *The Garden* and covers a range of topics from dinner-table arrangements to wreaths, buttonhole bouquets, and religious items (see IIc). It includes more than 50 small wood engravings, many of which she attributes to William Robinson (the magazine's editor), suggesting that they were created in the *Garden*'s office; they did not necessarily feature in the original articles. Note that the 'Asplenium flabellifolium' currently refers to the Necklace fern native to Australia and New Zealand.

Hassard (1851–1930) was a journalist, whose contributions to the *Garden* were generally signed by 'A. Hassard', an indication that her name was valued by its mixed readership of professionals and amateurs that included women. The use of initials, a practice employed by the majority of her male colleagues in the press overall, was starting to decline from the 1870s; the naming of contributors was, however, more customary in 'technical' journals.

John Wills, 'Plants for house decoration', *Journal of the Royal Horticultural Society* (1893), reports on the amounts spent by individuals on floral displays and on the benefits of the expanding trade in plants to horticulture and to wider society. Wills (1832–1895) was a London nurseryman who ran a profitable floristry business, Wills and Segar (c. 1870–present). He refers to a number of nurserymen, some of whom feature elsewhere in this collection. His own company, however, is mentioned obliquely, possibly because much of his business was associated with public events (see Volume 2).

The difficulties of maintaining a healthy display of plants indoors are satirised in a French source of 1891 by the pseudonymous 'Member of the Society for the Protection of Plants'. 'A drama in five acts' ('Un drame en cinq actes'), *Le Jardin*, features a series of illustrations with captions prefiguring comic strips of the next century. The final illustration is signed, but we have been unable to identify its owner. The strip chronicles the suffering of a Latania, a palm native to Mauritius which was then a French colony. The reference to 'Paul et Virginie' recalls the 1788 novel by Bernardin de Saint-Pierre (possibly the French novel with the greatest number of editions and reprints) set in Mauritius and the 1876 opera by Victor Massé in which a natural way of life is corrupted by the artificiality of contemporary French culture. The servants, 'Joseph' and 'Baptiste', have traditional Christian names. The anonymous 'Madame' is depicted as being incapable of tending her plant.

The final source, extracts from Josiah Conder's *Flowers of Japan and The Art of Floral Arrangement*, was published in the same year and offered a revolutionary aesthetic for indoor plants. A large-format publication (27 x 35.5 cm), the plates are credited to Japanese artists Tsukioka Yoshitoshi (1839–1892) and Kawanabe Kyosui, a woman (1868–1935). Conder (1852–1920), a trained architect who worked in Japan, was among the first to favour the apparent simplicity of Japanese flower arrangements; a trend that entered some aristocratic or royal households from around the 1900s and may be seen today in 21st century minimalist homes.

42

'MANAGEMENT OF TOWN GREEN-HOUSES, AND OF PLANTS IN CHAMBERS AT ROUTS'

J. C. Loudon

Source: J. C. Loudon, 'Management of Town Green-houses, and of Plants in Chambers at Routs', *The Green-House Companion* (1824), sect. 8 [241–251] excerpt, pp. 246–251

The management of plants in chambers can only be understood to apply to the short time in which they are kept there. This time should be as short as possible, if it be intended that the plants should live and thrive afterwards. The only true way to have a fine display on the chamber-stage, is, never to bring the plants there till they are just coming into flower, and to remove them when the flowers first show indications of decay, unless the plant appears to be growing sickly before, which with heaths, geraniums, and camellias, is very often the case. During the time they are kept in the chamber, the surface of the pot should either be covered with fine fresh moss or coarse sand, to lessen the evaporation from the moist earth of the pot, and to prevent the earth from caking with the heat of the chamber: the water which passes through the bottom of the pot into the saucer, should be instantly extracted with a sponge to prevent evaporation, and none should be spilt on the leaves or the stage. As much air should be admitted by opening the windows, as is consistent with the use of the room, and the stage should, in the day time at least, be kept as near the windows as possible. It is not essential that the sun should shine on the plants, so as they have abundance of reflected light. It is almost needless to observe, that they should before being brought there be tied up, or otherwise arranged with the greatest neatness, and that while in the chamber, all decayed or injured parts should be removed, and any dirt or dust carefully wiped off with a moistened sponge. When the plants are taken back to the green-house, a little extra heat and moisture will in general recover them.

Plants at routs require little management while there, but must be tastefully arranged individually by rods and threads, and well syringed, and also watered at the root. The soil should be covered with moss, and the pot either cleaned and painted in any appropriate body colour, or chalked, or covered with coloured paper. An earthen brown with black and gray lines is among the most suitable colours, whether for the temporary painting, chalking, or papering. Where the plants are

DOI: 10.4324/9780429198809-50

only to remain one night, they need not be set in saucers, but only on paper or small carpets of the size of the bottom of the pot; but where they are to decorate the apartments for two or three days in succession, they should be set in saucers on a little gravel, and over the gravel the saucer filled brimful with moss or fine green turf. This looks well, and the space occupied by the gravel admits of giving the plants daily a little water, which greatly refreshes them in this state of trial.

The arrangement of the plants in the rooms is various, and depends on the kind of rout or entertainment. In common cases they are placed in recesses, and on side-tables, and near glasses which may reflect them; and a few choice specimens are scattered over the floor as single objects. But in more select entertainments, a proportionate attention is paid to their arrangement. During dinner a few pots of fruit-bearing shrubs, or trees with their fruit ripe, are ranged along the centre of the table, from which, during the dessert, the fruit is gathered by the company. Sometimes a row of orange trees, or standard peach trees, or cherries, or all of them, in fruit, surround the table of the guests; one plant being placed exactly behind each chair, leaving room for the servants to approach between. Sometimes only one tall handsome tree is placed behind the master, and another behind the mistress; and sometimes only a few pots of lesser articles are placed on the side-board, or here and there round the room.

The drawing-room is sometimes laid out like an orange-grove by distributing tall orange trees all over it in regular quincunx, so that the heads of the trees may be higher than those of the company: seats are also neatly made over the pots and boxes, to conceal them, and serve the purpose of chairs. One or two cages with nightingales and canary-birds are distributed among the branches, and where there is a want of real fruit, that is supplied by art. Sometimes also art supplies the entire tree, which during artificial illumination is hardly recognised as a work of art, and a very few real trees and flowers interspersed with these made ones, will keep up the odour and the illusion to nature.

Sometimes large picture galleries are laid out in imitation of parks in the ancient or modern style, with avenues or with groups and scattered trees. At masqued routs, caves and grottos are formed under conical stages, and covered with moss and pots of trees, in imitation of wooded hills. In short, there is no end to the arrangement of plants at routs: and the reader is not to suppose that only real plants with roots are necessary for this purpose; for, provided a few of these be judiciously introduced, all the rest can be effected by branches of box, laurustinus, laurel, juniper, holly, &c., decorated with artificial flowers and fruits, and fitted to stems or trunks to answer either as trees or shrubs; and besides these, whole pine and fir trees, the spruce especially, can be cut over, and thus admirable groves formed in a short time. Artificial supplies of odour of the rose, the orange, or the jasmine, are readily supplied. Much romantic splendour may be produced in this way with little expense of green-house plants.

Next to the common domestic flower-stage, and a handsome plant placed here and there in spare places in the lobby, hall, staircase, &c., are a few choice specimens of tall plants in fruit or flower, distributed in the drawing-room: the orange, the camellia, the acacia, and tree heaths, are well adapted for this purpose.

43

'DECORATION FOR BALLS AND SOIRÉES'

Jules Lachaume

Source: Jules Lachaume, 'Decoration de bals et soirées' ('Decoration for Balls and Soirées'), *Les Fleurs Naturelles* (1847), pp. 75–78

In meetings which have pleasure as their goal, flowers, the emblem of pleasure, must be spread everywhere, with both hands: from vestibule to boudoir, an apartment must be carpeted with flowers on the day of a reception, and become a temple of Flora.

Natural flowers not only adorn and perfume an apartment, but they infuse it with that vague poetry which transpires from a countryside enamelled with flowers; they establish a general harmony between an apartment and all the young women, animated flowers, intelligent and loving flowers, crowned and adorned with other flowers, with which they would dispute the prize of grace and beauty.

In a well-organised ball, one should arrive at the salon only through a double hedge of camellias, heathers, rhododendrons, myrtles, and rare plants, tropical or showy.

In houses where several living rooms adjoin, and where two-way mirrors make it possible to see from one to the other, it is necessary, on both sides of the mirror, to put large planters, or clumps of plants which merge their flowers and their leaves. One must take care, however, to clear the middle of the mirror, so that one can enjoy the fantastic look of these charming women, crowned with flowers and diamonds, standing out against a background of flowering foliage.

Fireplaces should also be covered with camellias, azaleas, rhododendrons, and thick evergreen foliage to great effect. Care must be taken to ensure that the flowers stand out against their foliage. The pots should be covered with a dense green moss.

A dining room alcove will also be adorned with a bed of plants at eye level. Sometimes lights are interspersed with the plants in the bed, so that they cast shadows and produce a more beautiful effect. At the front should be placed plants with light and serrated foliage, such as mimosas, heather, etc., and behind, plants with large and thick foliage, to create a vault.

A console leaning against a mirror should be lined with low, light plants to emphasise the mirror. On each side we will place two tall plants of equal height, to hide the frame of the mirror; the pots will be lined with moss.

DOI: 10.4324/9780429198809-51

The setting and tiers of an orchestra will be converted into a floral grove arranged in an amphitheatre so that all the plants can be seen distinctly. This massif must be treated boldly, with plants with thick foliage, which can produce large shadows, contrasting with the lights of the orchestra.

In the decoration of an apartment, the same rules of agreement and harmony must be observed as in crowns and ornaments. The colour of the flowers must be matched to the shade of each room; thus, in a white living room, we will put pink, purple, red or blue flowers, with a beautiful background of greenery; in a red salon, pink, hot-pink, blue, white flowers, etc.

Finally, let harmony reign everywhere in the decorations; for harmony is always the essential condition of beauty.

44

'MOVABLE TRELLIS'

Louis Van Houtte

Source: Louis Van Houtte, 'Treillis Mobile' ('Movable Trellis') *Flore des Serres* (August 1848), p. 380

DOI: 10.4324/9780429198809-52

A simple glance, cast on the vignette opposite, will immediately understand the utility and pleasure that can be derived from light constructions of this kind, which, fitted on rolling boxes, can be transported everywhere where caprice will want to take them, even in apartments.

All species of climbing plants, passionflowers, nasturtiums, *Echites, Ipomoea, Convolvulus, Maurandya, Bignonia, Thunbergia*, etc., will flourish there and by their free display on a double trellis, whose shape is left to the taste of the builder, will make the richest, most graceful effect.

The mobile trellis, the model of which we give here, consists of a box placed on two wooden rolls *a a*, fixed on each side, by a strong iron ring screwed onto the uprights *b b*, which control its movement; the box, about one metre long by 60 cm wide, is lined inside with a row of bricks: a material that absorbs and retains humidity. The trellis, properly speaking, is double, 3 or 4 metres high and the width of the box.

The expense of these light and graceful constructions is absolutely minimal; and even if they were more expensive, the pleasure to be derived from this kind of work amply compensates for the expense.

L VH

45

'A PLEA FOR PYRAMIDAL PELARGONIUMS'

George M'Ewen

Source: George M'Ewen, 'A Plea for Pyramidal Pelargoniums', *Gardener's Magazine of Botany, Horticulture, Floristry and Natural Science*, vol. 1 (January to June 1850), pp. 68–70

PRESUMING that we shall have gained an important point, if we can satisfy our readers that the form of training which we recommend is worthy of imitation, we deem it expedient, before entering upon the practicability of the subject, to show that variety of form is desirable; not that we wish, or expect, to see the present dwarf form abandoned. No; but only that another form, viz., the pyramidal should be introduced. We are prepared to show that it is not less natural, practicable, nor is it less appropriate, than the dwarf, squat form; and that it will much increase the beauty and interest of a collection of these plants as a whole, and would add largely to the number of admirers and growers of the Pelargonium, if this form of growth were to be generally introduced.

Because of the delight which *variety* of form produces, pyramidal training is desirable. Who is not fond of variety? Let us go and look upon yon forest – scan the fields – glance at the wild-flowers and plants by the highways and byways – take a grand survey of the vegetable creation; – or, if you choose, let us confine our range, and take the same species or varieties, and have them each shorn of their now varied forms, and then say if they have not lost more than half their beauty and interest by the change. Now, all we plead for is, that we should not pay exclusive attention to one – the dwarf – form; but that we should grow the tall form also.

Another inducement we would urge in favour of the pyramidal training, is *novelty*. To tell gardeners that anything *new* gives increase of interest, and oft-times pleasurable excitement, is to repeat what they must have constantly observed. Indeed, much of the success of gardening may be traced to, if it does not altogether, depend upon thus giving an agreeable excitement.

Another and yet more important inducement, we would urge, is their utility, or adaptation to almost any situation. Yes; you can place them with *effect*, whether in or out of bloom, on the centre, or side stages, where the sun, light, and air, may play freely upon them; or you can lodge them on circular hoops of iron or

DOI: 10.4324/9780429198809-53

brackets, in a hall or staircase, as represented in the woodcut. How much, too, would they enliven the tameness of the Pelargonium house as it is; or, where they are still more at home – yet greater favourites – in those large conservatories, where you can walk past, and around them, and feel as if in a garden. Again; decorative objects are perchance wanted for special occasions – none will be found so suitable as the pyramids for centre-pins in the group of flower-stands, or baskets; whilst as pillars in the entrance-hall, or by the corridor, or as beautiful plants for the table, they are equally appropriate.

Again; they will "fit in" to any style of gardening, whether geometrical or gardenesque; in rows by the straight walk – singly on the smooth lawn and in the dressed border at intervals, or as a centre plant in the mass of colour; – by turns relieving, surprising, and always greatly pleasing. And who can doubt, that, in the time coming, – when a yet higher standard, and a purer taste, is exemplified in the arrangements of our exhibitions, – our pyramidal-formed plants will be there, and so managed as to bear a thorough inspection – when no stakes nor glaring artificial props need be theirs – while the admitted dull uniformity of the collections in the Geranium tents will have passed away, to be no longer tolerated.

If we have done justice to our subject, we have succeeded in showing that our pyramidal Pelargoniums are, or may be, objects of great beauty and interest, as well as great novelty and utility, and will largely compensate for the labour bestowed on them; we now proceed, therefore, to treat on the not least important part – the *practicability* – of the subject.

Methinks I hear some one argue that the strength of the plant will be sure to go to its top, leaving the bottom and body of the plant naked. Another says that it will be *so long* before a Geranium is grown to the height and girth you require (say six feet by three – not more). And another won't think of it, because they occupy so much room. To the first objection it is replied, that the tendencies spoken of can be successfully counteracted, on the same principle that we train our wall-trees wheresoever we will. To the second, we submit that one season is not long; and, when once the plant is trained, it lasts for years. And to the third objection we reply, that the stage-room required is not more than that occupied at present, the extra height requisite being now unoccupied. It may, however, be well to remark here, that it is necessary to have in "the mind's eye" the model at which to aim in producing. We suggest that the height of the plant should be twice its diameter; that is, if the breadth of the plant at the base (which shall be just at the rim of the pot) is three feet, the height may be six feet, or thereabout. We must also remark, that in ordinary circumstances this height should not be exceeded; for this reason, amongst others which could be adduced: – in this class of plants the flowers are so placed as to display themselves better when looked upon, than when looked up to; and, for a similar reason, there should be no swell on the body of the plant. With such plants as the Rose, Fuchsia, Scarlet Geranium, Heliotrope, &c., there need be no limit to the height, if due proportions are maintained. In a

Staircase and hall, with geraniums in vases and on brackets.

future paper we may give a few suggestions on the *fitness* and adaptability of these subjects for pyramidal training.

[We append to the above very suggestive remarks some notes on the same subject, showing the effect of pyramidal plants as ornaments to the hall and staircase; with the accompanying practically illustrative sketches, which are furnished by Mr. H. Noel Humphreys: – "When last in Paris," writes Mr. Humphreys, "I was much struck with a beautiful effect produced by plants placed outside the rich iron balustrade of a principal staircase, at the hotel of the Count Auguste de B – . In that instance, the pots containing the plants were lodged in circular loops of iron, connected by a strong bar fixed firmly to the exterior of the staircase, at and between each loop. By this inexpensive means the odour, the freshness, and the floral beauty of the plants accompanied the visitor to the very door of the principal saloon, without encumbering the staircase, or encroaching in any way on the space devoted to free ingress and egress, which is seldom, in modern buildings, sufficiently spacious to bear curtailment. Geraniums grown in the pyramidal form would be peculiarly suited to place in such a situation, as their height would bring them above the balustrade, and leave their best flowering portion to the unimpeded view of the visitor, while a plant grown in a lower form would not possess this advantage. But instead of the iron loops, which can only be regarded as a cheap temporary contrivance, I would suggest a more decorative, and at the same time more permanent mode of placing flowering plants outside the balustrade of a staircase, and, at the same time, one which might be made subservient to other purposes. I propose making the principal moulding beneath the exterior of the balustrade jut out at certain distances, and by describing three-fourths of a projecting circle, form a series of brackets, upon which ornamental vases containing plants might be placed. Thus, on other occasions, when a profusion of light might be more desirable than flowers, bronze candelabra, each surmounted by a single light, with a ground glass globe, might be substituted for the vases containing flowers, or they might alternate with them, which would produce a very rich and pleasing effect. In the accompanying sketch I have shown the Geraniums placed upon the circular brackets as described, and, in the hall beneath, I have arranged a row of vases, exactly beneath the Geraniums, which I suppose to contain finely-formed Hydrangeas, grown with a single stem like the Geraniums, but with bushy round tops, in the orange-tree form, to contrast with the pyramidal figure of the plants above. Between each Hydrangea I would place a climbing-plant, neatly trained to a tall rod; they might be Maurandya Barclayana, Thunbergias, or Loasa lateritia; but they should be all the same, and very trimly trained, as this *symmetrical repetition* is necessary wherever irregular forms have to be grouped with architecture."

Pedestal.

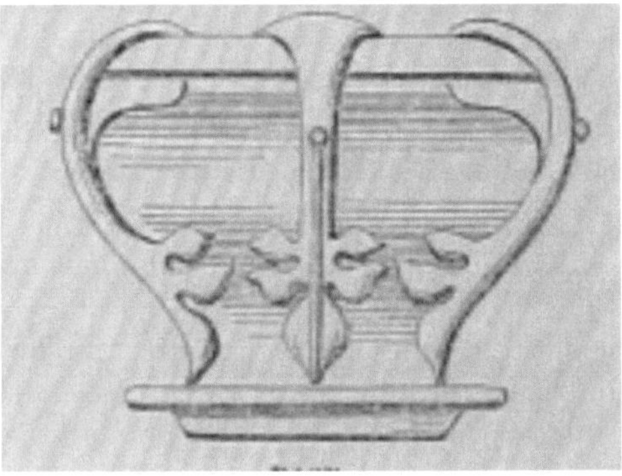

Vase.

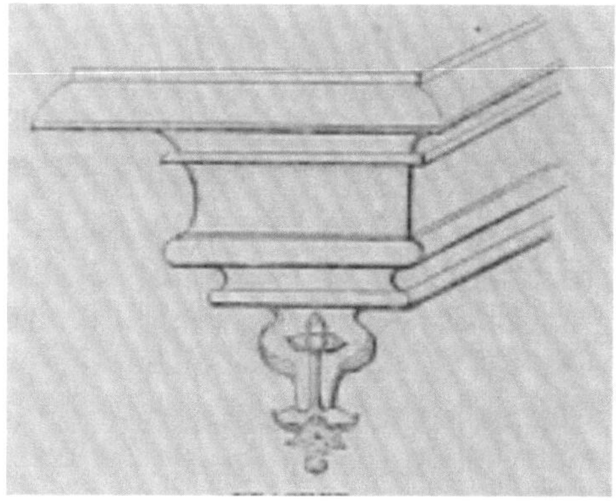

Bracket.

46

FLORAL DECORATIONS FOR THE DWELLING HOUSE

(Excerpts)

Annie Hassard

Source: Annie Hassard, ch. 10, 'Screens for the Fire-place', ch. 17 'Pot Plants in Rooms', *Floral Decorations for the Dwelling House* (1875), [60–66] excerpt, pp. 63–64, 109–114

Ivy Screen for Fire-place.

Screens for the Fire-place.

During the summer season, when there are no fires in the drawing-room, tastefully-decorated screens fitted into the fire-place have a charming effect. These often consist of looking-glass and specimens of dried Ferns; but, as they do not

 DOI: 10.4324/9780429198809-54

come within my range, being dried, I shall pass them over, merely remarking that, though handsome, they are expensive. The best plants with which to cover screens are the common or variegated Ivies. First, a box should be procured, the width of the fire-place, to stand inside the fender; it should be made either of zinc or wood, and should be ornamented with coloured tiles – in fact, a box such as one sees on hundreds of window-ledges; at each end, in the back corners, an upright iron rod should be fixed sufficiently high to meet the ornamental marble over the grate; between these rods a piece of fine wire netting should be strained, so as to form a screen on which to train the Ivy; this wire back should completely cover the iron or steel grate; over the holes in the bottom of the box some broken crocks should be placed, and over the crocks should be put a layer of Cocoa-nut fibre; then the box should be filled in with a mixture of rotten turf and sharp gritty sand. Some plants of Ivy should next be procured and planted firmly and rather thickly in the box, so as at once to cover the screen. Along the front of the box, set on the soil, may be pot plants, or the surface of the soil might be covered with Selaginella denticulata intermixed with cut blooms of large-sized flowers.

POT PLANTS IN ROOMS.

NO floral ornaments look better in sitting-rooms than well-grown pot plants, which have a great advantage over cut flowers, inasmuch as they last longer in perfection. To townspeople who have no gardens or glasshouses they are a real boon. Plants of all kinds, suitable for rooms, may be obtained in Covent Garden Market, but, being, as a rule, reared tenderly in heat, they soon drop their blossoms when placed in their new quarters. Avoid, therefore, plants brought direct to the market from a stove, or other house, much heated, and select such as have been brought up hardily, for no care will keep plants just out of a hothouse fresh or healthy-looking; the sudden change from heat to cold is certain to make them flag. Having made sure that the right kinds of plants have been obtained, the next thing to be considered is, how the pots are to be concealed, as common garden pots are anything but ornamental. To remedy this evil, some set their pots in paper covers of different colours; others slip them into china pots. Rustic covers are also sold for hiding pots, while some conceal them in the following manner: – A bundle of common Brake Fern is obtained, and the ends of the Fern fronds, cut in a slanting direction with a sharp knife, are inserted in the soil with which the pot is filled, close to the rim, over which they are broken, so as to make them droop and hide the pot. For those who may not like this plan, and who will not go to the expense of china pots, a new improvement on the plain rustic expanding covers has lately been introduced, differing only from the old style in being ornamented with artificial Ivy, Vine, and other varieties of leaves which tend to give it a pretty appearance. Such pot-covers are well suited for plants in halls, or for ordinary use; but, for the dining-room and dinner-table, they should be of china, than which nothing is more effective. Some are made to stand on a table or bracket, painted in pairs to match; while others are mounted singly, on tripods of iron or wood. I have seen

Ornamental Flower-pots.

a design made of bronze to stand on a table; at its base were two china pots, and raised about a foot or 18 inches, supported on a pillar of ornamental bronze, was another china pot. Being of bronze this stand was probably intended for the hall or dining-room, as I also saw several in ormolu, which would look lighter for drawing-room decoration. After plants have been placed in such ornamental pots or covers, the surface of the soil in which they are growing should be covered either with silver sand or Wood Moss; if with the latter, and any of the plants happen to be Ferns, care should be taken that the Moss does not cover any of the young

Plant Bracket.

fronds. All plants should be removed to the conservatory at night, if there be one; if not, into some room in which no gas has been burning. It is bad enough to keep plants in a gas-lighted room for any length of time, but, after the gas is turned out, they should be removed, as the atmosphere, which is still charged with its fumes sadly injures them. In the morning water should be given them if they require it. Palms and other fine-foliaged plants should have their leaves sponged over with water every morning, and the flowers of Pelargoniums and similar plants may be made to last for a long time if a drop of gum, such as is used for cut flowers, is dropped into the centre of each bloom. By observing these little items, plants may be kept in good health just double the time they otherwise could be. Plants look well set on brackets, though this style of decoration is not often resorted to; I am certain, however, that if tried, a good effect would be the result. Brackets may be purchased made of different materials; some are of terra cotta, others of china; some are made of wrought iron, painted to imitate bronze, others of Swiss carved wood, and others, again, are gilded. Some brackets are fixed flat against the wall, others are made to fit into corners. Those made with a hollow, into which the pot can be dropped out of sight, are much the best for upright-growing plants; but where an Ivy-leaved Pelargonium, or any drooping plant, is employed, it does not matter, as the foliage of the plant, if a good specimen, quite conceals the pot from view. One of the most effective plants for this style of decoration is a large and handsome specimen of Asplenium flabellifolium (Fan-leaved Spleenwort). No one who has not seen a plant of this so placed can have any idea of its beauty.

47

'PLANTS FOR HOUSE DECORATION'

By Mr. John Wills, F.R.H.S

Source: John Wills, 'Plants for House Decoration', *Journal of the Royal Horticultural Society*, vol. 15 (1893), pp. 84–88

No one who has visited the pleasure gardens of the rich or noticed the humble garden-plots of our country cottagers, or even glanced at the window-boxes and pot-plants in the artisans' dwellings in many of our smoke-begrimed towns, can for a moment doubt that the love of flowers generally, and the taste displayed in their arrangement, have increased tenfold amongst us during the last twenty years. And that this increased love of flowers has immensely benefited horticulture and those engaged in its most elevating pursuit, is also a well-known fact. And as the amount of money spent on any particular luxury or amusement is to some extent a measure of its popularity and progress, it will perhaps be interesting first to trace the rapid strides which have been made of late years in the floral decoration of houses and public buildings, as indicated by the sums of money spent upon it, and then to consider some of the plants most suited to the purpose.

Twenty-five years ago £25 or £50 would have been considered – was considered – an extravagant and altogether unheard-of sum for any nobleman or gentleman, however wealthy, to spend upon the decoration of his house for a dinner party or a ball, or any other entertainment. But in 1871 the rage for elaborate floral decorations came in, if one may say so, with a rush. It was initiated by the late Sir Edward Scott,[1] whose love of flowers and whose liberality were both unbounded. He gave his house up into his florist's hands for three full days, with *carte blanche* orders, regardless of expense, the only stipulation being that the handsomest decorations possible should be produced, and the result was both so novel and so beautiful that almost everyone who gave great entertainments during that London season of 1871 followed in the track which he had pointed out.

During the following years many and much larger decorations took place, amongst which may be instanced what at the time was called the "hanging gardens of Babylon," on the occasion of a ball given by the Marquis (sic) of Bristol,[2] when six tons of cut Ivy alone was used, to give a castellated effect to the bare walls of an improvised ball-room. A few days afterwards one gentleman

DOI: 10.4324/9780429198809-55

gave a magnificent entertainment, the flowers for which cost over £500. Various other similarly decorated entertainments followed, the result being that more than £3,000 was paid to one single firm for floral decorations only, in less than one month, and Messrs. Veitch, Turner, Paul, Bull, Lane, and many other great plant-growers were very largely drawn upon, nothing being considered too expensive or too rare. Magnificent Orchids, Roses by the ten thousand in a single day, as well as innumerable Ferns and other decorative plants were used.

On July 21, 1873, the first large public entertainment was given in the conservatory of the Royal Horticultural Gardens, in connection with a ball given in honour of H.R.H. the Prince of Wales. Here for the first time ice was largely used for cooling the heated atmosphere of the ball-room, and after this initiative by the Society, as much as forty tons of ice was used on different occasions. At one gentleman's mansion in Belgrave Square the floral decorations more than once cost over £1,500. Many other similar decorations followed in which Orchids and ice were largely used. These few instances are mentioned out of many, in order to show what has caused the enormous increase of late years in the cultivation of plants for house and window decoration.

In a letter to me, Mr. Assbee,[3] of Covent Garden Market office, says that he roughly estimates the number of vanloads of English-grown plants sent into Covent Garden Market last year (1891) at 20,000, and in addition to this, immense quantities of boxes of flowers were brought into the market by hand, and much larger quantities still were sold by auction in the Floral Hall from all parts of the Continent, the Channel Islands, &c. Mr. Assbee further tells me that the number of growers who attended and sold plants in Covent Garden Market twenty years ago was only about thirty persons, whilst at the present time they number over three hundred.

As an instance of the immense quantities of plants, &c., supplied by the various growers, I may just mention that the firm with which I am connected has during the past year bought from the market at Covent Garden, and from other sources, over a quarter of a million of plants, consisting of Ferns, Palms, and all kinds of flowering plants. And when we consider the vast quantity of plants and flowers sent to London and the different provincial markets, it will at once be seen how *enormously* the cultivation of plants and cut flowers has increased, and to what an extent it has benefited horticulture commercially, providing additional employment for large numbers of men, women, and children. You cannot even go into any back street in London without seeing eight out of every twelve window-ledges covered with plants; and this not only shows that a great elevating influence is being developed amongst the working classes, but that it is a benefit to vast numbers of working people who are employed in the cultivation of plants.

It will be within the recollection of most horticulturists, and of those who love flowers and can appreciate their effect when judiciously arranged, what magnificent groups of plants were arranged for effect some few years ago by the leading London nurserymen and florists at the Royal Horticultural Society's Gardens, and at Regent's Park, Manchester, and various other places; and how much it did towards infusing and developing a taste for a more natural style in the arrangement and adornment of conservatories, dwelling-houses, and garden decoration in general. Nature is the greatest schoolmaster, and the only guide the floral artist should take lessons from. He should "consider the lilies[4] how they grow," and watch our meadows and the banks of our meandering streams clothed with their simple flowers, and see how all is arranged for good effect, and watch how the juxtaposition of colour is displayed. Nature never arranges two things alike, only in cases of trees or shrubs, when seen in woodland scenery. This, therefore, is the great lesson to all who would aspire to arranging plants for effect, never to arrange any *match-pairs* of plants in their groups, but make them as dissimilar as possible. The wonderful display made by the Covent Garden growers[5] on recent occasions at the Royal Horticultural Society's great spring shows at the Temple, is a sufficient proof of what the energetic and plodding industry of horticulturists can attain.

I will now give a list of a few of the many plants suitable for house and window-box decoration. For either purpose the following may be used, and can in all cases be depended on if kept properly watered and clean. Palms are to all intents and purposes the most useful, by reason of their graceful habit and long-lasting qualities. Amongst these may be recommended *Corypha australis, Latania borbonica,* and *Cocos Weddelliana*. This last is decidedly the most graceful and best lasting Palm that can be used for house decoration. Up to a recent date it was generally thought that the plant required a much warmer temperature, and that it would only last a few days in an ordinary living-room. I have proved, however, that with ordinary care it will last for more than two years in even a cold and draughty position. I have had some plants in my house for nearly two years, which are looking at the present day as well as they did the first day they were brought in. They are also very pretty when grown in small thimble-pots, which can then be easily put into the smallest drawing-room vase. *Phœnix rupicola* is also a very graceful and useful Palm, but has not the long-enduring qualities of *Cocos Weddelliana. Kentia Belmoreana*, like *Cocos Weddelliana*, is one of the most useful for all purposes. Plants of this lovely Palm can be grown to a height of 12 to 15 inches in a 2½-inch pot, and will last in a drawing-room six months or more if kept well watered and clean. We have some specimens from 30 to 35 feet in height which are simply grand, and although they have been subjected to all sorts of hard treatment for many years they are very little the worse for it. Any of the following are also available for room or flower-box decoration, namely, *Areca Baueri* and *A. lutescens, Cocos flexuosa, Chamærops excelsa, Geonoma gracilis, Phœnix reclinata* and *P. tenuis, Rhapis flabelliformis*, and *Thrinax elegans*.

Amongst ornamental and coloured foliage plants the following may safely be relied upon as being very useful and satisfactory, and possessing long-lasting qualities: *Ananassa sativa, Asparagus plumosa* and *A. procumbens*. Plants of the latter are most graceful and useful by reason of their pretty feathery and procumbent habit, and their adaptability for decorating vases and mantelpieces. When used with Orchids or other flowers they produce an exceedingly charming effect.

Bambusa nigra and *B. falcata* are two most graceful and useful plants. Many handsome Crotons may be used with impunity for house and window-box decoration. The following Dracænas are also very pretty, and may be used to great advantage: *Dracæna australis, D. fragrans Lindeni, D. Linita, D. Goldiana*, and many other beautiful varieties. *Nidularium fulgens* and other ornamental Bromeliads may be freely used, and will retain their beauty for a long time. *Pandanus Veitchii, Phormium tenax, Pourrettia mexicana* and *P. argentea*, Tillandsias, *Aspidistra lurida*, and its variegated form, are most useful and never-failing plants. Several of the Fittonias are also pretty. *Hydrangea hortensis variegata* is a beautiful plant when grown in small pots, being prettily variegated, and is very effective in any decoration. The never-dying Ophiopogon, any number of Ferns, and various other plants too numerous to mention, are available for all purposes.

Editorial Endnotes

1 '[T]he late Sir Edward Scott' – Sir Edward Dolman Scott, 2nd Baronet (1793–1852), was an English landowner and a Whig politician.
2 'Marquis of Bristol' [sic] – Frederick William John Hervey, 3rd Marquess of Bristol (1834–1907), was a Tory politician.
3 Mr. Assbee – 'market superintendant' (London metropolitan archives, Bedford) E/BER/CG/E/05/01/015).
4 'He should "consider the lilies how they grow"', Luke 12:27, King James Version.
5 The Covent Garden growers showing at the RHS spring shows took place at the Temple (City of London) from 1888 to 1912.

48

'A DRAMA IN FIVE ACTS'

Anon.

Source: Anon., 'Un drame en cinq actes' ('A Drama in Five Acts'), *Le Jardin* (1891), p. 223

A Drama in Five Acts: The Odyssey of a Plant

First scene. 2.00 in the afternoon. Delivery.

Second scene. Midnight. Duo of Paul and Virginie in the shade of a Latania. 25 degrees centigrade.

DOI: 10.4324/9780429198809-56

Third scene. 9.00 in the morning. Joseph and Baptiste giving an air bath at 10 degrees below. Long discussion about the quality of refreshment.

Fourth scene. Noon. Effects of the refreshment on the Latania. Bewilderment of Madame! . . .

Fifth scene. . . . who arrives at the Garden with the Latania, whose appearance surprises the Director, who cannot help.

This little story is that of all the house plants whose owners do not look after the plants themselves. Whoever has gone walking in Paris on a morning with a temperature of six degrees, has noticed that windows are open for the health of the household, and that the plants are placed on the balconies, probably to keep them free of dust. No plant can resist this treatment; perhaps the series of little sketches we publish here will save our beloved plants from the tortures which the negligent reserve for them.

A Member of the Society for the Protection of Plants

49

FLOWERS OF JAPAN AND THE ART OF FLORAL ARRANGEMENT

(Excerpts)

Josiah Conder

Source: Josiah Conder, 'Arrangement of Flowers, Introduction', 'Lineal Distribution', *Flowers of Japan and the Art of Floral Arrangement* (1891), pp. 21–23, [43–50] excerpt, 45–50

ARRANGEMENT OF FLOWERS

INTRODUCTION.

THE arrangement of cut flowers in vessels of various kinds has become, with the Japanese, a decorative art of considerable refinement, compared with which the Western methods of floral composition appear but haphazard combinations. The bouquet, the wreath, and the garland, all depending for their beauty upon the close massing of blossoms and greenery in luxurious confusion, bear no resemblance whatever to the more austere and open compositions of the Japanese. The fact that many of the most charming flowers of the country are those of trees, the blossom-clad twigs of which it is impossible to arrange in closed and rounded masses, may in some manner explain the open lineal character given to floral designs; but the same treatment is applied equally to flowering plants and grasses which lend themselves far more easily to the European method of grouping. The peculiarity of treatment noticeable in these flower arrangements is closely connected with the Japanese manner of observing and enjoying floral nature. Whereas the Western amateur devotes his attention mainly to the blossoms, the Japanese lover of flowers bestows his admiration on the whole character of the plant or tree producing them. The rugged nature of the plum trunk, with its straight, stiff shoots, or the graceful sweep of the branches of the weeping cherry, are to him inseparably associated with any beauty which the blossoms themselves possess. The lines of branch and stem, the form and surfaces of leaves, and the distribution of buds and blossoms, all receive their full share of attention. The loveliest buds and blossoms torn from their stems and crushed together in a mass, with ferns or other greenery between them, convey to the Japanese mind no idea of floral art or beauty. The art under consideration is, in fact, based upon a representation, more or less

DOI: 10.4324/9780429198809-57

conventional, of floral *growth;* and, principally for this reason, the compositions are made to assume an open character in which the individual forms of branches, stems, leaves, and flowers are all clearly expressed.

The Japanese term *Hana*, translated as *Flower*, has, in the art of floral arrangement, a much wider signification than its English equivalent. Among the so-called *Flowers* of the seasons are included certain evergreens and other flowerless shrubs and trees, some of which hold very high floral rank. The Pine and Bamboo for example, both occupy a very important place in what are called *Flower Arrangements*, and the Maple with its reddening leaves is used as one of the principal *Flowers* of Autumn.

Seasonableness in the choice of material is one of the leading principles which guide the designer. The luxurious taste for choiceness, as associated with the idea of rarity, is diametrically opposed to the rules of the art under consideration. Flowers blooming out of their proper season are, with very few exceptions, rejected for floral compositions, which are intended, in a manner, to be expressive of the particular period of the year. April blossoms, used in any other month, would appear to the flower-artist as incongruous and out of place as would be winter clothing worn in summer-time. It therefore naturally follows that a proper cultivation of the floral art requires a thorough acquaintance with the nature and growth of all trees and plants employed; and even a close observation of the varying characteristics of the same plant during different seasons, in the case of those which are common to several months. The Flag or Iris, for example, which is common alike to certain months of Spring, Summer, and Autumn, has a peculiar bend and vitality in its leaves, and a different length and vigour in its flower stems, at the various periods of its growth; and these distinctions are all kept in view when employing this flower in compositions.

The natural locality of production of trees and plants, whether it be mountain, moor, or lake or river bank, greatly influences the character of the design used. To arrange a water plant in the same manner and with the same feeling and surroundings as a land plant would be considered a great violation of the rules of appropriateness. Not only are blossoming trees and flowering plants treated as perfectly distinct in character, but minor divisions as to locality of production are often observed in both. Among plants a distinction is made between ordinary land plants, forest plants, mountain plants, and water plants; and among trees, land trees, forest trees, and mountain trees are distinguished in some cases.

The imperative necessity for a proper familiarity with the nature of all flowers used in compositions is one reason why it is forbidden to employ those of rare or little known plants, however beautiful they may be. The use of wild flowers, only known to the botanist, as well as rare or foreign flowers with the names of which ordinary folk are not familiar, is prohibited, unless the arranger has previously made himself perfectly acquainted with all the natural characteristics of such flowers. As one exponent of the art has quaintly expressed it, the artist must be thoroughly imbued with a sympathetic feeling for the character, habits, virtues, and weaknesses of the members of the floral kingdom from which he seeks his material, till he possesses almost the same love and tenderness for their qualities as for those of living beings

. . .

LINEAL DISTRIBUTION.

AN analysis of Japanese flower arrangements shows that the lines or directions taken by the different stems or branches form the basis of all compositions. While European floral decorations are merely combinations of masses of colour, in which blossoms and leaves alone play a part, those of Japan are synthetic designs in line, in which every individual stem, flower, and leaf stands out distinctly silhouetted. The treatment followed may be likened somewhat to the methods employed of distributing carved foliage in architectural panels.

The surface of the water in which the flowers are placed is technically considered to be the soil from which the floral growth springs; and the designer must here convey the impression of stability and strength. However good the upper lines of the composition may be, a weak *springing* at the base deprives it of life and vigour; for it must be remembered that not flowers alone, but floral growth and vitality, are to be expressed in the designs. The direction of the stems at starting need not be strictly vertical, but, if curved, the curves employed should be strong ones, and all weak bends and angles should be avoided. As a composition consists of several main lines, there will be severall ines of *springing*, or origin. In some cases the *springing* lines are all united from the surface of the water to some distance above, after which they separate in tangental curves in different directions; in other cases, each stem-line is kept distinct, being separated from the others from the point of origin.

In the distribution of the principal lines of the composition from the point of their separation, the artist studiously avoids an equal-sided or symmetrical arrangement, but he obtains a balance of a more subtle kind, which is at the same time productive of a pleasing variety of form. Balance and harmony without repetition is a governing principle in this as well as in other Japanese arts. The lines of each stem, or, in cases where numerous slender stems are combined, the central lines of each group of stems, receive first attention. The triple arrangement, by which is meant that governed by three prevailing lines, may be taken as the original model for all arrangements (see Plate XI. A).

The three lines of such compositions may be called, with sufficient fidelity to the more quaint native nomenclature, *Principal, Secondary*, and *Tertiary*. The *Principal* line, as the name implies, is the central and longest line of the design, and this is made to form a double curve with the upper and lower extremities nearly vertical and in a continuous line, the general shape thus assumed being that of an archer's bow. The *Secondary* line should be about half, and the *Tertiary* line about one quarter, of the length of the *Principal*, supposing all to be straightened out; and these two lines are arranged on different sides of the *Principal* in graceful double curves of varied character. As a general rule, the *Secondary* line has a more vertical, and the *Tertiary* line a more lateral tendency; the former being on the outside of the arched bow formed by the *Principal*, and the latter making a counterpoise on its hollow side. According as the hollow of the *Principal* faces

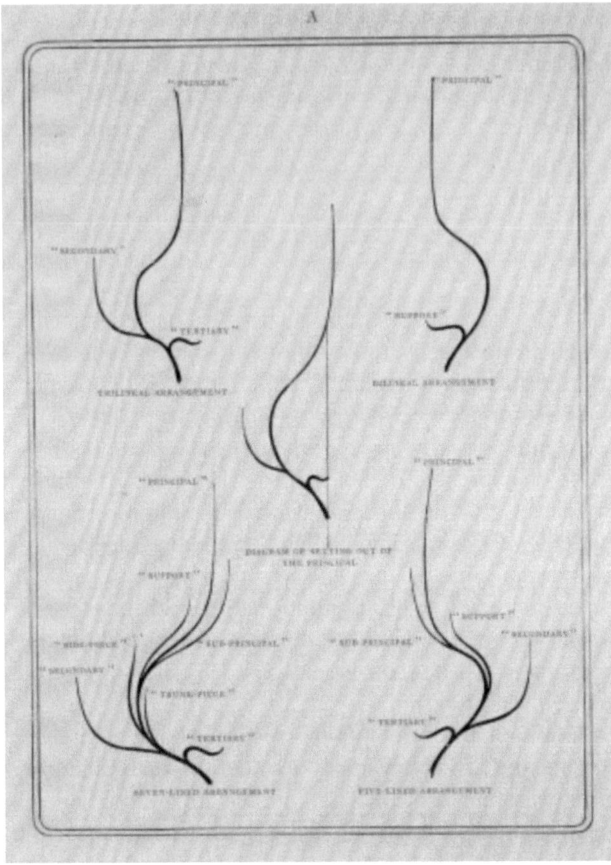

Lineal arrangement of stems in flower design

right or left, the arrangement is called a right or left composition. By changing the direction and giving a different character to the curves of these three lines a great variety of designs is produced, some of which are shown in Plate XI. B.

To produce a five-lined arrangement, two additional lines are introduced between the three previously named. The one placed between the *Principal* and the *Secondary* is called the *Support*, and the one between the *Principal* and *Tertiary* is called the *Sub-principal*. The *Support*, in length and importance, approaches more to the *Secondary* than to the *Principal;* while the *Sub-principal*, as its name would imply, in length and importance, approaches more the *Principal* than the *Tertiary*. In this way, it may be observed, a proper lineal balance and harmony is obtained; for supposing the three elementary lines to be valued 4, 2, and 1, according to their respective lengths and importance, it is plain that to preserve balance with the addition of two extra lines, that introduced between 4 and 1 should be longer

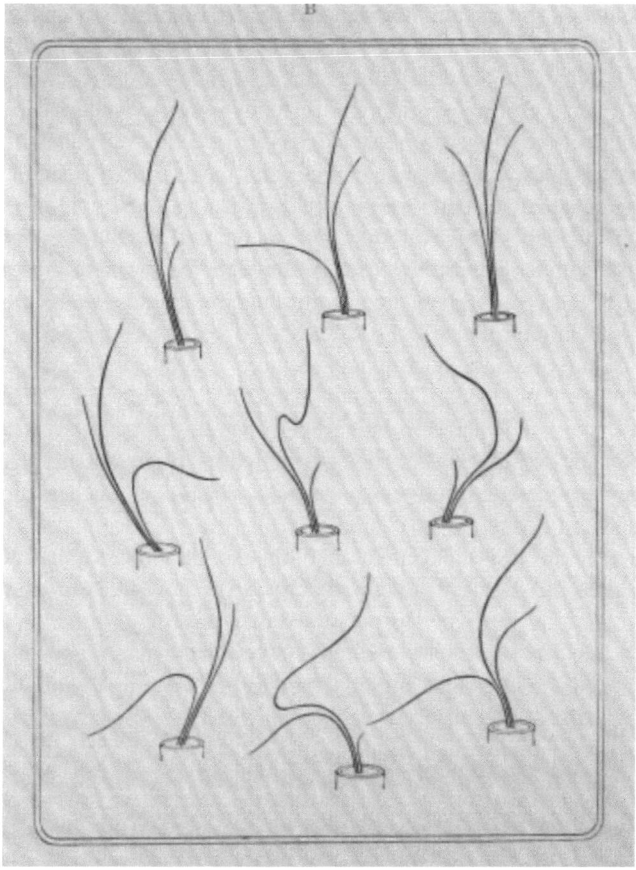

Various Trilineal Arrangements of Stems

and more approaching in value to 4 than that between 4 and 2, which should more approach the dimensions and character of 2. These additional lines, besides having different intermediary lengths, have other curves given to them, and are arranged so as best to fill the intervals with due regard to variety and balance. In the seven-lined arrangements two more extra members are added, one called the *Side-line*, and the other the *Trunk-line*. Their lengths are about intermediary, the *Side-line* being placed between the *Support* and the *Tertiary*, and the *Trunk-line* between the *Sub-principal* and the *Secondary*.

The different members of the above lineal arrangements have curious fancy names bestowed on them by the different Schools. For the triple style such classifications as that of *Father, Mother*, and *Self;* or that of *Heaven, Earth*, and *Mankind*, are used; and for the five-lined style the terms *Centre, North, South, East, West;* or *Earth, Fire, Water, Metal, Wood;* and sometimes *Yellow, Red, Black, White*, and

Blue are employed. The Enshiu School apply the following curious ideas to the different lines in an arrangement, namely: *The Heart, Help, The Guest, Skill, The Finishing touch;* referring to the different sentiments which should inspire the designer.

The general form of the above groups of three, five, and seven lines depends mainly upon the amount of curvature given to the *Principal* or centre-most line. In the simpler and less affected styles, this bow-like curvature of the *Principal* is slight and strong, but in the more exaggerated compositions, this governing line is boldly bent sideways from a point a few inches above the *springing*, and turned in a quick curve back again so as to bring the upper extremity vertically above the base, and to preserve the centre of gravity of the whole. The general form thus imparted is, as mentioned above, somewhat like that of a strung bow. From such a shape the transition is not great to that produced in the arc of a bow by drawing it; a more violent lateral curvature with less vertical height being produced. This latter character is given to the *Principal* chiefly when used in positions where too much height would meet with actual or optical obstructions and be detrimental to a good effect. In floral arrangements placed below a shelf, or in front of a hanging picture which must not be hidden, such treatment is often followed. This modification in the character of the *Principal* necessitates corresponding changes in the directions and curvature of the auxiliary lines of the composition.

Up to the present, slight or violent as may be the bow-like curvature imparted to the *Principal*, it has been always presumed that its general direction – that is the direction of a line joining its base and top – is vertical. Thus, supposing a tri-lineal composition enclosed in a right-angled tri-angle, such tri-angle would be standing on its point, with the hypotenuse, corresponding with the *Principal*, placed vertically (see Fig. 1). There is, however, another style of design applied to a large class of

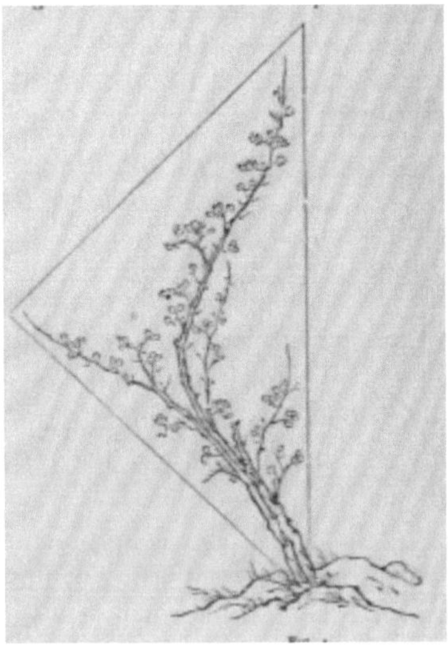

Fig. 1.

Fig. 2.

flower arrangements in which the *Principal* line of the composition has a horizontal or almost horizontal direction, and supposing a tri-lineal group of this style enclosed in a right-angled tri-angle the hypotenuse of the tri-angle would lie horizontally (see Fig. 2). This latter style is chiefly applied to flowers arranged in hanging vessels, or in standing vases placed on raised shelves, the intention of such compositions being to suggest floral growths on the edges of cliffs or banks, which lean over laterally. In such elevated arrangements it often happens that one of the auxiliary lines on the side towards which the *Principal* bends over, receives a decided droop and proportionate lengthening in order to emphasize the droop. Such drooping lines are technically called *Streamers*. The *Streamer* character may be imparted to either of the auxiliary lines, but redundancy must be avoided, and it is regarded as a fatal error to introduce *Streamers* on both sides of the *Principal*. The *Streamer* is very much used for floral arrangements in suspended vessels, in tall bamboo vases having side mouths, or in receptacles placed upon the edges of raised shelves. The prevailing idea in each of these cases is that the composition must suggest the wild growth on the edge of a bank or precipice over which the *Streamer* droops (see Plate XII. A).

The above description has been confined to three, five, and seven-lined designs. Single line and double line compositions, as well as those of nine, eleven, or more lines, are sometimes made, but their use is very rare. The double-line form is employed occasionally for the simplest of all flower arrangements, namely that of one flower and one leaf.

The different lines of a composition have hitherto been spoken of as if existing in one vertical plane parallel to the spectator, but actually, in addition to the directions mentioned to the right, left, upwards, and downwards, these lines have also directions of varying degrees forward or backward. In other words, the extreme points of these lines would require a solid and not a plane figure for their enclosure. These directions are best explained by supposing a bundle of stems placed in a vase of octagonal plan and designating the eight different facets of the vase respectively as North, North East, East, South East, South, South West, West, North West: then, imagining the South face of the vessel immediately in front

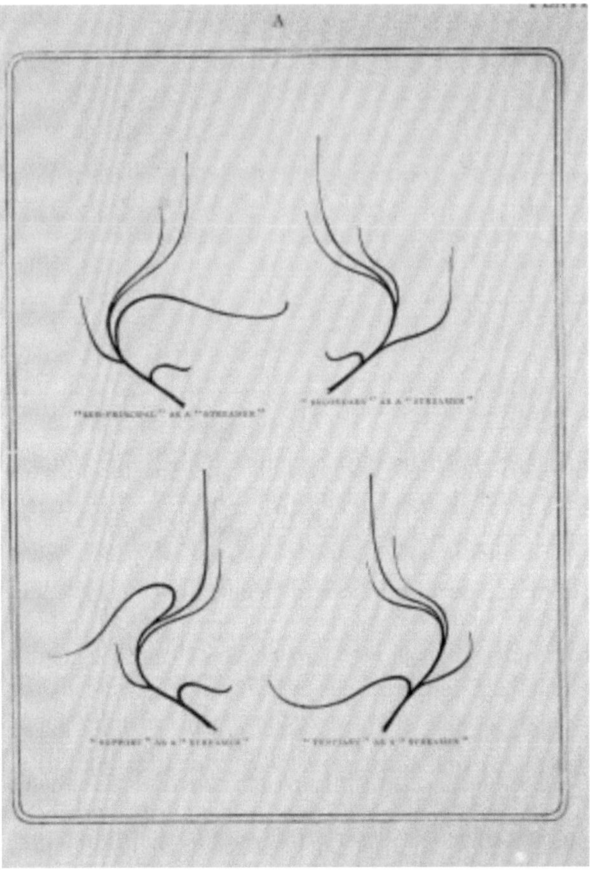

Lineal diagram of stems with "streamers"

of the spectator, and bearing in mind that all the stems coincide for some little distance from their origin, the cardinal directions which they take after separation would be as follows (see Plate XII. B): – The *Principal* bends North East, the *Secondary* South East, the *Tertiary* South West, the *Sub-principal*, which is between the *Principal* and the *Tertiary*, turns East, and the *Support*, which lies between the *Principal* and *Secondary*, terminates centrally over the vase. Of the two extra lines used in seven-lined arrangements the *Trunk-line* bends North East, and the *Side-line* bends West. In this manner a pleasing variety back and front, as well as laterally, is given to the compositions, and they are not the flat arrangements which might at first sight be supposed from explanatory drawings. Though designed principally with a view to being seen from a point of view immediately in front, they obtain hereby solidity and variety, and their effect when regarded from points to the right or the left are also taken into consideration to some extent.

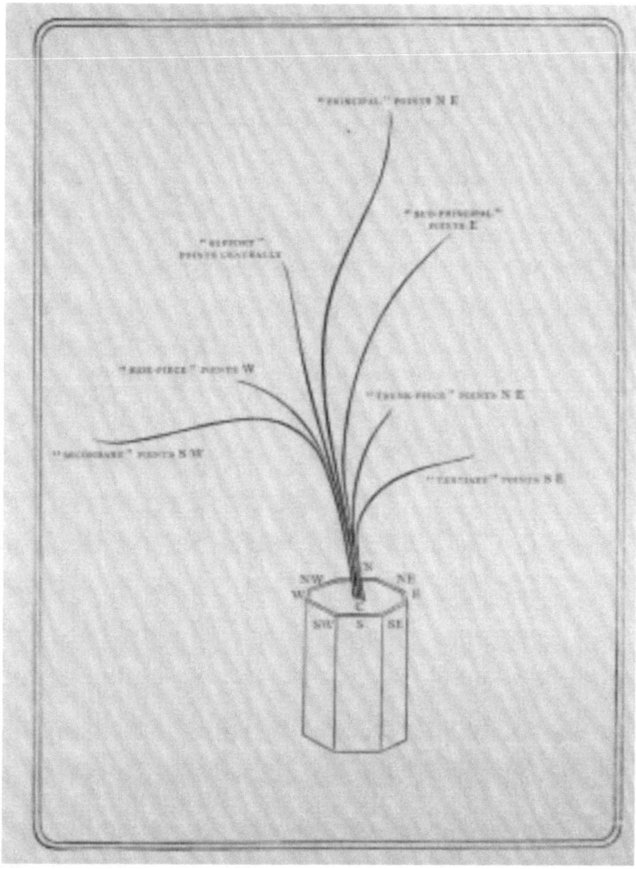

Diagram shewing cardinal direction of lines

Certain errors in arranging the lines of floral designs are pointed out to be strictly avoided. The first is that of *Cross-cutting*, which is produced by allowing two or more lines visually to intersect one another, so as to form angles.

A similar fault, called *View-cutting*, exists when a stem or twig cuts across its parent stem or branch; but this is allowed for certain trees, as for example the Plum, in which such intercrossing is characteristic of the growth.

Another error, called *Parallelism*, occurs when two or more adjacent stems or branches are exactly parallel to one another, a fault much aggravated when the lines are besides of equal or nearly equal length.

Window-cutting, so named after the curved or cusped windows of the temples of the country, is the name given to an error occasioned by allowing stems to cross and recross in curves so as to form looped openings.

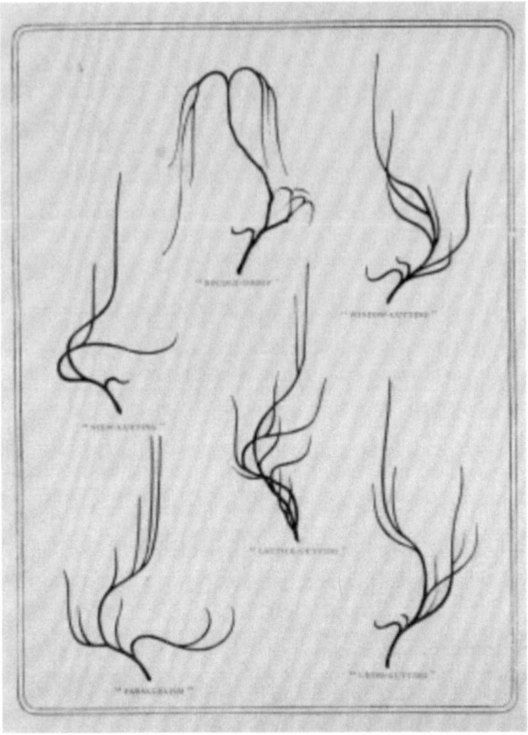

Errors to be avoided in floral arrangements

Lattice-cutting is an exaggerated form of *Cross-cutting*, and this error appears when several stems cross in parallel directions so as to suggest lattice work.

Another fatal error is that of the *Double streamer*, or drooped branch, used on both sides of the same composition. The above rules of prohibition, it will be observed, are for the most part similar to those followed in floral and other ornamental designs in the best styles of the architectural arts, being intended to prevent harsh, conflicting, or redundant effects and to be productive of harmony of line. They are illustrated in Plate XIII. A.

The various directions imparted to plants and branches of trees on the above principles are obtained, first, by a careful selection of suitable material; then, by twisting, bending, building together, and fixing at the base; and, lastly, by means of cutting and clipping off defective and superfluous parts.

c)

FLORAL DECORATION FOR THE TABLE

Editorial Headnote

50. E. A. Maling, 'Preface', ch. 9, 'Flowers for Dinner Tables', *Flowers for Ornament, and Decoration and How to Arrange them* (1862), iii–v, 57–69
51. Thomas C. March, 'Glass as a Material for Displaying Flowers', 'Arrangements of the Flowers etc.', *Flower and Fruit Decoration* (1862), pp. 7–14, 32–43
52. John Lindley, unsigned report on table decoration, *Gardeners' Chronicle* (13 July 1861), pp. 646–647
53. 'H', 'The Competitive Table Decorations at the Royal Botanic Gardens, by Gaslight', *The Garden* (20 July 1872), p. 52
54. Annie Hassard, 'Artificial Lighting of Dining-rooms', 'Plants Placed through Dinner Tables', *The Garden* (18 July 1874), pp. 55–56
55. John Perkins, 'Introduction', 'Breakfast or Luncheon Table', 'Dinner Table', *Floral Designs for the Table: Being Directions for its Ornamentation with Leaves, Flowers and Fruit* . . . (1877), pp. 9–10, 24, 31
56. Edward Luckhurst, 'The Arrangement of Cut Flowers', *Journal of Horticulture* (15 September 1881), pp. 237–238 and (22 September 1881), pp. 270–271
57. Shirley Hibberd, 'A Cheap Table Decoration', *Amateur Gardening* (8 May 1884), p. 6

This selection charts the development of decoration for the dining table through the changes of shape and material of vessels which contain plants and of the kinds of plants – choices which reflected trends in garden design. Two factors in particular contributed to the expansion of floral displays at the table: firstly, the 'French service' in which dishes were placed on the dining table for guests to help themselves was replaced by the Russian dining style, whereby dishes were served out to guests from a side table and floral decorations added to fill the space. By the 1860s, the Russian style predominated in fashionable households. Secondly, the innovative competition for table decoration of the newly named Royal Horticultural Society in 1861, which was judged by aristocratic lady patrons and created to raise funds, fostered these developments. However complex or expensive some of these floral decorations might be, a shared trope is that there are ways in which similar effects can be achieved at relatively low cost (and publication sales might increase).

The first two sources were published in 1862. The authors thank the businesses who have made their innovations possible, a form of advertising that is characteristic of domestic advice and gardening books of the period. They cover similar topics, both advocating greater use of wild or local plants (Robinson had not yet started on his journalistic career). E. A. Maling's *Flowers for Ornament, and Decoration and How to Arrange them* was issued as part of a series following her *Indoor Plants and how to grow them* and *Songbirds and How to keep them* (see IIa). The single, coloured frontispiece represents her design for a trumpet-shaped container (possibly of pewter) which regrettably she was too late to enter for the first RHS competition for floral decoration. The prize instead was won by Thomas C. March (1819–1898), a government official of the lord Chamberlain's office, whose floral arrangements have been attributed to his younger sisters. The design (and patent) for an elevated flower stand constituted a shift from the epergne, elaborate dishes that had become fashionable after the Great Exhibition of 1851 and the pyramid arrangements referred to by earlier commentators (IIb). March's larger-sized *Flower and Fruit Decoration* (price 3s.6d) explaining the uses of the stand offered ten additional colour plates costing 1s. each (the full price of Maling's book was 2s.6d). His arrangement for bands of coloured flowers indicated in three pages of diagrams at the end of the book (not included here) were influenced by the latest style in bedding plants (see Volume 5), as well as single- or two-coloured bouquets sold in Paris at the time. Although floral arrangements for a dinner of up to 20 are included, March claims that this is a practical book intended for staff or the master or mistress of the house, rather than the elite households of orchid owners. John Lindley was enthusiastic about the competition in a review that followed in the *Gardeners' Chronicle*, in terms of both design and the resulting improvements in the 'taste' of the wider public. The full-page illustration (uncredited) representing the prize winners graphically demonstrates the changes in shape and material of dining table ornament.

Over a decade later, the interest in table decorations continued unabated: 'H' reports on 'the competitive table decorations at the Royal Botanic Gardens, by gaslight' (*Garden*, 1872), favouring the winner of the third prize (Miss E. Harris) with her uncluttered display. Annie Hassard ('Miss E. Hassard' [sic]), winner of the third prize, contributed to the series 'The garden in the house' (*The Garden*, 1874). In 'Artificial light of dining-rooms', she suggests a combination of permanent and temporary arrangements of plants for the table under difficult lighting conditions. In 'Plants placed through dinner tables', an apparently simple method is represented almost five years after the striking appearance of a palm tree as the centrepiece of table decoration at an exhibition at the Crystal Palace, Sydenham.

Two head gardeners based at estates in Suffolk made further contributions. John Perkins (1824–1905) of Thornham Hall dedicated *Floral Designs for the Table* (1877) to his former employer Lady Henniker, Anna Kerrison (1812–1889). The large landscape format (folio, 335mm x 215mm) was the ideal vehicle for Perkins's 24 designs which were lavishly reproduced as chromolithographic plates (a recently developed process of colour printing based on the lithography). Their resemblance to garden bedding patterns (see Volumes 5 and 6) was noted by Hassard. She suggested that the disappearance of almost all vessels to contain flowers 'may be termed a kind of dinner-table bedding out' *Gardeners' Chronicle* (22 December 1877) (not included). The price of 30s. however, (and settings for up to a maximum of 100 people) may have precluded middle-class purchase despite the opening reference to the poet of 'everyday' rural life, William Cowper. Edward Luckhurst (1838–1901) was head gardener at Oldlands Hall, Sussex. In the first of two articles, 'The arrangement of cut flowers', *Journal of horticulture* (1881), he argues that floral decoration is a skill and should be taught, and he recommends a large glass fish bowl as ideal for floral arrangement for the table. Official courses in floristry did not start until the 1930s, although competitions for floral decoration spread to other societies from 1861 and remained a feature of horticultural shows until after the first world war. In the subsequent article he advises on the installation of a flower-arranging room – hinting at the vast costs that elite households were spending on flowers for the home in the latter decades of the 19th century. The extent to which head gardeners were involved in these arrangements is an indication of their multiple responsibilities in the period (and reflected throughout these volumes).

Finally, Hibberd's short piece 'A cheap table decoration', *Amateur Gardening* (1884), features an unsigned illustration of a relatively elaborate display that looks back to those of the mid-19th century.

50

FLOWERS FOR ORNAMENT, AND DECORATION

(Excerpts)

E. A. Maling

Source: E. A. Maling, 'Preface', ch. 9 'Flowers for Dinner Tables', *Flowers for Ornament, and Decoration and How to Arrange them* (1862), iii–v, 57–69

PREFACE.

AMONG the many books that we already have upon the growth of plants, there seem to be few, if any, on the after arrangement of what Florists call "Cut-Flowers."

This has, however, become of late no unimportant matter, both as a question of tasteful arrangement and skilful manipulation, and as a rather serious item of household expenditure; the *diner Russe* requiring so very many flowers, and the taste for having them increasing with such rapidity.

For myself, I have always been so very fond of everything connected in any way with flowers, as to have lost none of the many opportunities that have occurred to me, of obtaining good instructions in the art of their arrangement; and thus I hope that a little Manual, embodying these hints, and describing accurately many such designs, may not be unwelcome.

I have endeavoured in this little volume to explain, in the most clear and simple manner, everything that is required in arranging flowers; and I hope that those who adopt my hints for German wreaths and Russian dinners, French bouquets and Italian pavements, dishes of growing flowers, and ornaments for slabs and tables – not forgetting also the more massive wreaths and more elaborate fabrics for architectural decorations – may find that they afford them some means, which they had not before, of using flowers to the best advantage; economizing their number, when, as in towns, they are rare and costly luxuries; classing their colours rightly, and choosing the kinds correctly, so as to avoid a useless and harmful waste of their much-prized brightness.

I must not close this preface without acknowledging my great obligations to Messrs. Veitch, of Chelsea, Messrs. Hooper, of Covent Garden, and many other florists, who have given me information as to correct names and good varieties of flowers, and often shown me specimens I might not have seen otherwise.

DOI: 10.4324/9780429198809-59

In regard to the vases, fountains, flower-stands, &c., mentioned throughout this book, I have also to acknowledge the advantages afforded me by Messrs. Phillips and Co., of Oxford Street, and Messrs. Minton, in often showing me beautiful new designs.

The Frontispiece represents a Group of Flowers and Fruit designed by myself, the arrangement of which was so much liked by Messrs. Phillips and Co., that they have manufactured the Vase for sale, and it may be seen at their establishments in Oxford Street and Bond Street.

E. A. M.

FLOWERS FOR DINNER TABLES.

1. HAVING already given such very small details as to frame-work, mounting, making up, &c., for flower-vases in general, there is now little left to do but just to name the special flowers, and describe the designs, for some dinner-vases. I shall generally name a whole set together; but any one requiring only three or so, can very easily select some to suit their purpose, and adapted for the kind of flowers which they may have to use.

I will suppose that there are generally one large centre, two or four next size, and perhaps some smaller vases, to be filled with flowers, and will first describe an arrangement of Roses which I know to be very beautiful, and which is also most easily arranged. One large wide central vase is required here – a low-spreading basket borne up by Parian[1] figures, or supported on a stand, looks very well in this way; four smaller vases should, in a measure, correspond with the centre; and six, ten, or fourteen smaller ones, may be used as well.

2. In some cases – as in impromptu dinners in seaside or country sojourns – where, perhaps, the flowers are the only materials really in abundance, it may be amusing to know how to make them useful; and a wonderfully beautiful display may, indeed, be wrought at a little expense of most pleasant labour, and by the exercise of a little taste, or – as the people "there" would probably call it – *gumption*. A range of glass milk-pans – price from sixpence upwards – or an array of soup plates supported on finger-glasses, have been known to represent a splendid dessert service in a most effective manner – Sycamore leaves, and Plane leaves, the spreading foliage of beautiful Acacias, leaves of Water Lilies, Ferns, gathered on the hill-sides, and many other beautiful shapes of green, utterly concealing the poverty of their supports.

I hope that this may prove a useful hint, and lead to further attempts to beautify common things; for whatever may be thought of cheap and vulgar *finery*, there is no such thing existing as vulgar *beauty*, though it may be cheap.

3. I will now proceed to the Rose arrangement, only premising that in such a case as that above alluded to the green must be looked upon as a simple addition, just as if it were a part of the vase itself. The central vase, of course, requires the largest and most *prononcés* flowers, the four smaller ones ranking next to it in size. For the larger vase, then, the Géant des Batailles Rose is one of the very best,

and of the most brilliant crimson. Placed three or four together in the centre of the centre vase, it might be surrounded with white Moss, or Provence, or some other pure white Rose; long sprays of small white-clustered Roses, mixed with those of the lovely crimson Pæstum, or of some like briary kind, hanging about the edge and mingling with the white.

Pretty healthy foliage is far more important for this than flowers. Still some beautiful clusters might well be wired on stalks possessing better foliage than their own might be. Many of the very small briar Roses are so exquisitely lovely, that even the Wild Rose sprays would be far from out of place.

For the centre vase a pointed design, or a thick band of crimson and white alternately, where such designs are wished for, would also do very well, instead of the simply-mingled bouquet; with, in the former case, little foliage; in the latter, much. Two of the other larger vases should be in similar colours, and of a corresponding design, all circles, or all grouped; but if crimson is the ruling colour in the centre piece, white should be so with them – or the reverse, if otherwise. The Rose foliage, or Rose leaves and Ferns alone, should be also used exclusively for all.

The side vases need not match in any way, though all should contain either white or crimson as one of their colours. Crimson delicately shaded into pink would be very beautiful, placed alternately with white shaded into rose. The various flowers must be chosen, however, with great precision, as so many, especially of the year-or-two-old Roses, have so heavy a tinge of bluish lilac. In the centre of each side there might be a vase of the palest Tea Roses, arranged lightly amidst their foliage as for a drawing-room vase, and relieved by a border of the very palest pink.

Of course, the vases down each side of the table must match; if a different centre is therefore used, the number of pairs of side vases must be uneven.

4. I will next describe a very graceful and most cool looking arrangement for a dinner table in the hottest time of the London season, when cool and refreshing things are most to be desired.

In this design, both vases and arrangement conspire to the same end – the vases being made all of frosted glass and of crystal dew-drops, for the especial purpose of looking really ice-like.

These vases are of the shape of which the Frontispiece represents the largest: they are made by Philipps, in Oxford Street, and will, I think, become very popular on account not only of their perfect coolness, but also because of the ease and convenience of separating entirely the flowers and the fruit, while still obtaining the full embellishment that each gives to the table, the perfect facility afforded for the cleaning, and the useful attribute of holding much or little fruit and few or many flowers.

As I designed, or rather adapted, these vases, from one for flowers, bought, a short time ago, at Philipps's shop in Bond Street, I have, of course, tried their capabilities in many ways: my own object having been to compete for the prize at Kensington Gore for table decoration: for which attempt, unluckily, I was too late in the field last year.

5. My idea, in the design of these, was that while presenting a pile of fruit lying on cool thick leaves, such as might be disturbed as readily as that in any other fruit dish, the flowers or ferns above would make a gracefully waving shade, without interrupting at all the view across the table – which is such a great discomfort.

6. The accidental advantage of a few sprays of Fern, and wreaths of drooping Roses, being sufficient flowers; and the wide limit left for much or little fruit, while producing the same effect in either case, will, I think, make the arrrangement useful for either large or small requirements.

The vases, being simply made of glass, are also very pretty for the drawing-room table when entirely filled with flowers.

I will now proceed to describe some of the flowers that I have found best suited for these kinds of vases.

7. It seems to me that *height* should not be attempted. A good arrangement is to place first in the vase either a branch of Rose-leaves or a bushy piece of Myrtle, or such-like green, just to fill the centre without standing up, and to keep the flowers and leaves properly in their places.

The Ferns will always then wave widely enough around; and a few sprays of Heath or Epacris, graceful blue and white bell-shaped flowers, such as the Campanulas and beautiful Lily tribes, Wistarias, again, and the pale Rose Acacias, are most exquisite, both for leaves and flowers, if laid in large, fan-like layers all round the vase.

8. Few things would be actually more fresh and beautiful than vases filled with white and rose Acacias; the central and larger bouquet combining both amidst their own pale leaves. Wistarias and Laburnums even would be beautiful, where a pretty simple group was all that was required. Roses, Fuchsias, Lilies, and Passion-flowers; the lovely wreaths of the Peruvian climbing Lily, or rose-coloured Lapageria; sprays of drooping Orchids and of summer climbers; even the great white Bindweed wreathed around the stem – all these would be beautiful.

9. Sometimes, again, the places might be reversed; and grapes having been placed to hang from the upper vase with their long wreaths of leaves, the dish below might be filled with water-lilies floating in their own wave.

10. A design which always answers very well, is, however, the first mentioned Ferns; placed lightly and yet abundantly to droop around the vase, and to overshade the white and purple grapes in the dish below. Perhaps the loveliest vase that I ever saw, contained Ferns alone; but then that was in the very hottest weather, when green fresh foliage was so refreshing.

11. In colder weather, more red becomes desirable, and rose-coloured and white Camellias, Heaths and red Poinsettias, Hyacinths and red Tulips, are amongst the flowers to which we have to fly.

Then, again, we may take to the lower dish; and filling it with Hyacinths, with snowdrops, moss, and blue-bells, we may make such a dish as will last us half the winter.

A zinc pan made to fit into the dish, either in two pieces or passing overhead, would answer for this most perfectly.

12. I have, however, two special arrangements to recommend at present.

The first is all Roses.

The dishes, in all cases, being filled with fruit grouped around the stem, the Roses have to be arranged above in their own many shades.

White Banksia Rose is lovely for entwining part of the stem itself; and then the common shell-tinted Ruga Rose, the exquisitely shaped Noisettes, and the long Multifloras, the old sweet Pæstum, or the shaded Blaria, are at once very common and abundant everywhere, and first-rate for beauty.

Tea Roses, Moss Roses, the Provence and the various kinds of Chinese Roses, all give many flowers worthy of a place; but those which I named first deserve some pre-eminence for their dark healthy foliage, as well as for their beauty and their sweetness.

Suppose the central vase to be grouped with crimson Roses, or the sweet pink moss-roses, gathered in the centre – large clusters of the drooping Banksias all around the edge, and little bunches of pink flowers like the centre Rose here and there appearing – the dish below might be heaped with purple grapes, surrounded at the top by a wreath of their own leaves. The other vases might then well contain white Roses, and these might be broken up with red and crimson flowers; while bright red fruit beneath, peeping out through leaves, would continue the brilliant colour.

Smaller vases standing all round the table might be filled with fairy roses, the sweet little Rose de Meaux, and the white Noisettes, so exquisite for those baskets which white little china figures carry upon their heads.

13. Another, but a wintry design, for decking out these vases, looks, I know, most brilliant, and can be arranged most easily.

A thick fringe of green is laid in the centre vase, and all around it runs a ring of waving sprays of Heath or Hyacinths, of the palest rose and white.

Next to these comes a circle of white Camellias, or of Tulips – four or five are enough – with a fringe of Fern leaves made to stand up above them; and then again a group of Heaths or Hyacinths; while bright rose-coloured Camellias or Tulips form a sort of crown rising up above. In this design the Ferns keep up the wide impression, and take off the appearance of a too great height.

The vases meant to match can be made most charming, either with pink Hyacinths, or rose and white Azaleas: in the first case, it is better that the leaves should be chiefly at the edge; in the other they are far better placed in a group together, to overshade the flowers.

A drooping Orchid spray, or a wreath of some blue or rose-coloured climber, would be a great advantage; though I have known long sprays of creeping moss, or even of the little ivy-leaved trailing plants, look more artistic than any flowers would have done in their place.

14. Drooping flowers and foliage are especially to be desired in vases of the shape already mentioned; and from the kind of glass employed, the green is one of the most effective features. The smaller vases – those at the top and bottom of a table, or at the sides – should in a measure correspond with the centre one, having at least *one* of its colours for the ruling tint.

When the other vases are to be very gay, two, of Fern leaves only, on either side the centre, have a delightful effect.

The other larger groups may contain Azaleas, or Anemones, Camellias or Tulips, reversing the colours used in the largest vase; that is, placing the brightly coloured flowers lower in the vase, and the white ones over.

Of course these arrangements can be always infinitely varied; one thing, however, being observed, especially in cases where the white has the lower place; namely, to place a thick layer of widely spreading green between it and the vase. A most beautiful arrangement of white china and the beautiful Parian vases was formed of centre figures supporting baskets filled with the two coloured flowers, rose and white, matched with white and rose. A break on each side the table was made by a basket of beautiful mixed Azaleas, and then all round pairs of the little flower-baskets were filled with Primroses and Violets.

An exceedingly pretty and effective arrangement is that recommended in the "The Lady's Guide;"[2] baskets, namely, of gilded or enamelled wicker-work (which should be made very light and open), containing a dish of flowers planted in sand or moss.

This would be a most useful plan for the employment of the "growing bouquets," which I have elsewhere named. I should, however, venture (when the basket is very open) to suggest the use of a white china dish or bowl, rather than of a tin, as the gilded tracery of the basket seen against the china has a good appearance.

White small Hyacinths and rose and white Van Thol Tulips are first-rate flowers for these kind of baskets. The Tulips are most exquisitely tinted with a pale waxen rose colour, and being about six inches high they are more graceful-looking than the pretty and sweet little *red* Van Thols, though the latter do charmingly amongst low plants like Primulas. Heaths should always be seen by candle-light before they are selected for dinner-table use, as they vary much according to light.

White and pink Azaleas, blue Achimenes, white corollaed Fuchsias – such as Princess of Prussia and Madame Cornelissen – Roses and Geraniums, are also plants well suited to these baskets. A prettily shaped fountain-like plant looks extremely well with the basket covered with the little blue Lobelia, or some small creeping flower. It does not do for these baskets to mix too many plants, if plants are used in the place of cut flowers.

I cannot forbear from borrowing from the book alluded to[3] a hint which I am afraid my readers will think I ought already to have laid to heart. It certainly is an unfailing rule that nothing is worse taste than a too crowded table, especially one crowded with little trivial pieces. In most cases, however, where more than three or five groups of flowers or plants are used, I take for granted that they are mixed with fruit, as in the vase where a bouquet of Ferns surmounts a pile of grapes, or as in those pretty groups where a china figure stands by a large mass either of fruit or flowers, holding on its head a basket of appropriate leaves and blossoms.

I do not know if others will agree with me in delighting so much in seeing what may be done with our own common flowers to equal the brilliance of the foreign and most expensive plants. But so far from being ashamed of them, there is nothing I delight in more than to have people asking about flowers which they do not recognize as inhabitants of their hothouses, and to be able to tell them that they are common garden, or even hedge-row flowers.

Editorial Endnotes

1 Parian is a white unglazed porcelain. The introduction into Britain is credited to the Copeland & Garrett factory (formerly Spode) at Stoke-on-Trent from 1842.

2 'Baskets, namely of gilded or enamelled wicker-work', [attributed to] 'The Lady's Guide' – *The Lady's Guide to the Ordering of her Household, and the Economy of her Dinner Table,* by a lady (1861); a book of dinner plans and recipes throughout the year to guide the inexperienced mistress of the house. Precise wording not identified.

3 The book alluded to is *The Lady's Guide . . .*, p. 59: 'The effect of a handsome service may be entirely spoiled by covering the table with too many objects'.

51

FLOWER AND FRUIT DECORATION

(Excerpts)

Thomas C. March

Source: Thomas C. March, 'Glass as a Material for Displaying Flowers', 'Arrangements of the Flowers etc.', *Flower and Fruit Decoration* (1862), pp. 7–14, 32–43

GLASS AS A MATERIAL FOR DISPLAYING FLOWERS.

CLEAR and transparent glass harmonizes so admirably with the soft colouring of flowers and foliage, and lights up so brilliantly at night that on the whole it is preferable to every other material, though silver and porcelain have formidable claims, and, indeed, have hitherto been far more extensively used for table decoration.

Silver and plate were indispensable under the old regime when the tablecloth was removed after dinner to the great discomfort of all the guests; the white metal was a good contrast to the dark polished wood, and as some moving was necessary, solid candelabra and epergnes were preferred, and strong cut decanters and wine glasses. A few years ago, however, the Russian service, which for some time had been used abroad in most good houses, was imported into England, and after a time became thoroughly domesticated, putting the old service to flight, as the brown Hanoverian rat[1] is said to have annihilated the native black rat. A few lovers of old custom "still hug the chain,"[2] preferring the slavery of carving for twenty people on a hot July night to the luxury of ease and unrestrained conversation; but these dissentients are rare, and at all events the white cloth remains untouched to the last with its burden of fruit and flowers, and brilliant glass and lights. Since this change greater care has been bestowed upon the appointments of the table, – beautiful and delicate glass has succeeded to the more solid descriptions, and the merits of this material generally have been more prominently shewn. To my mind, nothing is equal to glass for the display of flowers, fruit, and lights, and for these reasons: – If the flowers are good, they are like diamonds, the setting is almost superfluous; the simpler the vessel the more beautiful they look, but a silver or plate epergne would be thrown away without carving and variety of form. Porcelain in the same way requires colour and exquisite painting, otherwise the manufacturer's skill and the charm of the material are useless. Take a handsome

DOI: 10.4324/9780429198809-60

centre-piece of silver, a group of cupids beautifully modelled, a stag-hunt, or a halt in the Desert; these may be exquisite things in themselves, but they demand so much admiration for themselves that their modest tenants in the upper story are quite overlooked. To indulge the skill of the artist, the cupids and scroll work occupy the whole of the ground-floor and the drawing-rooms, while the unfortunate flowers are forced into the garrets where they appear to the least advantage. I object to see them sitting "on the knifeboard," looking down from a perilous height upon the heathen mythology. So also with porcelain its very beauty interferes with the full effect of the flowers.

These remarks relate principally to ornaments for the dinner table; it is different in reception-rooms, where the colour and exquisite shapes in porcelain render that material also very eligible for vases, which it must be recollected, are frequently unfilled with flowers. There is room here both for glass and china. As a general rule, however, it will be found that all complication of form and brightness of colour are better avoided in the choice of vessels which are intended to contain flowers. It is these which are to be looked at, not the vase or scaffolding which holds them; and glass, from the difficulties of its manufacture, and by reason of its perfect transparency, will always be simpler and less obtrusive than any other material. The same argument applies to the earthenware flower boxes, which decorate so many London windows, since plate glass and white stucco have succeeded to prison bars and dull red bricks. These flower-boxes which should be kept down in tone, are frequently made so bright and varied in colour, that the plants are perfectly bewildered in them, and the effect is a mass of confusion. So also in furnishing houses; an extra bright carpet, or a paper of glaring colours, often destroy the whole effect of a room. When people marry they always mean to have the most charming house in the world; the carpet manufacturer rolls out before them in full magnificence a superb carpet with the most brilliant of flowers, which is so handsome in itself, that it defies resistance. The paperhanger in the meantime having hurried them through several volumes of patterns, till they are nearly blind and can only see the brightest, fixes them at last with a beautiful paper, all colours and gold; the effect of these irresistible patterns is, that the frames and pictures, the Dresden china, cabinets, and encoigneurs,[3] are completely absorbed, and for years the carpet keeps staring up at them, and the walls keep staring down at them, till they are glad to welcome the sober influence of London dirt, which like time equalises all things. I have seen papers and carpets which would neutralize the whole contents of Wardour-street or the South Kensington Museum.

It is this too important and self-sufficient "entourage," or setting which so often destroys the effect of flowers. Nothing equals them in beauty, for the short time they last, and their reign should be supreme and absolute.

Perhaps these opinions may not be universally endorsed, for silver and porcelain are really so handsome in themselves, and tastes are, luckily for manufacturers, so widely different; there are even people who would not "bemean themselves" by setting before their friends an epergne of so cheap a material as glass. There are others, however, upon whom parents have settled more taste than money, and economy will not be unacceptable to these, if it can be shewn that the less expensive taste is also the most effective. I will give way to my own preferences in these few remarks, and confine them to the consideration of glass vessels of simple form and even surface, undiversified by ornament of any kind, in which flowers, fruit, and foliage, may be grouped without any hindrance, according to the design which pleases the arranger the best.

FLOWERS SHOULD BE GROUPED ACCORDING TO SOME DESIGN.

A large and uninterrupted surface gives scope for form and design; pull down an old street, clear away the whole space, and the architect can carry out his cherished plan; but in crowded old cities, he can only do a little patchwork here and there. Descending to the more unimportant subject of Flower Arrangement, and turning to the diagram below, it will be observed that the circular dish or base of this epergne is extremely wide and very shallow. For the centre decoration of a long dinner-table, the base will be 15 inches across, while the depth is little more than an inch. As a circular border of leaves should invariably be added, in order to hide the rim and extend the surface, the diameter of the base will exceed two feet. The advantage of this large surface for flowers and foliage low upon the table, is that the whole beauty of the arrangement is taken in at a glance. You look down upon it and see every leaf and blossom; it must, however, be remembered, that every defect is equally visible, and that great care must consequently be taken to select handsome leaves without blemish, well shaped flowers, and graceful accessories. It is the wide shallow base which is the most effective part of the epergne. Flowers always look well in large circular dishes without any superstructure, but the upper part also gives variety to the form of these designs. The tall slender stem suggests a cluster of flowers and buds round its base, and invites you to twine around it some graceful branch of cobœa, jessamine, or passion flower, by way of introduction to the upper glass, where there are also flowers and drooping foliage, perhaps hanging fruits, or falling blossoms, the whole harmonizing and corresponding with the flowers and leaves below.

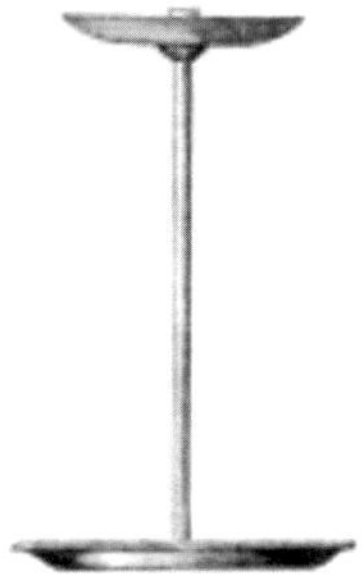

In these simple shapes, if there is little to help, there is at least nothing to obstruct. When the whole is arranged, the scaffolding will be entirely concealed, with the exception of the bright clear glass of the stem, partly hid with foliage, and glimpses of the glass rim above between the border-leaves. The surface to be filled is large, and it will not suffice to crowd in a quantity of different flowers with no relationship to each other except their bright colours. This may captivate the least discerning of the million – what Mr. Beaton calls the "good gracious"[4] people – but if the desire is to please those who have correct and refined taste, it will be necessary to adopt some definite design or leading idea, which must be adhered to more or less throughout the whole arrangement. Let us take an example, in which Water-lilies form the principal feature. It would be a barbarism here to intermix Fuchsias, Geraniums, or Roses, which have no kind of similitude to Water-lilies in habit or place of growth; but where this exquisite flower rises to the surface in some slow-running stream, you will find Forget-me-not fringing the bank, and small waving reeds and grasses, perhaps some wild Iris, and in the wood which borders the stream will be Ferns and Mountain-ash, Creeping Ivy, and moss which no velvet can rival, and here and there, amongst thousands of blue Hyacinths which colour the whole underwood, will be a few bells of slender white Hyacinth, if the season is not too late. With these and similar materials carefully selected, a group may be elaborated which, in its humble way, may be almost a work of art. So again with Primroses and Violets, you have an early spring group in which the leaves of those flowers, with Trefoil, wild Strawberry, Ivy, and wild Rose will be far more in keeping with the leading idea than the intermixture of Maidenhair fern, although, in its proper place with cultivated flowers, nothing can equal this fern in lightness and beauty.

If you wish to make a decoration of Grapes, it will be impossible to add effect to the simple beauty of vine-shoots, leaves, and tendrils, but ivy is so classically associated with the grape, that you may use for a border its dark, shining carved leaves, near which all flowers, fruits, and foliage appear to gain in freshness and brightness. In arranging other fruits the foliage should always be mixed; but when the leaves are less graceful, shoots of the Vine, Hop, or Currant may be added, or you may use the leaves and sometimes even the blossoms of the flowering varieties of the fruit-tribe, such as the double white Cherry, the Potentilla with its strawberry leaf, and the Pyrus Japonica.

Ferns are of such infinite variety, that some of them are in harmony with all arrangements; but less hardy varieties, together with Begonias, Maurantas, Cissus Discolor, &c., are more at home with Poinsettias, Gloxinias, Camellias, and other things which are grown under glass.

Most of the spring flowering shrubs, such as Lilac, Laburnum, and Wistaria, look best with a border of large dark ivy leaves, which are sure to be growing near them, and form their natural relief at a period when most deciduous leaves are not full-grown; delicate branches of Arbor Vitæ and Yew, and small varieties of the fir-tribe are also most useful and appropriate with all winter and spring flowers, and Berries of the tree Ivy and Arbutus will afford a variety of colour and outline. It is a curious fact that the Ivy and Oak-leaf are scarcely used at all in decorations of natural flowers, while they form the stock-in-trade of the wood-carver and metal-chaser, being, in fact indispensable in all works of art where nature is carefully followed in the form which most pleases the eye. Mr. Standish,[5] of Bagshot, first suggested to me to try Oak-leaves, and I have since used them constantly as borders for these decorations, the leaves, when fine and well-chosen, forming an excellent relief to all garden flowers in summer. The branches and leaves of the Vine are almost equally neglected, with the exception of a few leaves which are sometimes put in a dish with Grapes, and are pressed flat with the weight of the fruit; yet artists, from the Greeks downwards, have been right in selecting for ornament these simple and beautiful forms, and it is absurd to neglect in nature that which is so admirable in art. The reason for the omission may probably be found in the fact that flowers have hitherto been arranged almost exclusively in tall vessels filled with water only, and that in these there is no space for the leaves to assume their natural freedom, but it is different with wide, shallow vessels, where sand and plastic clay are used to give consistency and keep the stalks in any position required. Here such leaves are of the highest service and tend immensely to relieve the flowers. With a border of Ivy, Oak, or Vine, or a fringe of the common ferns (Filix mas and fœmina), which grow in every wood, you have a framework which will do justice to your flowers, however beautiful, and there is this advantage, that they cost but the trouble of gathering.

So much for design and for appropriate mixtures. These hints may be carried to affectation, and sometimes the choice of material is not at hand; but if you aim at good taste and refinement, fitness and propriety of admixture should never be lost sight of.

ARRANGEMENT OF THE FLOWERS, &C.

I. Large Flowers of a Rounded Form

These look well arranged in circles just within the leaves of the upper and lower borders, the flowers being at equal distances from each other, as near the edge of the zinc base and glass upper dish as possible, and, if anything, rather overhanging the edge.

The number of flowers to form each circle must depend upon their size and also upon the width of the epergne, but generally speaking, seven flowers in the lower circle, and five in the upper, will be found the best proportion; in addition to which, two or three flowers and buds will be required to form a cluster at the base of the glass stem, and the same for the centre of the upper glass, but these may be of another kind if you prefer it. The engraving below represents Geranium with Lily of the Valley arranged in the manner above described.

The stalks of the border flowers should be from two to four inches long, and they should be inserted in a slanting manner rather than quite upright, so that part of the flower may project a little beyond the rim. A quantity of flowers and foliage stuck in perpendicularly and looking up at the ceiling, will have a stiff ungainly appearance; they should therefore be made to incline outwards from the centre, and, arranged in that manner, will be seen in full by all who are sitting round. Much effect is lost by contracting the two circles, the flowers are brought too near to each other, insufficient space is left in the centre, and the leaf border looks long and straggling.

Large round flowers are very easy to arrange, and the effect is seen the moment they are placed in position within the border. Take, for instance, the common example of fine roses; seven of these are placed in a circle at equal distances, just within the lower border of fern, some rose leaves are placed under or round each flower, three or four other roses, one smaller and higher than the rest, together with some buds, are arranged in a cluster round the glass stem, and other foliage added whenever it is necessary to break the flat line of the moss. A spray of Cobœa or Jessamine is then twined round the stem and fastened with some light wire. This completes the lower part, and the upper glass is treated much in the same way, but with fewer flowers, adding, if you have them, four or five fronds of Maidenhair fern to bend over gracefully from above, while the same number may be added below, resting lightly over the Roses. There are few decorations of large flowers which are not improved by this beautiful fern, but its effect is lost when numerous small flowers and leaves are used.

Take another instance, that of Water-lilies with a border of Ivy. The same number of flowers should be used and in a similar manner, not forgetting to add some green buds between the open Lilies; a cluster of the same flowers and buds will form the centre, and at the base of the glass stem you may place some wild Hyacinths or blue Veronica, and some fronds of fern arranged like a growing plant, more Ivy, and if you like it some Forget-me-not between the stem and the circle of flowers. The same arrangement will be carried out in the upper glass with five Water-lilies and buds. If you use ferns above, they must not be too high or spiky, and a branch of wood ivy should be twined round the glass rod.

The leaf of the white Water-lily is very handsome, but will fade almost immediately if it does not actually rest upon water, so that it cannot well be used. The yellow Water-lily has a smaller and very well shaped leaf, which lasts longer, and can be used with greater convenience, but on the whole good ivy leaves are preferable.

II. SMALLER FLOWERS ON LONG SLENDER STALKS, SUCH AS CARNATIONS AND PINKS, ROSES DE MEAUX, &c.

These should be planted over the whole surface of the clay and sand, with foliage between the flowers. The diagram will give some idea of the mode of placing the flower stalks. The side represented by *a* shows the more correct plan. Those shown at *b* are too upright.

The number of flowers will be in proportion to their size. Supposing that an epergne of 12 inches diameter is used, two dozen Carnations, if they are at all fine, will be sufficient – say 15 below and 8 or 9 above. Three dozen would be required of Picotees, large Pinks, or Roses de Meaux, 22 below and 14 above; of small Pinks a greater number still must be used. In this kind of arrangement the flowers should have

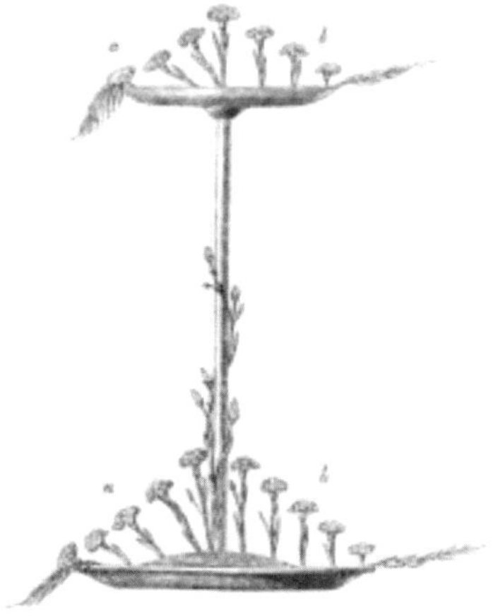

the appearance of growing, and the foliage of the plant is for the most part indispensable. There should be a good foundation of clay, extending over two-thirds of the base, because each stalk must be firmly and singly supported in it, sand alone being quite insufficient for tall slender stems. There is no flower which excels the Carnation for these decorations, and we will take it as an example. About three flowers should be arranged near the base of the glass stem, and another higher up the glass, with buds considerably higher again, but not quite to the top of the rod. Several other carnations lower in height should then be planted round the three first-mentioned, and, finally, an outer circle of eight or ten, not greatly higher than the moss. The flowers should thus rise gradually from the border to the stem, and the Carnation grass or piping should be placed round the base, up the stem, and between the flowers. In the upper glass less of this foliage should be used – one or two lower spikes only in the middle. Both the flowers and foliage should bend outwards from the centre as I have before explained. It is more important in this kind of arrangement than any other, and I have endeavoured to show the right and the wrong way in the foregoing diagram.

In buying flowers it is a rare thing to obtain them with stalks long enough to reach from the clay to the top of the stem. This is immaterial with Roses de Meaux, Pinks, and Carnations, for these will live a considerable time without being placed in moisture, and short pieces, with the proper sized buds and small flowers attached, can, therefore, be fastened on to the stem with wire just where they look best, the end of the stalk being easily concealed with a little of the foliage.

III. FLOWERS IN SOMEWHAT LONG AND POINTED BUNCHES, AS LILAC, DELPHINIUM &c.

These look remarkably well when arranged in circles, the bunches of flowers not over large, at equal distances from each other, and projecting considerably beyond the commencement of the border leaves. Ten of these bunches will be enough for the lower circle of the epergne, as shown in the diagram, but the bunches would be larger and the arrangement less formal. The upper will require eight or nine. The same flower should also be in the centre, both above and below, and foliage between. The bunches should be allowed to hang

over, or rather incline downwards, as much as their habit of growth will allow. Lilac may be taken as an excellent example of this mode of arrangement; its numerous half-hanging branches clustering gracefully round the stem and over the border leaves of ivy. Double blue Delphinium requires the same treatment, and the mass of bright blue colour is splendid. Syringa also and Flowering Currant belong to this class, which comprises many beautiful garden flowers accessible to all. As a variety, I may add the drooping blue Campanula, which, made into pointed branches alternating with bright pink Roses, is one of the prettiest possible arrangements; and this plant may be bought always during the month of July in Covent-garden market. Fuchsias are something between this division and the next, but they look very well arranged in points, allowing sufficient height for the flowers to hang gracefully. They are, however, best as an accessory to different groups, a few of the flowers being so placed as to droop from the upper glass, or round the handles of the baskets.

IV. DROOPING FLOWERS LIKE WISTERIA, OR FRUIT LIKE GRAPES.

The chief feature in this class is that the long pendent flowers, or heavy bunches of fruit, are made to hang down from the edge of the upper glass, and for this purpose are fastened with wire or ribband round the glass socket in the centre, as marked at *a* in the diagram overleaf. Each wire, twisted two or three times round this socket, will support a bunch of flowers or fruit upon each side, the stalk being pinned into the moss and sand at the edge of the glass so as to keep it steady. Any number of bunches can thus be attached, as shewn in the lower diagram, and the effect of pendent fruits and flowers alternating, with ivy or other border leaves, is extremely good. Three, four, or five bunches of Wistaria would be sufficient, but of Laburnum double the number may be used with advantage. In the lower part of the epergne the flowers or fruit should be arranged in an arched form, so as to show the shape of the bunch, and for this purpose it is desirable to make the foundation of clay higher in the centre than usual, or else to heap up moss into a curved surface. With Laburnum it is better to arrange a cluster of the flowers round the stem, and small bunches should fall from underneath the upper glass; but the branches of Wistaria are too large for this purpose, and ferns, or some pieces of ivy, can be substituted. If you are making a decoration of grapes, treat them in much the same manner as the Wistaria – placing three long pointed bunches of purple fruit alternately with three of green, the latter being the shortest; five or six smaller bunches of purple and green grapes will hang downwards from the upper glass, with more fruit above, vine leaves whenever they are wanted, and a vine-shoot round the glass stem. The shape of this decoration is shown in the frontispiece.[6]

Editorial Endnotes

1 'Brown Hanoverian rat is said to have annihilated the native black rat.' A joke about the change from traditional (French service) to Russian dining and derives from the

Protestant Hanoverian dynasty taking over from the (historically Catholic) Stuarts, as the ship the Hanoverians travelled in reputedly contained the brown or grey rat which decimated the natives.

2 "Still hug the chain" – conditions under which slaves were held captive.

3 Encoigneurs – 'Encoignure', furniture designed for corners of a room.

4 "Good gracious" people – A phrase coined by the head gardener and journalist Donald Beaton and seen, for example, in relation to the White Polyanthus, *Journal of Horticulture and Cottage Gardener* (21 May 1861).

5 Mr. John Standish nurseryman, of Bagshot – 'Messrs Standish & Co.', The Royal Nurseries, Ascot, Berkshire. John Standish was also known for the cultivation of a dessert apple.

6 'The shape of the decoration is shown in the frontispiece' – The frontispiece referred to is not included here. A version of the March frontispiece features instead the accompanying image to the Lindley report on 'Table decoration', *Gardeners' Chronicle* (1861).

52

UNSIGNED REPORT ON TABLE DECORATION

John Lindley

Source: John Lindley, unsigned report on table decoration, *Gardeners' Chronicle* (13 July 1861), pp. 646–647

IT is reported that the Council of the Royal Horticultural Society intend to renew next year, in some form, the prizes for TABLE DECORATION. It is as well therefore, before the peculiarities of the recent competition are forgotten, to place before the reader a faithful representation of the three combinations which obtained prizes on the 5th of June. In order that no error may have crept into these figures the art of the photographer has been called in; for he cannot add or omit anything, and thus we have an absolute certainty that the groups were neither more nor less than what is now shown – except, of course, as regards colour.

The merit of the skilful arrangements to which the first prize was awarded has been much discussed; and those who cannot perceive how superior graceful form is to mere heaps of colour, have not hesitated to condemn the judgment that conferred it. Surely the accompanying engravings will lead them to modify their opinion. To us it seems impossible not to see that, however handsome the other groups may have been, they were unmistakeably inferior to that produced by Mr. MARCH and his family. For how did he obtain his beautiful effect? Not by masses of flowers of priceless value, quite unfit for the purposes of a dinner table, to last for a few hours; but by some Ferns, some Lycopods, a few Roses, some Forget-Me-nots, Vine leaves, and small clusters of Grapes. Any body, with the most modest income, could imitate this. More resplendent decorations may be within the reach of the upper ten thousand, but of nobody else. The business of all great associations for the improvement of the social condition of the masses, and the Horticultural Society is nothing else, as its new name Royal implies, is to employ its means for the general benefit, and not for the advantage of a few; and if the taste displayed at the most modest board is improved that is in itself a considerable gain. Your Museums of Science and Art, with all their parliamentary grants, are for nothing else – in the intention of the public at least.

We need not describe the groups forming the subjects of the 2d and 3d prizes. With the clever combinations of colour which they possessed they were doubtless fine examples of table decoration. But it must be owned that with all its beauty, No. 2 was lumpish, and that No. 3 owed its effect to its stands, its costly Grapes and costlier Pine Apples.

DOI: 10.4324/9780429198809-61

Gardeners' Chronicle (13 July 1861) p. 647

53

'THE COMPETITIVE TABLE DECORATIONS AT THE ROYAL BOTANIC GARDENS, BY GASLIGHT'

'H.'

Source: 'H', 'The Competitive Table Decorations at the Royal Botanic Gardens, by Gaslight', *The Garden* (20 July 1872), p. 52

AT the recent horticultural show at Birmingham a darkened tent was prepared expressly for the exhibition of dinner-table decorations, as described in THE GARDEN of the 30th ult.; but at the Botanic Gardens, on the evening of July 11th, the table show was a genuine and legitimate night display. The decorations that obtained the first prize (as already known) were those exhibited by Mr. Buster; and, on the whole, they deserved that distinction. The central object was, as is so frequently the case, the very general tall vase of the trumpet form, at the mouth of which was a very elegant display of grasses, the feathery character of which is daily becoming more fully appreciated for decorative purposes. These were varied by maidenhair ferns of two or more kinds, as well as by some other slender-growing species, sparingly and lightly used, and relieved only by three or four sprays of a delicate small-flowered white Campanula; while a careless additional grace was added by a trailing spray or two of, I think, Maurandia. The base of the tall vase was surrounded by a series of semi-circular glass troughs, which formed a scalloped enclosure. These low troughs were filled with forget-me-nots, mingled with water Ranunculus and a few small ferns, very delicately and neatly used. The two secondary objects consisted of similar vases, of lower proportions, rising from their own low saucers or plateaus, the vases being furnished in a somewhat similar manner to the central one, but yet ingeniously varied; the saucers having sprays of white Campanula rising from ferns, richly but sparingly mingled with roses. The small vases with buttonhole bouquets were of the usual character, but very neat and elegant. On the whole, the general refinement of the arrangement of Mr. Buster's decorations was of a marked and very pleasing character.

The table which occupied the next place in excellence, in my estimation, was one decorated by the elegant taste and light artistic hand of Miss E. Harris, to which only a third prize was awarded. I was delighted with Miss Harris's composition; first, because it was not overdone; secondly, because it left broad spaces

DOI: 10.4324/9780429198809-62

of the "best damask cloth," which is the greatest and simplest glory of the dinner-table, undisturbed in snowy whiteness by extraneous objects; and, lastly, because the decorations, as thus kept within the strictest bounds of tasteful propriety, were in themselves light and elegant. The grasses and ferns of her principal vases were delicately assorted and displayed, and were simply varied by a few sprays of dwarf white harebell, and one or two other objects as chaste and simple; while in saucers beneath were roses – only a few – nestling among fresh, mossy-looking ferns, and partially concealed by a feathery veil of lacey grass-blooms. The general effect was full of elegant repose, which should always be aimed at, rather than that fussy, overdone profusion of decoration which is now so often seen.

A table decoration furnished by Miss E. Hassard, which obtained a second prize, was composed of six tall glass vases, from which rose graceful plumes of grasses, almost entirely unaided by the colour of flowers, being only just flecked here and there with a little starry blossom. At the bases were roses, white lilies, and geraniums, within a fringe of ferns. Altogether it was very pretty, and chiefly remarkable for the important part which common field grasses were made to play, with very elegant effect.

The adjoining table had a handsome silver epergne as a central object, well filled with the usual display of ferns, &c., the chief characteristic feature being the surrounding of the central epergne, and also the four corner ones, by favourite glass troughs, of pretty forms; these being filled with the best kinds of variegated geranium leaves, alternating with strongly-marked flowers, such as Pansies, blotched Pelargoniums, and Mimulus, with other kinds of flowers displaying two or more colours. The effect was original and striking, but perhaps wanting in repose. Mr. J. Mortlock's table, the vases and other ornaments of which were entirely of his white Dresden chinaware, must be considered (though well enough embellished with flowers) to be chiefly an exhibition of his elegant ceramic manufactures, which are remarkably well adapted for the successful display of floral decorations on the dinner-table. His gigantic dinner-table vase, furnished with palm leaves and a few large and striking flowers, looked remarkably handsome in the brilliantly lighted tent. Messrs. Daniels' table had a large mirror-plateau for a centre-piece, on which some small china swans looked very pretty with their forms reflected in the simulated lake, round which a border of water lilies did very successful duty. Messrs. Daniels can evidently exhibit very attractive table furniture. Messrs. Gardener's table, with a similar plateau, made much greater pretensions to importance. The swans, of real Dresden, were much larger, and of really artistic character, and instead of being placed on the mirror-lake, appeared as though forcing their way through the sedges and other foliage and flowers with which it was bordered. The vase of flowers rising from the mirror was of the usual kind. In two smaller plateaus of similar character, pretty Carrara groups of Venus and Cupid seemed floating in a graceful cockleshell on the mirror, surrounded by water lilies and suitable foliage. Small corner vases had Japan lilies, ferns, and drooping grasses of the great oat-grass kind, and in their saucers there were white sweet peas and delicate ferns. Altogether it was a very handsome table, but somewhat too profusely covered with decorations, which is a defect easily remedied on a future occasion. Mr. Charles Wood (floral artist) added a special feature to his floral decorations; richly plumaged birds being made to flutter round his table lamps, among the flowers with which they were wreathed; while he had for a centre-piece what he termed a "paradise fountain," the crowning feature of which was a bird of paradise, with spread wings drooping round, like a shower of pale amber, towards the flowers below. Messrs. Ward, also, exhibited "ornithological table decorations," as they style them, in which lovely birds nestled among the flowers in very picturesque fashion. But such novelties are, after all, of very questionable taste, and though pretty, and likely to be fashionable for a time, will scarcely retain a permanent place as legitimate dinner-table decorations. Miss Hassard obtained a prize for a table-decoration in which two large pots of growing ferns were the chief features, though the clay pots were entirely undecorated; for a dinner display such articles of the commonest kind of clay-ware ought surely to be clad in some kind of evening dress. The arrangement of table fruits by Miss Harris looked pretty and tempting in the glare of bright light, and the composition deserved the prize which it obtained. I have thought it worth while to notice these table decorations at some length, because they had previously been criticised by daylight, while their carefully prepared effects are intended to be seen, as a rule, by artificial light only.

H.

54

'ARTIFICIAL LIGHTING OF DINING-ROOMS'

Annie Hassard

Source: Annie Hassard, 'Artificial Lighting of Dining-rooms', 'Plants Placed through Dinner Tables', *The Garden* (18 July 1874), pp. 55–56

[Miss Hassard, so well known for her charming dinner-table and other floral decorations at our great flower shows, and who has so often contributed to The Garden articles embodying her experiences in that way, has, we are glad to say, acceded to our request that she would re-describe them in consecutive order. In the articles, of which this is the first, the whole question of table decorations will, therefore, be dealt with in detail. – Ed.]

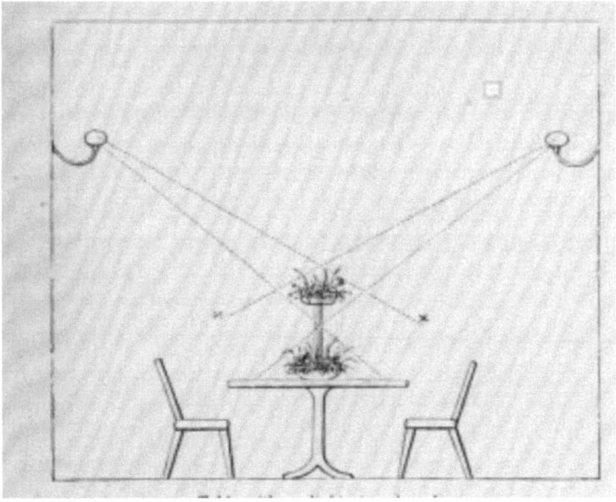

Table with gaslight at each end.

The selection of stands for the decoration of the dinner-table must depend (as has been well shown by Mr. W. Thomson, in The Garden, whose illustrations we reproduce) on the manner in which the room is illuminated by means of artificial

 DOI: 10.4324/9780429198809-63

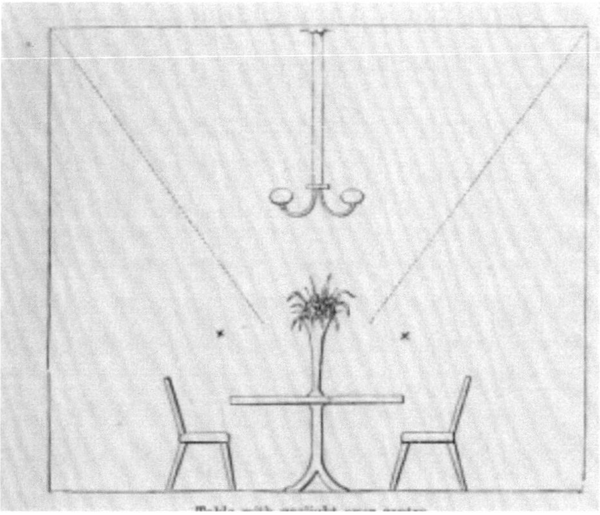

Table with gaslight over centre.

light. Generally speaking, there are but three ways of lighting rooms with gas or wax lights, viz., lights placed on the table, bracket or wall lights, and the chandelier or gasalier. The style of stand or stands selected for the table wholly depends upon which of these modes of lighting is adopted. For instance, if the room be lighted by two gas lights or chandeliers, one at each end of the table, a March stand should be chosen; but if illuminated by a gasalier placed over the centre of the table, the selection of such a stand would be a mistake, for, the light being above the stand, though the flowers arranged in the top tazza would be well shown up, those in the lower one would be put into the shade; the top tazza would catch all the illuminating rays, and keep them off the lower one as effectively as if a screen had been placed over it. By such an arrangement as that shown in the accompanying illustration, each tier has its proper share of light. Where the gasalier is placed over the table, the best form to use is a high single slender trumpet, or some such shape, so that when the light happens to fall on it, it will not throw any shadow on the arrangement which may be placed round the base of the vase. When the light is placed over the centre, flat arrangements can be used with advantage, though for my own part, where possible, I give preference to something high in the centre. Undoubtedly the best stand for a room thus lighted is the trumpet, or some modification of it – such a stand, in short, as that represented by the accompanying illustration. When about to purchase new stands, these remarks should be borne in mind, and also when arranging pot plants on the dinner-table; for the mode of lighting affects them quite as much as stands. If the lighting is not taken into consideration, large Fern fronds and similar foliage would throw heavy shadows on the table-cloth, which would spoil the look of any arrangement in their vicinity.

Plants Placed Through Dinner-tables.

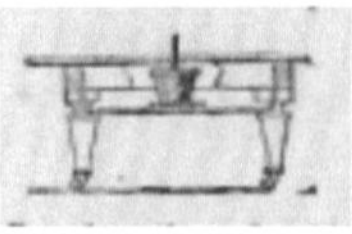

Pot plant through table.

Plant through table arranged with Ferns, &c.

Many will doubtless be ready to exclaim, "Who would cut holes in one's dinner-table or table-cloth for such a purpose?" Well, allow me to explain how this can be effected without damaging either the table or cloth. With the exception of those who have seen suitable plants put through a table, no one can have an idea of the elegant effect which they produce when arranged in that way. Imagine the striking effect which young Tree Ferns or Palms have in such positions; their elevated fronds shading from the blaze and glare of light the smaller arrangements of flowers and fruits laid here and there on the snowy damask below them, a result which cannot be produced unless the plants are put through the table. This may be done in several ways. Dinner-tables are generally of two kinds – the one the telescope, into which may be put as many or as few spare leaves as may be desirable; the other that with a large centre and spare ends, which may be added at pleasure. There are two ways of putting the plants through these tables. Now-a-days dinner-tables are never, I need hardly say, uncovered, therefore it is of no consequence of what material the top of the table is made, provided one leaf, reserved for a plant in the centre, or two if for a plant at either end, be made of common deal, either the same size as the mahogany one for which it is substituted, or only a foot or 18 inches wide, if the table is required to be made shorter. This spare leaf should be cut in two, and a half-circular piece cut out of each piece, which, when replaced in the table, will leave a circular hole in the centre. Before this deal leaf is lifted into the space from which the mahogany one has been removed, a wooden box, or anything which will form a stand for the plant, should be put into proper position; the plant should then be placed on it and carefully raised to a proper height, keeping always the surface of the pots just below the level of the table. Each half of the spare leaf may then be fitted into its place, when the stem of the plant will occupy the centre of the circular hole cut for it. If there is a plant to be placed at each end of the table, the best way of supporting them is to place a long plank underneath. Should Ferns be used, the holes cut out of the spare leaves must be of the same size as the top of the pots, as, in the case of such plants, the pots must be flush with the table. The other plan of arranging plants in this way is to have a small piece of brass or iron that may be drawn across and fastened when the table is within an inch or an inch and a half of closing, which will prevent its either closing or opening. This form is suitable only for Palms, and similar plants. It is in this way the table with extra ends must be managed, and about which there is no difficulty; for I constantly practice it. Let us now turn to the covering of the table. The way the table-cloths (for it takes two to each table when so decorated) are arranged, is as follows: – Each cloth is opened and spread along each side of the table, so as to meet in the centre or lap over each other about an inch. They should be kept as flat round the base of the plants as possible, and the cloths should be pinned together so as to prevent them from being open. A clean flat-iron should next be obtained with which the folds should be neatly pressed, so as to obliterate any appearance of a join. If half the width of the cloth is not found enough to allow it to droop gracefully at each side allow more; but always keep the middle fold flatly pressed, a trouble which will be well repaid by the effect produced. My tables both at South Kensington and Birmingham had plants put through them. At the former I had a graceful pair of Pteris tremula, and at the latter a pair of Chamædoreas. The accompanying illustration represents a plant of white Bouvardia, with Ferns and scarlet Pelargoniums arranged round the base. All plants put through tables should have some similar kind of arrangement round the base, otherwise they would look unfinished, and, indeed, anything but ornamental.

55

FLORAL DESIGNS FOR THE TABLE

(Excerpts)

John Perkins

Source: John Perkins, 'Introduction', 'Breakfast or Luncheon Table', 'Dinner Table', *Floral Designs for the Table: Being Directions for its Ornamentation with Leaves, Flowers and Fruit . . .* (1877), pp. 9–10, 24, 31

INTRODUCTION

I THINK it was Cowper who said, "He who tells a long story, should take care that it is not made a long story by his manner of telling it. His expression should be natural, and his method clear; the incidents should be interrupted by very few reflections, and parentheses should be entirely discarded." My story would be a long one were I to recapitulate all my experience in Table Decorations during the past twenty-nine years, but I will endeavour to tell it as shortly as possible, and my method shall be as clear as I can make it.

In this book-making age, writers generally give some reasons for their publications. I have two: one is, I think every man should contribute his mite for the benefit of his fellows; the other is, that I think a work on this subject, showing at a glance how a Dinner Table for from eight up to one hundred persons should be decorated, will be welcomed not only by those who are called upon to embellish the tables of the upper classes of society, but also by that important and numerous class who are in the habit of giving entertainments which aim to be distinguished by their simplicity and elegance, rather than their mere costliness.

Whilst I do not presume to teach those who may possibly know more than myself, still I am sanguine that this work will give some valuable hints to others who have but little knowledge and experience of this kind of decoration. Personally, I have never seen or even heard of any work like the present; it is, therefore, superfluous to add that nothing has been culled from other sources.

Most ladies love flowers and take a great interest in Floral Decoration, and by universal consent in matters of taste and refinement, none are better judges. The talented Lady who has kindly permitted her name to be associated with this work stands in the foremost rank as regards taste in dinner table and other decorations.

 DOI: 10.4324/9780429198809-64

The Dinner "à la Russe" has, within the last ten or fifteen years, become so fashionable and popular that to have a dinner in these days without the accompaniment of Cut Flowers, Plants, Ornamental Leaves, &c., would be like dessert without wine. Some years ago shallow receptacles made of zinc or glass in various shapes were much used. This was a step in the right direction, by bringing the flowers, &c., more directly under the eye. These trough-like-looking vessels were filled with sand, and a border of juniper shoots, ivy and other leaves, place outside, and the centres filled up with flowers. These after a time gave way to a more elegant and effective style of decoration, a style, venture to say, which will hold its own for many years to come. This plan is to have the principal part of the decoration on the tablecloth itself.

A flower garden, when looked upon from some distance above its own elevation, is far more pleasing to the eye than otherwise; so it is with table decoration. Some may perhaps think that laying the leaves, flowers, &c., on the cloth might soil the damask; but with proper care on the part of the gardener or person to whom the work of decorating the table is entrusted, not the slightest disfigurement need be made. Others may exclaim, What a waste of flowers! But here again it is not by the large quantity of flowers, but by "the way" in which they are arranged that the effect is produced. A few leaves and flowers elegantly combined are far more effective and pleasing than when put on the cloth in heaps, regardless of form or colour. Lightness and elegance should be the motto of every dinner-table decorator, and the view across the table should not be obstructed more than is absolutely necessary.

After the decorator has tried some of the designs which this work contains, he will soon find out that the decoration of a table is more easily done than he might imagine, and he will readily admit the superiority of the work when well done by this method to the plan formerly adopted with the old zinc trough.

I may mention that in large establishments about five designs will be required for each week, and it will of course be advisable to have them as different in character as possible each night. The materials for the one decided upon may often be gathered two or three hours before wanted, and well damped with clean water (flowers excepted). Fern fronds, lycopodiums, and such things, are best immersed in water for a few minutes to cleanse them from all grit, &c., and vine leaves will often require sponging: when all are ready, they should be placed carefully in a flat basket or box, and be put in a cool place until wanted. About half an hour or an hour before the time for dinner – according to the size of the table to be decorated – operations should be commenced. It is necessary to see that all lamps, plants, dishes, glasses, spoons, knives, forks, &c., are first properly placed on the table, as any irregularities in this matter would probably upset the whole arrangements.

When all is ready, the centre of the table should, if possible, be done first, finishing off at the outsides. Each piece of fern, ivy, &c., should have a slight shake to relieve it of any superfluous water which might otherwise remain about it.

As each design will speak for itself, and a brief description is given of the materials employed, it will be only necessary to state that after the company have left the dining-room, the groom of the chambers or the under butler should place any fern fronds, leaves, or anything else worth preserving, into water, whereby many useful pieces may be saved, and be made available for other purposes the following day.

I may here observe that those who prefer to have fountains on the table, with jets of water playing at dinner-time, have only to consult the persons who fix such contrivances; and those who wish to have pots of plants sunk through the table can have extra leaves made of common deal, in which circular holes can be pierced at the required distances.

In concluding these introductory remarks, I would add that I do not presume to have fully exhausted this interesting subject, but I hope that I have succeeded in adding "one more link to the chain," and therefore beg the forbearance of those critics who may think the volume incomplete.

JOHN PERKINS.

THE GARDENS, THORNHAM HALL, SUFFOLK.

October, 1877.

XI.

BREAKFAST OR LUNCHEON TABLE.

IF the materials are not laid on too profusely, and with ordinary care, this Table has a very pretty effect. The Wreaths round Nos. 1 and 5 are of Myosotis (Forget-Me-Not). Dishes Nos. 2 are surrounded with the scarlet-flowering Cratægus; Nos. 3 with borders of Apple-Blossom; and White Roses are placed round Nos. 4. Nos. 1 and 5 are places for plants; and Nos. 2 are Low Dishes to be filled with Mixed Cut Flowers prettily arranged. Nos. 6 are of the Lycopodium apoda: this should be grown in Saucers for the purpose. Nos. 4 are intended for Fruit; and Nos. 3 to contain Mixed Cut Flowers.

PLATE 11.

XVIII.

Dinner Table.

THIS Design, when carefully worked out and displayed under a good light, is very effective. Round each Lamp, Nos. 1, are placed Six Fern Fronds, Pteris aquilina, or any similar Fern, with a straight piece of Amaranthus caudatus (Love-lies-bleeding) down the centre of each frond. Nos. 2 may be Gold or Silver Cups; Nos. 3 are intended for Fruit; Nos. 4, Confectionery; and Nos. 5 indicate Plants. The Outside Wreaths are composed of Fuchsia (Sunray), and flowers of the same, placed upright, about four inches apart. In this, as in all the other Designs, when the lamps are suspended from the ceiling, Plants or Cut Flowers may take their place on the Table.

PLATE 18.

56

'THE ARRANGEMENT OF CUT FLOWERS'

Edward Luckhurst

Source: Edward Luckhurst, 'The Arrangement of Cut Flowers', *Journal of Horticulture* (15 September 1881), pp. 237–238 and (22 September 1881), pp. 270–271

IT is quite an exception to my ordinary experience to find a gardener able to arrange cut flowers really well," was the extraordinary expression that fell quite recently from the lips of a lady whose exquisite taste in the treatment of cut flowers is indisputable, who, moreover, is the owner of a large garden, and whose experience of the doings and capabilities of gardeners is sufficient to impart great weight to words not lightly spoken. A feeling of regret to which they gave rise prompts the inquiry, Is this so? and if it is, What can be done to effect a general improvement in a matter of such importance?

That the prizes given at flower shows for dinner-table decorations and for bouquets have done something in this direction is undoubtedly true, but ought we to rest content with this? The inspection of finished work is instructive by conveying a general idea of how it should look when done, but how to do it is the question to which the sight of it so frequently gives rise. "Can I ever attain to such excellence as this?" is the unspoken thought of many a puzzled but aspiring beholder. It would be misleading to assert that it is possible to render proficient all or even many of those who aspire to the mastery of an art for which, just as in landscape gardening, there must be a natural taste – an intuition that readily adapts itself to the refining influence of artistic training, and originates and brings to perfection conceptions of its own different from others, and yet devoid of all that offends the highest taste and culture.

Let not those be discouraged, however, who are sensible of the non-possession of such natural gifts, and who can never hope to achieve great things in this direction, for all flowers are so beautiful and lend themselves so readily to graceful combinations, that a knowledge of what colours may best be used together, and when, with a few lessons – actual practical demonstration by an accomplished artist – would suffice for ordinary purposes. But where are such lessons to be had? Machinery in motion, working dairies, bee-driving, and many similar methods of popular education are established facts at public exhibitions, but floral decoration

 DOI: 10.4324/9780429198809-65

has advanced no further than a mere exhibition of finished work. If the managers of horticultural societies wish to add a feature to future exhibitions that will attract the public and afford greater gratification than anything that has yet been tried, let them secure the services of a good floral decorator – a man who cannot blunder at his work, of fluent speech and polished address – a lecturer, in point of fact, and let him have a large table with vases and flowers of his own selection, so placed in an enclosure that as many persons as possible may look on. Then let him show how cut flowers can be made to look best in vases and other vessels suitable for sitting-rooms, corridors, and wherever they are required, in a variety of ways for dinner-tables, and also in hand and buttonhole bouquets, explaining the reason of all he does, and answering the very numerous questions with which he is quite certain to be assailed. It may very safely be predicted that this would be a step onwards so highly appreciated that the lecturer and not the public would tire first. It would, moreover, tend more than anything else could do to instruct and aid many a willing anxious learner, rendering his duties in floral arrangement easier to himself and satisfactory to his employers.

Tasteful combinations of flowers and foliage that harmonise not only in colour but in form, are our chief aim in this work. Avoid mixtures of gaudy colours: if scarlet is the chief colour use no yellow shades, but rather choose white for contrast. For example, just now the bold scarlet flowers of Vallota are plentiful, so also are the lovely white flowers of Anemone Honorine Jobert; try them together, using the Vallota blooms singly for a groundwork, out of which springs the white Anemone, also single, yet with the flowers gracefully clustering. This is a charming mixture for a dinner-table, and serves to illustrate my meaning both in its application to form and colour. In so treating a Marsh stand lately the Vallota flowers in the bottom saucer had a bold fringe of Fern fronds, a few small sprays of Maidenhair Fern rose among the mingled white and scarlet flowers, and around the glass stem was twined a spray of Selaginella cæsia with the lovely bluish metallic hue well developed, partly hidden by tufts of half-opened Anemones, dried Quaking Grass, and a couple of leaves of Geranium pratense which clustered around the base. The same flowers were repeated in the tops with the addition of bold clusters of scarlet-flowered Begonia fuchsioides. Had this stand been for a daylight decoration some spikes of single blue branching Larkspur might have been used with excellent effect. But blue never answers for lamp light, and is never used because it looks black, just as yellow becomes an unsightly shade of white under artificial light, and is, therefore, always avoided for dinner-table work.

Before turning from the blue flowers a recent success with some may be noteworthy. A large glass fish bowl had its edge boldly but not heavily fringed with Fern fronds, through and between which the clear, flashing, transparent water was visible. Upon the Ferns rested broad circles of Gloire de Dijon Roses in clusters; between the Rose clusters the white flower spikes of Ligustrum japonicum projected irregularly, two or three being made to droop downwards upon the green fringe, and the centre was entirely filled with spikes of single branching

Larkspur, principally of blue shades, with a mixture of pink, a few curved deep blue spikes bending charmingly over the Roses, while a dozen or two clusters of dried Quaking Grass (Briza minima), springing irregularly out of the central flowers imparted a pleasing air of sprightliness and finish to the whole. This was for a daylight repast.

Pink and carmine are excellent colours for the groundwork of lamp-light decoration, and it should be added they are equally effective by day. Coleus Sensation from the rich carmine centres of its foliage answers admirably for the dinner-table; Maidenhair Fern, Anemone Honorine Jobert, and white Begonia flowers being all that are required with it to form a chaste combination that lights up well. Some shades of pink are not liked so well as others, the deep purplish-pink of Anemone japonica being avoided for brighter shades, such as we have in the flowers of the old variegated bedding Pelargonium Manglesii. Insignificant as these are in a bed, yet they are of so much use for all arrangements of cut flowers as to be eagerly sought after whenever they are in bloom. Nothing can be more lovely than a groundwork of pure white Roses with Pelargonium Manglesii flowers mingled with and springing out of it. For the dinner-table it, perhaps, is most liked in tall slender glasses managed in this way: Each glass first has four sprays of Selaginella cæsia put in so as to be gracefully pendant, and most of it long enough to rest upon the tablecloth, then come six or eight of the small pink Pelargonium trusses, two or three very choice little fronds of Maidenhair, and half a dozen clusters of dried Briza. Sufficient slender glasses must be so dressed to form a double chain around a cup or whatever is used for the centre of the table. Care is taken not to crowd the glasses, and the pendant chain is formed by twining two sprays of Selaginella together, so as to gracefully connect the glasses. The last time I so used them was for a small table, and only about a dozen glasses were required, but above them five small transparent fish-bowls were suspended by very fine wire from a chandelier, four of them forming a circle, with the fifth suspended a little lower from the centre. Much longer streamers of Selaginella were used in the bowls, a pendant chain being made from bowl to bowl just as from the glasses below, only it was much bolder in character. White Fuchsias with bright pink corollas were used with Maidenhair Fern in the four outer bowls, and another Fuchsia with large rosy pink flowers in the central one. Briza maxima was used in the suspended globes instead of the smaller variety, to give due importance to the central group.

Of white flowers there are many which are found most useful whenever they are to be had. Lily of the Valley is so exquisitely graceful that it would seem impossible to arrange it badly, and yet it may sometimes be seen woefully crowded. Roman Hyacinths, too, are very lovely and answer admirably for slender glasses. Stephanotis reigns supreme from spring till autumn, and may be used in a variety of ways. What can be more suitable for a boudoir-table than a basketful of its sweet-scented flowers mingled with delicate Fern fronds? Not long ago I used it for a small dinner-table successfully in a somewhat novel manner. Three small transparent glass bowls were put in the form of a triangle upon a flat stand covered with crimson plush, a fourth glass was placed on the top of the others in the centre, and then they were dressed with Selaginella cæsia, Stephanotis, and dried spikes of Briza minima. Four flowers only of Stephanotis were required for each of the three lower glasses, and six flowers for the upper one. Four of these little groups were prepared, each upon its crimson stand, and they were so placed upon the table that the Selaginella sprays met upon the cloth. The effect was very chaste and lovely.

THE ARRANGEMENT OF CUT FLOWERS.

In all large establishments where floral decoration is valued there should be a "flower room" especially devoted to the work. It should contain a table close to the window, another large table or two, a sink with a water tap, a large lofty cupboard with plenty of broad shelves, and a broad shelf or two along any convenient part of the walls. The window table should have a drawer for scissors, knives, wire, string, dusters, cotton wool, a packet of gum arabic, and a bundle of flower sticks. Some white sand, charcoal, and moss should also be kept in boxes under one of the tables, care being taken to pass the sand through a fine sieve before it is put in the box, and there should be a large iron spoon for placing the sand into the flower stands. A small piece of charcoal is placed in every vase to keep the water sweet. A single drop of gum dissolved in water and allowed to fall carefully in the centre of a flower prevents its shedding its petals; it is not often that this has to be done, but it is worth knowing when flowers become scarce. A set of water jugs, a very small waterpot with a finely perforated rose for moistening sand, a sponge and set of brushes for cleaning vases, and a clock, renders our list of the flower-room furniture complete.

Flower stands, vases, and china flower pots are kept in the cupboard and upon the wall shelves. Perhaps there are few things more difficult than the selection of these indispensable adjuncts to floral decoration by the uninitiated; so varied are the forms and so charming the colours set forth in tempting array at the china shop, that even experienced hands are not unfrequently led to purchase articles that prove unsuitable for a given purpose. Coloured flower stands should if possible always be avoided, for if they are used flowers must be chosen to harmonise with them, and so our selection would always be unduly restricted. White stands are of course admissible; but the most useful are undoubtedly those of transparent glass, which, although colourless, yet when filled with clear water impart an air of brightness that mingles most pleasantly with the varied charms of the flowers. Of all forms fish globes are the most useful for the dinner table, and the flower-room cupboard should contain a complete set of them, consisting of a dozen of each of the larger sizes and five or six dozen of each of the smaller ones downward to those that will not contain more water than a teacup. A single globe of the largest size answers admirably for the centre of a circular table, just as three or more do for a long table. The smaller sizes all prove useful in turn, being used either singly in circles around the large ones, or in groups in a variety of ways. My meaning will perhaps be most readily understood by those whose resources are frequently severely taxed to impart a distinct character to a series of consecutive table decorations. A considerable number of circular wooden blocks of various thicknesses and diameters to suit the different globes, all of them covered with dark crimson plush, are used to elevate the globes above the tablecloth, salvers of glass mirrors set in silver being sometimes used for the large central globes.

Of other table stands what are known as Marsh stands are still used in the plain forms in which they were originally brought before the public. Various modifications have found favour according to the promptings of individual taste or fancy; a favourite one of medium size, consisting of a bottom saucer for sand, from the centre of which springs a single water vase, is very useful. Glass baskets are also useful; so, too, are the small slender glass vases suitable for a single Rose, or two or three trusses of smaller flowers, of which mention was made on page 237. They are very elegant and are usually frosted. Iridescent and crackled glass vases are very beautiful – so beautiful that they are usually found in sitting-rooms, but there should be one or two sets for dinner-table work, and they are not expensive. For corridors, vestibules, landings of wide roomy stairs, and various nooks and corners, gigantic china vases 2 or 3 feet high are best for standing upon the floor; but if there are tables or brackets, then there is nothing better than large china bowls filled if possible with fragrant flowers. What is more liked than a bowl of Sweet Pea blossom, or Mignonette, Honeysuckle, Lily of the Valley, Harebell, Roses, Violets, or Primroses? All are welcomed in their seasons. Depend upon it, it is not the costly Orchid or other rare flowers requiring expensive glass structures for their culture that are most often in favour even in a palace. There is a craving for the "old-fashioned flowers" of our childhood, and only a selection of those of the most easy culture are needful to carry on the work of decoration well. Of Roses we cannot have too many, especially of Moss Roses, Damask, Cabbage, Gloire de Dijon, and Maréchal Niel. Let there be plenty of Sweet Peas – not small clumps, but long rows in succession throughout summer and autumn. I have now two rows, one on each side of a path some 200 feet long, just bursting into bloom, forming an avenue of sweetness and affording a bountiful supply of cut flowers. Anemones, Zinnias, Asters, Scabious, pink and white Mallow, Love-in-a-Mist, pink and white Canterbury Bells, Columbines, Cornflowers, Larkspurs, Lupins, Pinks, Carnations, Picotees, Myosotis are all highly valued for the rich supply of cut flowers which they afford, every one of which can be turned to account for decorative purposes. – EDWARD LUCKHURST.

57

'A CHEAP TABLE DECORATION'

Shirley Hibberd

Source: Shirley Hibberd, 'A Cheap Table Decoration', *Amateur Gardening* (8 May 1884), p. 6

A Cheap Table Decoration is here figured for the delight of many who love changes in such things, and have no particular taste for toys that cost more than they are worth. Messrs. Stevens and Williams, of the Brierly Hill Glass Works

(2, Old Bailey, London, E.C.), have lately brought out a series of vases and jardinets in coloured glass, fitted with moveable Minton pots, in which flowers or plants may be placed, with a view to their lasting long in a perfect condition, while there need be no slopping of water about, to the injury of the furniture of the chamber decorated. The jardiniere figured is fashioned of coloured glass, but the colour does not "kill" the flowers if a reasonable amount of foliage is employed. As the reader may perceive at a glance, the furnishing of a decoration of this kind is a matter of taste and means; it is a mere holder, but being made of glass, it may be placed anywhere without fear of communicating damp to the cloth or table it rests upon. Very many people fail to keep flowers in a receptacle of this kind, and they ask about the use of charcoal and daily renewal of the water and so forth. If they will permit us to give them a word of advice worth a thousand pounds, they shall have it for nothing, or for some infinitesimal part of a penny. Fit up a thing of this kind to your own taste and supply the flowers with as much water as the thing will hold without running over. Then, *do nothing more*. Do not disturb the affair in any way until its beauty is past and gone, when, as a matter of course, the flowers may be turned out and the glass have a cleaning and be again furnished with another display. You want no charcoal, no change of water, no clipping of stems, "no nothing." Be at peace. You have built your flowery castle; let it stand until it falls to pieces. It will last longer and keep its looks better by this trustful plan than by any amount of fidgetting and managing within the compass of human invention. Should a philosopher appear and say this must be done, and that must be done, sweetly decoy him away give him a comfortable seat at the dinner table and discourage all talk about table flowers.

d)

LANGUAGE OF FLOWERS, WAX MODELLING AND LEAF PRINTS

Editorial Headnote

58. Henry Phillips, 'Preface', 'Introduction', *Floral Emblems* (1825), [v–xvi], excerpt, pp. v–ix, [1–26], 23–26
59. Jules Lachaume, pt. 2.,'Langage Emblématique des Fleurs' ('Emblematic Language of Flowers), *Les Fleurs Naturelles* (1847), pp. 79–89
60. Ann Pratt and Thomas Miller, 'Preface', 'A', *The Language of Flowers, the Association of Flowers, Popular Tales of Flowers* (1870s), pp. 5–6, 15–16
61. Sarah Carter Edgarton Mayo, 'Preface', 'Key to the Fortune Teller', 'White Flowers', *The Floral Fortune-Teller: a Game for the Season of Flowers* (1847), pp. iii–vi, 8–22
62. Emma Peachey, ('The Editor', 'Preface'), 'Roses – Instructions', *The Royal Guide to Wax Flower Modelling* (1851), pp. ix–xiv, 4–8, 30–35
63. Annie M. Williams, 'Introductory Remarks', 'To Make Leaves and Stems for Wax Flowers', 'Arrangement of Flowers for Baskets, Vases, etc.', 'To Make Stamens, Pistils etc., for Wax Flowers', *Wax-Flower Modelling Made Easy* (1871), pp. 1–12, 17–25
64. Edward Parrish, 'The Leaf a Type of the Tree', 'How and What to Collect', 'Maceration', 'Seed-vessels', 'Bleaching', 'Mounting the Specimens', *The Phantom Bouquet . . . on the art of skeletonising leaves . . . and adapting them to embellish homes of taste* (1862), pp. 13–24, 30–41
65. James F. Robinson, 'Leaf Printing from Nature', *The Garden* (29 November 1873), pp. 432–433
66. F. W. Burbidge, 'How to Skeletonise Leaves and Seed-vessels for Winter Decorations', *Domestic Floriculture* (1874), pp. 195–199

During the 19th century, a number of experts on the language of flowers and floral decoration were men; these areas of knowledge, however, also provided middle-class women with a respectable space of work and leisure in terms of authorship and in the creation of the variety of media representing plants and flowers that were particular to the period.

The sale of books on the language of flowers was a publishing phenomenon which took place over several decades of the 19th century, initially within Europe and becoming a transatlantic trade from the 1850s. The 'language' they describe is rather a vocabulary of flowers, matched to meanings or associations that varied considerably. The main influence on English language writers was Charlotte de la Tour, whose *Language des Fleurs* was first published in 1819. It did not appear in translation until 1834 but it had already been partially absorbed into Henry Phillips' *Floral Emblems* in 1825. Phillips (1779–1840) was a well-established horticultural writer with scientific credentials, known for his landscape gardening in Brighton when it was rapidly expanding. He toned down de la Tour's references to religion and any double entendre and directly translated almost half of her 300 definitions. He lists Scottish and Gaelic flower names and their meanings alongside the English (pp. 13–16 not included) and acknowledges the 'Eastern' origins of the language to re-figure them within an English or Western poetic tradition; a process of forging national identity through history that is evident in subsequent writing on the subject.

In France, Jules Lachaume, *Les Fleurs Naturelles* (1847), acknowledges the diverse sources of the language of flowers and, like Phillips, aims to simplify. His interpretation of named flowers beginning with the letter 'A' is, in many cases, identical to the meanings attached in *The Language of Flowers* (1870s) by Ann Pratt (1806–1893), botanical artist and author (see Volume 2), and Thomas Miller (1807–74), a poet and writer on rural matters. There are differences; the aloe, for example, means 'caquet' or 'cackle' (with the implication of a 'big mouth', 'prattler', or 'gossip') according to Lachaume, while Pratt and Miller suggest affliction or grief. Their list of plants is considerably longer than Lachaumes', partly because they include fruits and trees and partly because of the 'discovery' by Europeans of additional plants from beyond the continent – the azalea, for example.

In the US, a number of women authored books on the language of flowers. *The Floral Fortune-Teller* (1847) was the last published work by author and editor Sarah Carter Edgarton Mayo (1819–1848). Her book's commercial success is suggested by the publication of a second edition in 1855. Within its pages she flaunts her literary knowledge in the form of a game to be played with a bouquet of flowers. The numerous quotes range from the plays of Shakespeare (*Othello*, *Hamlet*, *Julius Caesar*, *Twelfth Night*) to the poetry of Wordsworth (*Michael, A Pastoral Poem*) to Spenser's *Faerie Queen*, amongst others.

Emma Peachey, *The Royal Guide to Wax Flower Modelling* (1851), included the literary 'meaning' of each flower to be modelled. The diverse strands of the history of wax-modelling in the reproduction of flowers for religious purposes, as an activity for elite women, and in scientific display, as seen in the reproductions of the human body and for the identification of plants, came together in 19th-century Britain (and in the US from the 1850s). Peachey (circa 1807–1875), who took a commercial risk in publishing her book herself, appears to have taken the precaution of employing an editor (male, identity unknown) to introduce her *Guide*. Peachey, the daughter of an army surgeon, outlines her products and methods in her introduction. In the excerpt she emphasises the femininity and safety of her particular art and its educational or scientific value. She contests claims, made in an article from the *Manchester Examiner,* that wax modelling was dangerous; the article was widely circulated in the press throughout Britain and parts of the Empire.

Two decades later, religious and literary quotations and the language of flowers are absent from Annie M. Williams's *Wax-Flower Modelling Made Easy* (1871). Williams (1848–1922), who married Henry Carter in 1891, was the daughter of B. S. Williams, nurseryman and florist; advertisements for his business predominate in her small-sized publication. Both she and Peachey offer botanical terms for the structure of plants. The protection of wax flowers under glass (as with living plants) which had been observed since the 1820s features in the single, coloured frontispiece.

A Phantom Bouquet was a funerary souvenir made out of wax flowers and is referenced in the title of *The Phantom Bouquet . . . on the art of skeletonising leaves . . .* (1862) by Edward Parrish. The predominance of black in funereal settings is echoed in images within the publication. Parrish (1822–1872), a chemist, dedicated the book to his wife who produced several of the prints. The illustration included here is credited to 'August Wilhem' (n/d). To skeletonise a leaf, the pulp is removed from leaves through 'bleaching', a process that was first practised by naturalists and botanists to investigate leaf structure. The resulting prints were kept in scrap books or framed and displayed.

More than ten years after the publication of the Parrish pamphlet in the US, photographer James F. Robinson (n/d), who was based in Dublin and was a member of the Dublin Photographic Society, contributed an article titled 'Leaf printing from Nature' to *The Garden* (1873). He briefly discusses leaf printing before elaborating on a series of photographic techniques. His article was reprinted almost verbatim in the first edition of William Burbidge's *Domestic Floriculture* (1874) and in Annie Hassard's *Floral Decoration for the dwelling house* (1875) (see IIb). Both authors thank William Robinson, the editor of *The Garden*, for allowing them to use the article, but they do not credit the original author. Burbidge (1847–1905), who trained at the gardens of the Royal Horticultural Society and at Kew Gardens, was a skilled draughtsman and a member of staff at the *Garden* until 1877. In the source included from the second 'enlarged and revised' edition of *Domestic Floriculture* (1875), the passages on photographic processes have been removed and an alternative account of leaf printing – 'skeletonisation' – was inserted and credited to Dr. Dickson, a medical man based in Edinburgh. The presence of the original account (by James Robinson) of different photographic processes in Hassard's book (not included here) suggests that women were active in this field, as exemplified by Anne Atkins' use of the cyanotype in *Photographs of British and Foreign Algae* (1843–1853). The quantity and nature of some of the chemical substances required for photographic printing, including silver nitrate and copper sulphate, implied a recklessness in regards to personal safety comparable to the risks taken by 19th-century gardeners in general (see Volume 4).

Further Reading

Bea Howe, *Antiques from the Victorian Home* (1973)

Ann B. Shteir, '"Fac-Similes of Nature": Victorian Wax Flower Modelling', *Victorian Literature and Culture*, vol. 35 (2007)

John Whiteknight, 'Capturing an era under glass', *19th century*, vol. 33, no. 2 (2013), pp. 28–35

58

FLORAL EMBLEMS

(Excerpts)

Henry Phillips

Source: Henry Phillips, 'Preface', 'Introduction', *Floral Emblems* (1825), [v–xvi], excerpt, pp. v–ix, [1–26] excerpt, 23–26

PREFACE.

In forming this collection of Floral Emblems, the first care was to avoid perplexity, by creating new symbols, with such flowers as have been previously used in the hieroglyphics of the ancients or described in the verses of the poets.

Therefore, when the compiler found the same plant made to represent more than one design, he selected the emblem of the greatest antiquity, or the one most established by custom, and in some instances he has noticed their various interpretations.

Amongst the numerous species of plants which Europe has received from China, America, and New Holland, but few have hitherto found their way into the language of flowers, and some of these he has ventured to present, as emblematical of such characters or passions, as appeared most appropriate.

The numerical emblems, as well as those for the days of the week, and the months of the year, were devised by him in order to render the work more complete, so as to enable the emblematists to record dates by floral symbols.

As the invention of symbolical language appears to have originated in eastern nations, the author's endeavours have been to collect the emblems of those countries, through the aid of their poets, and the assistance of various travellers.

The most ancient works on heraldry have been consulted as well as the sculpture of classical nations and the paintings of modern ages. The assistance of the poets will be seen in every page, and it would be unjust not to mention the author's obligation to Madame la Comptesse de Genlis, to Madame de Latour[1] and also to the author of "*Parterre de Flore*." In this symbolical assemblage the author has carefully avoided all indelicate allusions or *double-entendre* that could be offensive to modesty, his object has been to establish a settled collection of floral emblems, and to render them as amusing as the decorative dress of the poet, and the sparkling garb of the wit would allow. And although he presents a flower to fit

DOI: 10.4324/9780429198809-68

every cap, none are personally intended, but the whole are offered for the selection of the wearer, and should weeds be discovered where flowers are expected, he flatters himself they will be few, and that those few will be found inoffensive, for although the work may be considered more adapted for amusement than for utility, he would not willingly offer entertainment through the assistance of immorality.

The author trusts that his work will not be condemned because it allows a communication of sentiment.

> "By all those token flowers, that tell
> What words can never speak so well."
>
> *Byron.*[2]

It is observed by Lady Mary Wortley[3] Montagu, that in Turkey, you may through the assistance of these emblems, either quarrel, reproach, or send letters of passion, friendship, or civility, or even news, without ever inking your fingers, for she says, there is no colour, no weed, no flower, no fruit, herb, nor feather, that has not a verse belonging to it.

When a Turkish lady sends a congratulatory message, or a ceremonious invitation, it is generally accompanied with some emblematical flowers carefully wrapt up in an embroidered handkerchief. The freshness of these flowers show the speed of the messenger, whilst the selection speaks in silent language the sentiments of the party from whom they are sent, and whose rank is also distinguished by the costliness or beauty of the embroidered envelop. These are merely the remains of the ancient customs of the eastern nations, where all was symbol, emblem, and allegory, and it must not be supposed that emblems were invented for the purposes of intrigue, since we find the scriptures full of the most beautiful parables, moral symbols, instructive allegories, and poetical emblems, that have ever been composed or collected.

INTRODUCTION.

In the gallant days of knight-errantry, respectful and faithful love often had recourse to the symbolical language of flowers, and the bashful swains of our rural scenes still frequently explain their first passion by the assistance of these emblems. Although unpractised in the rules of the floral grammar, they are instructed by nature to seek the earliest or the fairest flower of the season, scarce knowing why they wait at the stile, or wherefore they enter the wicket to present it. Their wish is to please, the flower proves it, and a smile rewards them.

Cottage friendship is frequently commenced through the aid of flowers, for where the tongue is disposed, but reluctant to say to its neighbour, come let us live in friendship and kindness, a nosegay will explain the wish, and the offer of a few flowering plants expresses more powerfully a desire for good neighbourhood, than all the words of village dialect.

But to express ourselves more readily by floral emblems, it is necessary that we should lay down grammatical rules for the government of this silent language before we proceed to the dictionary of emblems.

The first principle is to observe that the pronoun *I*, or *me*, is expressed by inclining the flower to the left, and the pronoun *thou*, or *thee*, by sloping it to the right, but when represented by drawings on paper, these positions should be reversed, as the flower should lean to the heart of the person whom it is to signify.

The articles *a, an*, and *the*, may be expressed by a tendril, the first by a single line, thus , the second by a double tendril , and the third by one with three branches .

The second rule is, that if a flower presented upright expresses a particular sentiment, when reversed it has a contrary meaning. Thus, for example, a rose bud upright with its thorns and its leaves means, "I fear, but I hope;" if the same bud is returned held downwards, it signifies, "you must neither hope nor fear;" but if the thorns be stripped off, it expresses, "there is every thing to hope;" deprived of its leaves, it signifies, "there is every thing to fear." Thus the expression may be varied of almost all the flowers by changing their position. The flower of the marigold, for example, placed on the head, signifies "trouble of spirits;" on the heart, "trouble of love;" on the bosom, "weariness."

The pansy held upright denotes "hearts' ease;" reversed it speaks the contrary; when presented upright it is understood to say, "think of me;" but when offered pendant, it means "forget me." And thus the amaryllis, which is the emblem of pride, may be made to express "my pride is humbled," or "your pride is checked," by holding it downwards either to the left or the right, as the sense requires.

In the same manner the wallflower, which is made the emblem of fidelity in misfortune, if presented with the stalk upwards, would insinuate that the person was considered no friend to the unfortunate. Thus various sentiments may be expressed by all the flowers noticed in these emblems, but which are principally intended as floral mottos.

Editorial Endnotes

1 Madame La Comptesse de Genlis (1746–1830) was a French writer known for her novels and theories of education for children. She is now best remembered for her journals and the historical perspective they provide on her life and times.

'Madame de Latour', generally known as Charlotte de la Tour (pseudonym of Louise Cortambert (1782–187)). 'Bouillet' is the name given to the author of *Parterre de Flore*, (Paris, Chez Marcilly, 1817).

2 'By all those token flowers' – 'The Maid of Athens, ere we part' (1810), Lord Byron (1788–1824), English poet.

3 Lady Mary Wortley Montagu (1689–1872) lived in Turkey when her husband was ambassador (1716–1718). Her account of her life there, *Turkish Embassy Letters* (1763), disseminated a view of the East to a wider public.

59

'EMBLEMATIC LANGUAGE OF FLOWERS'

Jules Lachaume

Source: Jules Lachaume, pt. 2.,'Langage Emblématique des Fleurs' ('Emblematic Language of Flowers'), *Les Fleurs Naturelles* (1847), pp. 79–89

From the earliest antiquity, especially in the East, flowers have had the privilege of symbolically expressing the thoughts and feelings of mankind. They were first struck by the analogy between their existence and human existence, of which flowers reproduce the different phases. Then, by studying them in their manners and habits, they soon found a large number of new analogies, and flowers became a veritable language, the charming and sweet language of the heart.

Only, it is to be regretted that this beautiful language was not always formed with a more rigorous logic: often the emblems are purely conventional, or at least we no longer know how to understand the relationships that the ancients were able to discover between certain flowers and the ideas or feelings they still symbolise today.

Son of a horticulturist, a horticulturist myself, I lived among flowers from my early years, and I cultivated and studied them with love. Today I am trying to publish, if not a new language of flowers, at least a more complete language, with more probable and more rigorous analogies. Instead of going to look for the words of this gracious language in distant countries, in inaccessible mountains, I take those which are constantly at our fingertips, in our gardens, in our flowerbeds, and in the countryside. Thus, instead of giving the mandrake, whose juice is a poison, as a symbol of rarity, I express the same idea using plants which flower rarely. The tremella, a species of fetid moss, which grows in the North, and which symbolises resistance, is here replaced by thorny plants, which express the same idea just as well.

Datura expresses disguise not, to my way of thinking, because the ancients disguised themselves with this plant, but because of the poison it hides under the beauty and scent of its flowers.

The rose, the violet, the carnation could, in the multitude of their varieties, offer a large number of synonyms to the language of flowers, so I took the liberty of giving an emblem to each species according to its qualities and its habits. Up to

 DOI: 10.4324/9780429198809-69

now we have only known the analogies of three species of roses: the *centifolia*, the *Four Seasons*, and the *white*; it was not wise to give a symbol to the *noisette rose*, the *tea rose*, the *Bourbon rose*, etc., etc., etc. We have filled this gap.

The language of flowers, to become fashionable and popular, needs to be simplified. Each person must find his plant dictionary by his side. Nature has spread flowers on the earth so profusely that even the smallest village has around it a number of flowers sufficient to express all the feelings of the heart!

It is this simplification that I set out to achieve. I would consider myself very happy if I could flatter myself that I had succeeded.

A

ABSINTHE, *Absence*. Absinthe comes from the Greek and signifies *without sweetness*. It is one of the bitterest plants. Is there a bitterness of heart comparable to the sorrow of absence?

ABUTILON, *Fragility*. The abutilon fears the slightest frost, its wood is spindly, its flowers hang like bells, because the peduncle is too weak to support them. This shrub needs support.

WHITE ACACIA, *Platonic love*. The savages of America have consecrated the acacia to the genius of chaste love and find its expression in a flowering branch.

ROSE ACACIA [*Robinia hispida*], *Elegance*. Nothing is as gracious and pretty as its clusters of pink flowers standing out against cheerful green foliage. One would call it an elegant evening dress.

ACANTHUS, *Symbol of the arts*. Since the Greek architect Callimachus chose the leaf of this plant to adorn the capital of the Corinthian column, modern architecture has retained this charming ornament.

ACHILLEA MILLEFOLIUM, *War*. Takes its name from Achilles. Ovid claims that Achilles used this plant to heal the son of Hercules whom he had wounded at the siege of Troy.

ACONITE, *Remorse*. Because aconite is a poison that has often been used in crime, and crime brings remorse after it.

ADONIS, *Sad memory*. According to the fable, Adonis was killed by a boar. Venus, who loved him passionately, shed tears over his fate. From these tears was born a small plant covered with flowers resembling drops of blood and which were called adonis.

MUSKY ADOXA or MUSCATELLINE, *Weakness*. Adoxa comes from the Greek and signifies *without glory*. It is a creeping plant with an odour of musk.

AGAVE, *Safety*. The Mexicans make impenetrable hedges around their houses with this long-spined plant.

AIRELLE or MYRTLE, *Treachery*. Myrtillus was the name of a treacherous horseman who, according to the fable, was changed by Mercury into the shrub of this name.

AGNUS CASTUS, *Frigidity*. The priestesses of Ceres made their virginal beds from the branches of this shrub, which they considered the *palladium* of their chastity. In later times several religious orders used a water distilled from *agnus castus* to calm the ardour of their senses.

ALISIER [service tree], *Harmony*. Its wood was used to make musical instruments.

ALOES (Parrot's beak), *Cackle*. Because its leaf imitates [the shape of] a parrot's beak.

ALYSSUM SAXATILE or Golden basket, *Tranquillity*. Because this plant, low and overgrown, has no fear of the wind or of the encroachments of other plants.

ALMOND, *Thoughtlessness*. The almond rashly blooms before the last frosts which often come to destroy all the hopes of its white and fragrant flowers.

AMARANTHUS, *Immortality, Love*. In Greek, this word means indefatigable. The amaranthus blooms in the autumn of the year the way that friendship blooms in the autumn of life. This plant remains in flower a long time.

AMARYLLIS, *Pride*. There is something of pride in the sharp tone of its startling colours.

AMETHYST or FENNEL, *Credulity*. The Romans deemed invincible in combat were crowned with this flower.

ANANAS [Pineapple], *Perfection*. This plant has a noble bearing; its fruit has a gracious form which would one say was sculpted by the hand of an artist, and one of the sweetest scents.

ANCOLIE [Columbine], *Madness*. The ancients crowned the statue of Madness with these flowers.

ANEMONE, *Abandon*. Anemone was a nymph beloved by Zephyr. Flora, jealous, banned her from her court and changed her to a plant which flowered before spring. So, abandoned by Zephyr, the anemone is delivered to the hard caresses of Boreas, who, having failed to make her love him, shakes her, half-opens her, and quickly withers her.

FIELD ANEMONE, *Sickness*. This plant makes animals that eat it sick, when it doesn't kill them.

ANEMONE HEPATICA, *Confidence*. It flowers in April, confident of the fine weather it heralds. Its flowers precede its leaves, so much does it count on the gentleness of spring.

ANGELICA, *Inspiration*, *Spirit*. The Lapps believe that by crowning himself with angelica the devil will electrify their lyres and inspire them to make beautiful verses.

ANTHEMIS, *Despair*. The anthemis blooms at the onset of winter and lifts our hopes for fine weather.

ARISTOLOCHIA, *Embrace*. This shrub embraces with its twining branches everything growing in its vicinity and smothers other plants under its broad leaves. The ancients crowned themselves with Aristolochia.

ARGENTINE [Potentilla], *Naivete*. This plant comes almost carelessly and blooms with a kind of abandon and naivety. It does very well in a border.

ARTEMISIA, *Happiness*. In some countries mothers crown their young children with this plant to protect them from all evil influence.

ARUM [Fly-catcher arum, Calla], *Trap*. This plant bears a flower which closes up, imprisoning at the base of its corolla the winged insect which comes to suck the sweet liquid it contains.

ARISARUM, *Ardour*. Monsieur Hubert claims to have observed on Mauritius that this species of arum gives off so much heat at the time of fertilisation that it is difficult to hold it in the hand.

SERPENTINE ARUM [Dracunculus], *Horror*. Because its trunk and tuberous roots give it the appearance of a snake.

ASPHODEL, *Regret*. The ancients placed this plant around tombs as a sign of regret.

ASTRAGALUS, *Hidden blessing*. The flowers of the astragalus are almost hidden under the down that surrounds them. [NB. So that's where the word duvet comes from! – it means "downy".]

ASTER GRANDIFLORUS, *Afterthought*. Because it blooms so late in the season. [NB. Arrière-pensée can also mean an ulterior motive, but I don't see quite how that fits.]

ARISTAEA, *Rigour*. Because of the severity of its dark blue flower.

AUBEPINE [Hawthorn], *Hope* and *Prudence*. The sweet smell of flowering hawthorn heralds the end of winter and gives hope for fine weather. One must be careful in detach a flowering branch of hawthorn, in order to avoid the burning sting of its thorns.

60

THE LANGUAGE OF FLOWERS, THE ASSOCIATION OF FLOWERS, POPULAR TALES OF FLOWERS

(Excerpts)

Ann Pratt and Thomas Miller

Source: Ann Pratt and Thomas Miller, 'Preface', 'A', *The Language of Flowers, the Association of Flowers, Popular Tales of Flowers* (1870s), pp. 5–6, 15–16

PREFACE.

IN a rude and primitive state, the words are few and simple that are required to express the thoughts and desires of an untutored race of men. But with every increasing want and every new desire, names and forms of thought must be created; until the brief vocabulary of the savage tribe swells into the complex dictionary of the intelligent and civilised people. In this enlargement and amplification of expression, Flowers and their associations have had a large share.

For instance, "Lovely as the Rose," "Fair as the Lily," or, "Modest as the Violet," are phrases that seem to have come universally into use. And no less do the Flowers prefigure hope and frailty. We strew them over the shroud of departed love, and plant them to bloom brightly above the grave, that they may speak in spring and summer of an eternal hope. To such and many other simple expressions of natural feeling, have been added gradually for centuries, many complex fancies from early classic poetry and the later writers.

Yes, flowers have their language. Theirs is an oratory that speaks in perfumed silence; and there is tenderness, and passion, and even the light-heartedness of mirth, in the variegated beauty of their vocabulary. To the poetical mind they are not mute to each other; to the pious they are not mute to their Creator; and ours shall be the office, in this little volume, to translate their pleasing language, and to show that no spoken word can approach to the delicacy of sentiment to be inferred from a timeously-offered flower. That the softest impressions may be thus conveyed without offence, and even a profound grief alleviated at a moment when the most tuneful voice would grate harshly on the ear, and the stricken soul can be soothed only by an act of grateful silence.

 DOI: 10.4324/9780429198809-70

We will not, upon a subject so varied, parade our learning by telling our fair readers what fine things Pliny has said upon it; or, with the spirit of prosing upon us, write a crabbed treatise upon the Egyptian hieroglyphics. We will even spare them a dissertation upon the floral alphabet of the effeminate Chinese; they had, and have, their flowers and their feelings, their emblems and their ecstasies. Let them enjoy them.

We shall do no more than rove through the European garden, to cull from it its beauties, to arrange them into odoriferous eloquence, and teach our refined and purifying science to those fair beings. The symbols of whose mortal beauty are but inadequately found in the most glorious flowers, and of whom the mental charms can be typified, only when we shall have reached those courts where the spring is eternal and the idea of decay unknown.

But little study will be requisite for the science which we teach. Nature has been before us. We must, however, premise two or three rules. When a flower is presented in its natural position, the sentiment is to be understood affirmatively; when reversed, negatively. For instance, a rose-bud, with its leaves and thorns, indicates *fear* with *hope;* but if reversed, it must be construed as saying, "You may neither *fear* nor *hope*." Again, divest the same rose-bud of its thorns, and it permits the most sanguine hope; deprive it of its petals and retain the thorns, and the worst fears are to be apprehended. The expression of every flower may be thus varied by varying its state or position. The Marygold is emblematical of pain; place it on the head, and it signifies trouble of mind; on the heart, the pangs of love; on the bosom, the disgusts of "ennui." The pronoun *I* is expressed by inclining the symbol to the right, and the pronoun *thou* by inclining it to the left.

These are a few of the rudiments of our significant language. We call upon Friendship and Love to unite their discoveries to ours; for it is in the power only of these sweetest sentiments of our nature to bring to perfection what they have so beautifully invented, the mystical yet pleasing links of intelligence that bind soul to soul, in the tender and quiet harmony of the one, or in the more impassioned felicity of the other.

THE LANGUAGE OF FLOWERS.

Part I.

Flowers, and the Sentiments which they Represent.

Abatina	*Fickleness.*
Abecedary	*Volubility.*
Acacia	*Chaste love.*
Acacia, Pink	*Elegance.*
Acacia, Rose	*Platonic love. Friendship.*
Acacia, Yellow	*Secret love.*
Acalia	*Temperance.*
Acanthus	*The Fine Arts. Artifice.*
Achillea Millefolia	*War.*

Aconite (Wolfsbane)	*Misanthrophy.*
Aconite, Crowfoot	*Lustre.*
Adonis	*Sorrowful remembrances.*
African Marigold	*Vulgar minded.*
Agnus Castus	*Indifference. Coldness.*
Agrimony	*Thankfulness.*
Almond, Common	*Stupidity, Indiscretion.*
Almond, Flowering	*Hope.*
Almond, Laurel	*Perfidy.*
Allspice	*Compassion.*
Aloe	*Affliction. Grief.*
Althæa Frutex	*Persuasion.*
Alyssum Sweet	*Worth beyond beauty.*
Amaranth	*Immortality.*
Amaranth	*Unchangeable.*
Amaryllis	*Timidity. Pride.*
Ambrosia	*Love returned.*
Amethyst	*Admiration.*
Andromeda	*Self sacrifice.*
Apricot Blossom	*Doubt.*
Anemone, Field	*Sickness.*
Anemone, Garden	*Forsaken.*
Angelica	*Inspiration.*
Angrec	*Royalty.*
Apocynum	*Deceit.*
Apple	*Temptation.*
Apple Blossom	*Preference.*
Apple, Pine	*Perfection.*
Apple, Thorn	*Deceitful charms.*
Arbor Vitæ	*Unchanging friendship.*
Arbutus	*Thee only do I love.*
Arum, (Wake Robin)	*Ardour.*
Ash, Mountain	*Prudence.*
Ash Tree	*Grandeur.*
Aspen Tree	*Lamentation.*
Aster (China)	*Variety. Afterthought.*
Asphodel	*My regrets follow you to the grave.*
Auricula	*Painting.*
Auricula Scarlet	*Avarice.*
Austurtium	*Splendour.*
Autumnal Leaves	*Melancholy.*
Azalea	*Temperance.*

61

THE FLORAL FORTUNE-TELLER: A GAME FOR THE SEASON OF FLOWERS

(Excerpts)

Sarah Carter Edgarton Mayo

Source: Sarah Carter Edgarton Mayo, 'Preface', 'Key to the Fortune Teller', 'White Flowers', *The Floral Fortune-Teller: a Game for the Season of Flowers* (1847), pp. iii–vi, 8–22

PREFACE.

THE custom of fortune-telling is nearly as old as time. Almost everything has been turned into an oracle. Even the dregs of a teacup have been supposed to possess the "gift of prophecy." Why, then, should not flowers, that are forever rising up, clothed with the mantles of prophets, and speaking in a language that has become as familiar as household words – why should not these little "floral apostles" be consulted respecting the mysteries of our earthly destiny?

At all events, they are pure and beautiful playthings for the fancy; and if any diversion on earth be innocent, it must be one whose instruments are poetry and flowers.

In selecting our oracles, we have drawn from the purest wells of English poetry; for who like Shakspeare can describe character, or like Wordsworth, Coleridge and Tennyson, picture a scene? We have also turned over the pages of the great German dramatists, and here and there culled from them a word of wisdom, or of prophecy. Our aim has been to condense into our brief oracles as much point, wit and variety, as the interest of the game will permit.

How well we have succeeded in our attempt to please, will be best determined by the future success of this little book. God speed it on a harmless way!

KEY TO THE FORTUNE-TELLER.

THE following five questions are answered, respecting the character and future fortunes of an individual.

1. What is my character?
2. What is, or will be, the state or quality of my love?
3. What is, or will be, my worldly fortune?

DOI: 10.4324/9780429198809-71

4. What is the scene in which most of my life will be spent?
5. What is the character of my future companion?

The person consulting our little Fortune-Teller, must select a bouquet of five different colored flowers, of whatever kind he chooses. The *White* shall describe to him his character; the *Blue*, the state or quality of his love; the *Purple*, his worldly fortune; the *Yellow*, the scene in which his life shall be chiefly passed; and the *Red*, the character of his future wife.

For example; if a gentleman bring a white Lily, a blue Violet, a Foxglove, a Marigold and a red Rose, he will first turn to the catalogue of *White* flowers, where, under the head of Lily, he will find this description of his character:

> "Of a free and open nature,
> That thinks men honest that but seem to be so,
> And will as tenderly be led by the nose
> As asses are."
>
> SHAKSPEARE.

The Violet, found in the catalogue of Blue flowers, will unfold the state of his "love matters" as follows:

> "You are now sailed into the north of my lady's opinion; where you will hang like an icicle on a Dutchman's beard, – unless you do redeem it by some laudable attempt."
>
> SHAKSPEARE.

The Foxglove will thus declare his worldly condition:

> "Poor, but content, is rich, and rich enough."
>
> SHAKSPEARE.

To learn the scene of his future abode he will consult the Marigold.

> "A cottage on a plot of ground,
> With large prospect north and south."
>
> WORDSWORTH.

The Rose will reveal to him the character of his wife:

> "A woman in her freshest age,
> Of wondrous beauty and of bounty rare,
> Full of great love."
>
> SPENSER.

By different combinations of flowers, a great variety of *fates* may be unfolded. Nor is the purpose of the game wholly confined to the reading of fortunes. Any gentleman wishing to compliment a lady, may do it in a very delicate and poetical manner, by selecting a bouquet of flowers that shall express the good fortune he wishes her.

The first paragraph upon each page is for the Lady, the second for the Gentleman.

PART I.

WHITE FLOWERS.

Describing your Character.

APPLE-BLOSSOM.

A most fresh and delicate creature.
A most exquisite lady.

SHAKSPEARE.

Thou hast been
As one, in suffering all, that suffers nothing;
A man that fortune's buffets and rewards
Hast ta'en with equal thanks; and bless'd are those
Whose blood and judgment are so well commingled,
That they are not a pipe for fortune's finger
To sound what stop she pleases.

SHAKSPEARE.

APRICOT-BLOSSOM.

The world hath not a sweeter creature.

* * * *

What an eye! – Methinks it sounds a parley of provocation. – An inviting eye, and yet, methinks, right modest.

SHAKSPEARE.

An eye, like Mars, to threaten and command,
A combination and a form, indeed,
Where every god did seem to set his seal
To give the world assurance of a man.

SHAKSPEARE.

ANEMONE, OR WIND FLOWER.

Your eye is like the star of eve,
And sweet your voice as seraph's song.

* * * *

Within your *soul* a voice there lives!
It bids you hear the tale of woe.

COLERIDGE.

Thou art not for the fashion of these times,
Where none will sweat but for promotion,
And having that, do choke their service up,
Even with the having; it is not so with thee.

SHAKSPEARE.

BACHELOR'S BUTTON.

Sell when you can; you are not for all markets.

SHAKSPEARE.

Jealous in honor, sudden and quick in quarrel,
Seeking the bubble reputation.

SHAKSPEARE.

BALSAMINE.

Thou hast always been gentle and good.

BEAUMONT.

A butterfly, a lord
Of flowers, garlands, love-knots, silly posies,
Groves, meadows, melodies and arbor-roses.

KEATS.

BLOOD ROOT.

You have deserved high commendation, true applause, and love.

SHAKSPEARE.

You have a nimble wit.
Your spirits are too bold for your years.

SHAKSPEARE.

CALLA.

Oh thou art fairer than the evening air,
Clad in the beauty of a thousand stars.

MARLOWE.

An ornament of nature, fine and polished,
A handsome youth, indeed!

FORD.

CANDY TUFT.

Handsome, young, and hast all those requisites that folly and green minds look after.

SHAKSPEARE.

You are one of those that will not serve God,
 if the devil bid you.

SHAKSPEARE.

CHINA ASTER.

Thou art as wise as thou art beautiful.

SHAKSPEARE.

Thy mind is a very opal! I would have men of such constancy put to sea, that their business might be everything, and their intent everywhere.

SHAKSPEARE.

CAMELLIA.

Of a constant, loving, noble nature.

SHAKSPEARE.

True of mind, and made of no such baseness
As jealous creatures are.

SHAKSPEARE.

CLEMATIS.

Of excellent discourse,
Pretty and witty; wild, and yet too, gentle.

SHAKSPEARE.

Believe me, in thy breast are thy destiny's stars. Trust in thyself. *Decision* is thy Venus. The Malignant, the only one that injures thee, is *irresolution*.

SCHILLER.

CANTERBURY BELL.

Faithful, gentle, good,
Wearing the rose of womanhood.

TENNYSON.

The gravity and stillness of your youth
The world hath noted, and your name is great
In mouths of wisest censure.

SHAKSPEARE.

COLUMBINE.

Ready in gibes, quick-answered, saucy, and
As quarrelsome as the weasel.

SHAKSPEARE.

You are rather point-device in your accoutrements; as loving yourself than seeming the lover of any other.

SHAKSPEARE.

CHRYSANTEMUM.

Thou art pleasant, gamesome, passing courteous,
But slow in speech, yet sweet as spring-time flowers.

SHAKSPEARE.

Thou art e'en as just a man,
As e'er my conversation coped withal.

SHAKSPEARE.

62

THE ROYAL GUIDE TO WAX FLOWER MODELLING

(Excerpts)

Emma Peachey

Source: Emma Peachey, ('The Editor', 'Preface'),'The Art of Modelling Flowers in Wax . . .', The Advantage of Wax Modelling . . .' 'Roses—Instructions', *The Royal Guide to Wax Flower Modelling* (1851), pp. ix–xiv, 4–8, 30–35

PREFACE.

THE Editor of this work, by Her Majesty's Artiste, MRS. PEACHEY, fairly entitled the ROYAL GUIDE TO WAX-FLOWER MODELLING, would fain leave the introduction, written by the same hand which rivals nature in her varying adornments, to unfold its historic, its poetic, its moral, and its suggestive graces – for it combines these; but having accepted the part, without which, since the days of Plato, no book is deemed complete, he essays a few prefatory observations and remarks.

Brevity, it has been said, is the soul of wit; but we may be brief when we know what is to follow, and for whom the following pages are designed.

Our fair readers will intuitively perceive that the scope of the instructive portion of this self-commending little volume is to facilitate their acquisition of an accomplishment at once royal and feminine in its origin and progress, and therefore worthy of their attention.

This elegant art requires but the fairy touch of a delicate hand to fill each available space in the chamber or drawing-room with the most perfect and beautiful imitations of the flower-garden.

"The morning flowers display their sweets,[1]
  And gay their silken leaves unfold,
As careless of the noontide heats,
  As fearless of the winter cold.

"Nipped by the wind's unkindly blast,
  Parched by the sun's directer ray,

DOI: 10.4324/9780429198809-72

The momentary glories waste,
The short-lived beauties die away."

Unaffected by change or climate, wax flower modelling perpetuates the transient glories of the floral seasons; places all the tender varieties under the immediate glance of the ever gratified eye of the artist, who can thus in the depth of winter exhibit to an admiring foreign guest the exotics of the far hemisphere, or the indigenous plants of her own loved land.

Who that has watched by the side of an invalid mother, would not feel an exalted pleasure in creating around her the magical representations of those flowerets and rosebuds her maternal hand was wont to rear? Who, in such a moment of ministering affection, would not feel how sweet the reward of a father's love, as his approving gaze spoke more than many words his thanks to the duteous child returing the early care of the fond partner of his griefs and joys? Contemplating such a scene as this, one cannot refrain from citing the language of the poet: –

"O! if there be a tear,[2]
From passion's dross refined and clear;
A tear so limpid and so meek,
It would not stain an angel's cheek;
'Tis that which pious fathers shed
Upon a duteous daughter's head."

The copious table of contents possesses great attraction for persons of refined taste, embracing every variety of flower usually modelled in wax: its arrangement is calculated to lead the learner, by easy steps, from the most simple to the most elaborate accomplishment of a very delightful task.

The sketch of her artistic life, with which the talented though unpretending authoress has favoured the public, cannot fail to prove useful and encouraging to the beginner, as it fully justifies the good old proverb, that "where there is a will, there is a way;" and that way is clearly and forcibly pointed out in the Royal Guide, so as to direct with perfect ease the willing fingers of the modeller to the attainment of her object, to excel in giving form and substance to her innate perceptions of the beautiful. Nor is this a selfish pleasure. These productions of skilled labour – if we may apply the word labour to an amusement – please the beholder, as they do the mind which calls forth the exquisite fancy which pencils these flowers.

The unanimous verdict of the Press will be found recorded at the end of the instructions. It is a remarkable fact, that so many Journals, giving in their separate awards, should have all concurred in opinion. This opinion is highly favourable to the Artiste and the art. The very language in which it is couched partakes of floweriness – if we may be pardoned for coining a word to express our meaning; indeed, we strongly commend for perusal these elegant notices of the Press; the writers evidently have been influenced by national considerations; for they speak of what they have seen as those convinced that, although there may be several wax flower modellers, there is but one MRS. PEACHEY – Her Majesty's Artiste, and an Englishwoman.

It is with no insular feeling that we express the same sentiment; but, nevertheless, we do feel it to be something to boast of, that our country-women will not have to learn the art of Wax Flower Modelling from foreigners, many of whom however have been amongst the now nearly 50,000 visitors attracted to the collection, by the notices of the Press, and who have expressed equal approbation.

The ROYAL GUIDE is essentially a domestic national instructor. But its teaching will not be bounded by our island shores. We venture to predict, that the ROYAL GUIDE TO WAX FLOWER MODELLING will, ere long, establish for itself a more than European fame.

The Editor would now conclude the task he has undertaken, and performed, as well as more immediate professional calls upon his time would permit, to the best of his abilities; but, beyond changing or transposing a word or term here and there, introducing some poetic gems, and correcting the press, he does not claim any merit for the work from his hands, that properly belongs to the Authoress, who has been called into the field, and to whom the reader is now fairly introduced, as to a pleasing and accomplished instructress in this Art.

THE ART OF MODELLING FLOWERS IN WAX WAS BROUGHT INTO ENGLAND.

I consequently subjoin an extract from Miss Strickland's[3] Life of Mary Beatrice, second consort of James II., A.D. 1686.

> "The beautiful imitations of natural flowers in wax which have lately afforded an attractive exercise for the taste and ingenuity of many of our youthful countrywomen, were first introduced into England by the mother of Mary Beatrice, as a present to her royal daughter; as we find by the following passage in a contemporary letter from a correspondent of the Lady Margaret Russell, which gives some information relative to the ornamental works then in vogue among ladies of rank, in the court of Mary Beatrice.
>
> " 'In gum flowers, Mrs. Booth tells me you and she is to doe something in that work, which I suppose must be extraordinary. I hope it will be as great perfection as the fine WAX WORK ye queen has, of nun's work, of fruit and flowers, that her mother did put up for her, and now she has 'em both for her chapel and her rooms. I do not know whether they be the four seasons of the year, but they say they are done so well, that they that see 'em can hardly think 'em other than the real.' "*

From the year last named until 1736 I have been unable to trace any knowledge of this elegant art.

When wax flowers were again introduced by an Italian, they were clumsily manufactured in comparison to those seen in the present age of improvement; for I had the opportunity of inspecting some of their "miserable remains" but a few years since. Still I must acknowledge I discovered some taste and much ingenuity in their construction, and am not too proud to own that I benefited even by examining these very inferior productions. I feel quite satisfied that the art of wax flower modelling is almost still in its infancy. It is no longer regarded as an amusement only. It is enumerated with other accomplishments essential to female education. It assists botanical studies, and promotes the views of flower painters, either in oil or water colours; even in drawing, wax flowers will be found excellent auxiliaries, far preferable as copies to the even surface of plates. I have myself been much gratified by furnishing flowers in wax to some of our first flower painters, who have assured me that they have proved of great utility, in cases where the evanescent properties of the flower of nature prevented the possibility of committing their similitude to canvas ere their beauty had faded. It affords me no small degree of satisfaction also, that my flowers were found useful as copies for some of the beautiful carved work in the late great and ever memorable Exhibition. I have also supplied them as illustrations to botanical lectures.

In thus referring to the utility of wax flowers, I am reminded of a partially unfavourable prejudice which has lately sprung up, from an article which first appeared in a Manchester paper, and which was subsequently copied into *The Times*, and other papers. It is possible ladies may be induced to abandon this delightful amusement, upon reading such a statement as the subjoined extract: –

> "The Danger of Modelling in Wax. – Few persons, especially, perhaps, of the many young ladies who are now practising the very pleasing art of modelling fruits, flowers, &c., in wax, at all suspect the great danger in which they are placed from the poisonous nature of the colouring matter of the wax which they handle so unsuspectingly. The white wax, for instance, contains white lead; the green, copper; the yellow, chrome yellow and vermilion – strong poisons all; while many other kinds of wax are equally poisonous, and, therefore dangerous. There are very many persons who are aware of the intense sufferings, for very many years past, of Mr. W. Bally, phrenologist and modeller in wax, in which latter branch he has laboured for 24 or 25 years, three of them as teacher of the art, at the Manchester Mechanics' Institution. Mr. Bally has been at times completely paralysed, and is now and has long been very nearly so, especially in the hands and arms; and he has also been afflicted with extensive ulceration of the throat, and has almost totally lost his voice. Both himself and his medical adviser, after a long attention to his symptoms, are satisfied that the primary cause of his affliction is the extent to which the subtle poisons in the wax with which he has worked have been absorbed into his system through the pores of his hands, while the

> disease has been generally strengthened, and one part of it accounted for, by the occasional application of his fingers to his lips while at work. Mr. Bally says, that he has known several cases in which young ladies have been attacked with partial paralysis of the hands and arms, after having devoted some time to the practice of modelling; but at the time he had no suspicion of the cause. As all the requisite colours can be obtained from vegetable matter, and as the use of mineral colouring seems to lead to such deplorable results, the subject should be carefully investigated by those working with coloured wax."
>
> – *Manchester Examiner*.[4]

It is not my intention to contradict an assertion so boldly set forth. I have no doubt the editor of the *Manchester Examiner* had some grounds for the article; but I think it right to state *that* which I can prove – namely, that the wax *artistically manufactured by me* is so perfectly harmless, that for the last fourteen years I have had it in my hands, upon an average from twelve to fourteen hours every day (Sunday excepted), and never in the slightest degree experienced any inconvenience or ill effects.

The small portion of colour I introduce undergoes a chemical process, which neutralizes entirely any deleterious properties appertaining to the few colours required to be used. It is quite unnecessary to introduce white lead at all. I was assisted by a practical German chemist to prepare borax, in such a manner, as to entirely supersede white lead. Now most of my readers will be able to testify how perfectly harmless must be borax, it being one of the drugs so constantly used with honey, and recommended by the faculty as an excellent remedy for canker in the mouth. I am, as I have previously stated, the daughter of a medical man, and am perfectly acquainted with the danger attending the absorption of mineral colours into the system: under these circumstances, it is not likely that I should myself use that which would be injurious. Ladies, who desire to enjoy the recreation of wax flower modelling, may indulge in the amusement with perfect safety, if they purchase the wax of me. At the same time, I wish it to be perfectly understood, that I do not insinuate, or attribute aught against any other person or persons who prepare wax for sale.

I DECIDEDLY OBJECT to the dry colours being rubbed into the wax with the fingers. I invariably apply the colours with a brush. It must be injurious to close the pores of the skin, even were the powders so used innocuous; but to say nothing of the danger of the method alluded to, it is a most dirty occupation, and ladies would not like to see their hands dyed with carmine, Prussian blue, or chromes. Such a method of tinting is likely to prejudice ladies against the work altogether; besides which, it renders the flowers much more fragile. The only time I ever use dry powder is in the form of bloom (peculiarly prepared arrowroot), which I throw on lightly, but never rub in. Having endeavoured to prove that there are no dangerous results likely to accrue from this pleasing occupation, I will proceed to shew

THE ADVANTAGE OF WAX MODELLING, OVER OTHER FANCY WORK.

And one great consideration is that the sight is not likely to be injured. The eye does not require to be fixed; it does not occupy so much attention as to prevent conversation, nor need the *body be bent*, – a matter of much importance with growing girls, many having suffered affections of the chest, and others disfigured for life, through continually stooping to frame work.

There is no monotony in this agreeable employment, for new varieties are continually springing up in nature; and a visit to the Botanical Gardens at Kew, or the Regent's Park, will at all seasons afford some fresh specimen. In referring to the former gardens, I cannot forbear expressing the deep sense of obligation I feel due from the public, and artists particularly – being myself one of them, – for the boon bestowed upon us by those powers who afford such facility for inspecting – free of charge – all that is lovely, choice, and rare. It is perfectly clear, according to my method, that the most elegant drawing room might be used, without suffering in its appearance during its operations. I would merely recommend that the table should be covered with paper, so that all small pieces might be kept together, and easily removed.

I shall proceed to give some instruction in Roses. The varieties now grown and named in our nurseries amount to the almost incredible number of two thousand. I shall therefore choose, for the information and improvement of those ladies who kindly think proper to place themselves under the guidance of my little book, about six kinds, such as I deem most useful, and as being different in form and colour. I think, when these are perfectly understood, any other kind can be copied easily from nature. I shall commence with the old favourite,

THE CABBAGE OR PROVENCE ROSE.

(ROSA CENTIFOLIA.) *Beauty*.

"The rose has one powerful virtue to boast
  Above all the flowers of the field –
When its leaves are all dead, and fine colours are lost,
  Still how sweet a perfume it will yield."

WATTS.[5]

Cut the petals from pale pink wax; colour the three smallest rather deep with the crimson powder. The split petals, marked on my pattern fifteen, are coloured the same, but rather a lighter hue. Each succeeding set are painted the same, but gradually diminish the colour until you arrive at the outer petals, which are the lightest of all. To form the petals, use a pin as little as possible; cupping them with the thumb or finger, according to its size. For the largest petals, use the thumb, so pressing each in the centre, while it is resting in the palm of the left hand, as to become perfectly round. The last two or outer rows are turned back with the head of a curling pin.

Prepare the foundation of solid wax, rolled round the end of a moderate size wire. It must be cone-shaped. The three smallest petals are crushed and placed at the point in a triangular form. The split petals, marked on my pattern fifteen, are united into clusters of five, and placed round immediately under the three that are crushed. Each succeeding row of petals are placed on in like manner, taking care that each petal is attached to the under part of the foundation, and *not* upon the side. I particularly name this, as I too frequently find ladies err in forming the roundness of the rose. The last two rows of petals, which are turned back, must be placed rather lower than those preceding. The calyx is cut from double light green wax.

The head of the curling pin is passed down the centre of each point previously to their being placed on. Pass a strip of double green wax close round the base of the flower, moulding it round and smooth with the thumb and finger. This is to represent the seed cup. When the calyx is affixed, it must rest against the back of the rose, and be so neatly moulded over the seed cup, as not to show any division or seam.

THE MOSS ROSE.

Beauty and Love.

Is similarly constructed (being one of the same class), but is rather smaller. It requires the addition of a little real moss. It must be the fine spray moss; and dried quickly, by placing a warm flat-iron upon it. It is affixed to the calyx and seed cup by pressing it on with the head of the curling pin.

THE WHITE ROSE.

(Rose Alba.) *Silence.*

Cut the petals from thin white wax. Tinge the lower part of the first three sets of petals with my lemon powder. Cup all the petals with the finger, turning the last or largest two rows back. Cut a few stamina in lemon wax, with the edge rolled to form the anthers; colour them with orange, and when quite dry, touch them occasionally with brown (crimson powder and cake sepia). These stamina are divided into clusters, eight or ten filaments in each, and about five in number. Make a small cone-shaped foundation; attach one cluster of stamina at the point. The ten small petals are affixed round, turning in various directions, and interspersed with the rest of the stamina. The whole of the remaining petals are placed on five in a row, the last two turning back. Finish off with calyx and seed cup, as in previous instruction. This rose is peculiarly adapted for bridal bouquets; and I must here mention, while alluding to the subject of bridal favours, that I made upwards of ten thousand of these roses upon the happy occasion of Her Majesty's marriage. It may afford some trivial amusement to my younger friends, to relate the following anecdote, in connection with the event just alluded to. About three years after Her Majesty was united to His Royal Highness Prince Albert, a gentleman visited my establishment, and inspected my specimens of flowers in wax with evident satisfaction. He represented himself as being a great admirer of wax-work generally; and stated, that he himself possessed a rare specimen, in fact, a perfect *bijou*. He should wish me to see it. I, of course, expressed some anxiety to behold such perfection of art; and accordingly, he sent his footman with a small box, charged with strict orders to be particularly careful in conveying the same. After removing sundry pieces of tissue paper, and as many of wadding, my surprise may be easily imagined, when I beheld one of the identical bouquets (white rose, orange blossom, and myrtle, tied with white satin ribbon) that I had myself manufactured upon the joyous occasion already alluded to. I am but human nature, therefore, I hope I may be pardoned for expressing and feeling a certain degree of vanity upon inspecting this Royal relic of my own hands; still, I am not blind to the fact, that the happy occasion for which the bouquet had been prepared, namely, the nuptials of our beloved Sovereign, had materially enhanced its value to the possessor; – but I will no longer digress from the leading feature of this work, but commence the description of the formation of

THE DAMASK ROSE.

(Rosa Damascena.) *Freshness of Complexion.*

"The rose, like ruddy youth, in beauty stands,
And would be cropped by none but fairest hands."[6]

Cut the petals from white wax, and paint them upon both sides (with my crimson), two-thirds down. Cup the petals as in the preceding rose. Cut two strips of stamina in lemon wax, tip them with my orange powder. Make a foundation of lemon wax, and pass round the strips already mentioned. Place in a triangular form the six small petals in clusters of sets of petals are placed on in clusters of five, the others singly in rows of five. The foundation is made similar to the damask rose, and the stamina show from the front of the flower. Calyx and seed cup formed as usual.

AUSTRIAN BRIAR ROSE.

(Rosa Lutea.) *Mingled Pleasure and Pain.*

This is a showy single flower, and very easily and quickly accomplished. There are but five petals, cut from bright yellow wax; colour them half-way down each with crimson: cup them a little in the palm of the hand, not using a pin at all. The foundation is rather small, and formed of green wax – one strip of stamina placed round, cut in lemon wax, tipped with orange powder. Calyx and seed cup formed according to previous instruction.

Editorial Endnotes

1 "The morning flowers display their sweets/" – verse from a collection of hymns (1840) of the Methodist Movement founded by Charles Wesley and attributed to his grandson Samuel Sebastien Wesley (1810–1876).

2 "Oh, if there be a tear/ From passion's dross" – Walter Scott, *The Lady of the Lake*, second canto (1827).

3 Agnes Strickland (1796–1874), historian

4 *Manchester Examiner* (1846–1894), a Liberal newspaper.

The article was reprinted in numerous newspapers in June and July 1851 across Britain and in India, but we have not been able to trace the original from the editions of the *Manchester Examiner* available, nor in other newspapers based in Manchester. The date of publication coincided with the Great Exhibition which failed (at short notice) to provide a suitable space for Peachey to exhibit a large-scale wax representation of a bouquet of flowers in a gold basket that included the Victoria Regia in all its stages of development, bud, blossom, and flower at its base (a 4-ft square mirror platform representing water). She chose to exhibit it instead at her own premises at 35 Rathbone Place.

5 "The rose has one powerful virtue to boast/Above all the flowers of the field/When it leaves are all dead" – Isaac Watts (1674–1748), *Divine songs: attempted in Easy Language for the Use of Children* (1777), Divine and Moral Songs, Song 3

6 "The rose, like ruddy youth, in beauty stands/ And would be cropped by not but fairest hands" – Not Identified

Source Endnote

* In the collection of private family letters of the Duke of Devonshire, at Chiswick Lodge: copied by courteous permission of his Grace.

63

WAX-FLOWER MODELLING MADE EASY

(Excerpts)

Annie M. Williams

Source: Annie M. Williams, 'Introductory Remarks', 'To Make Leaves and Stems for Wax Flowers', 'Arrangement of Flowers for Baskets, Vases, etc.', 'To Make Stamens, Pistils etc., for Wax Flowers', *Wax-Flower Modelling Made Easy* (1871), pp. 1–12, 17–25

 DOI: 10.4324/9780429198809-73

THE ART OF MODELLING FLOWERS IN WAX.

INTRODUCTORY REMARKS.

THE art of modelling flowers in wax has lately become a favourite study, from the fact of there being no other substance equal to it by which nature can be so closely imitated; all the little natural curves and lines that tend to make a flower so beautiful in its appearance, can be produced, owing to the flexible character of the wax.

In order to obtain good imitations, great attention must be given to the art. I have seen some wax flowers so very unnatural in their appearance and so indifferently made, that they were perfect monstrosities. Others again I have seen which were so closely imitated, as scarcely to be distinguished from the natural flowers; rules for these will be as far as possible laid down here, and if adhered to will doubtless, with a little patience and artistic arrangement, together with well-formed flowers, produce a most satisfactory result. Some people imagine it a very tedious and difficult task to produce these artificial but natural-looking flowers, nothing can however be farther from the truth, and these ideas arise simply because they have never tried, and are therefore in perfect ignorance of any facts concerning its accomplishment. Wax flowers form beautiful ornaments for any room, and are therefore exceedingly useful, as well as pleasing to the eye. They are exquisite for the mantel-piece, on tables in a recess, or indeed in almost any place or position: the glass shades can be procured to fit any sized shelf; and even on a dinner table when flowers are very expensive or out of season, a pretty group of flowers and foliage, with the glass shade removed, would look very charming. Objects that have not been treated upon in a work of this kind, will be brought before the reader's notice here, and I trust will be very pleasing to those interested in this art. A short list of botanical terms, simply those necessary in explaining the making of a flower, will be described before any definite explanation of the art of modelling is given, for everything connected with this art requires minute care, in order to make it well and true to nature. The flowers must not be allowed to get cracked, broken, or in any way disfigured, or the effect will be bad, and the trouble bestowed upon them useless. Wax-flowers can be made not only an interesting study to those who possess large gardens and grand collections of plants, but to those who have none in their possession, for what can be more delightful and invigorating than a ramble along the hedge-rows of the country, for the purpose of gathering subjects for modelling from our native wild plants, which, though not able to vie with those from tropical countries, yet produce gems of various kinds, such as the wild Rose and Honeysuckle, wild Hop, Anemones, Violets, Primroses, &c., &c.; and again, how charming to stroll by the streams and gather from their banks the blue Myosotis, and the various aquatics to be found luxuriating in their watery beds, and in some of the vales where the Lily of the Valley may be found thriving. What enchantment to roam and gather all these charming things, and these pleasures may be kept fresh and vivid in the mind by being able to copy them in wax, and thus cheer the heart through the dull and dreary months of winter, when rude winds and chilling frosts have killed both leaf and blossom of that fairy queen the white Water Lily, and the sparkling bright Forget-me-not is buried in fallen leaves and snow.

There are many different articles required in order to make a fair imitation of flowers in wax, viz.: pins for moulding, brushes for colouring, wire for stems, colours, and wax. The large pins, as shown in diagrams Nos. 1 and 2, can be obtained in box-wood or bone – the latter is far preferable, as they require frequent wetting, and are easier to use than wood; the other three are steel pins with china heads. Altogether, about five sizes are required, two bone, and three steel pins; the moulding process is performed by placing the wax petal of a flower in the hand and rolling it with the head of the pin to the shape, or as near that of the natural one as possible. The brushes must be sable, but they are only used when any spots or stripes of colour are wanted. The wire must be in three different sizes and covered with silk or cotton, or the wax would not adhere to it; the thickest wire is sold in coils, and is used principally for the footstalks of large flowers and of large leaves; the next two sizes are sold by the reel, and are used for the footstalks of small flowers, the wire having been previously covered with green wax. The colours to be used consist of powder, and are sold in bottles of different sizes; many of these can be mixed and will form others – for instance, blue and yellow will form green; blue and red, purple, and with an addition of white, mauve; yellow, with a small portion of scarlet, orange; the only really requisite colours to get, are blue, carmine, pink, yellow, brown, violet, scarlet, and some white, silvery looking powder, called frost, or bloom, for taking the otherwise too smooth surface of the wax off. These colours are rubbed on lightly with the

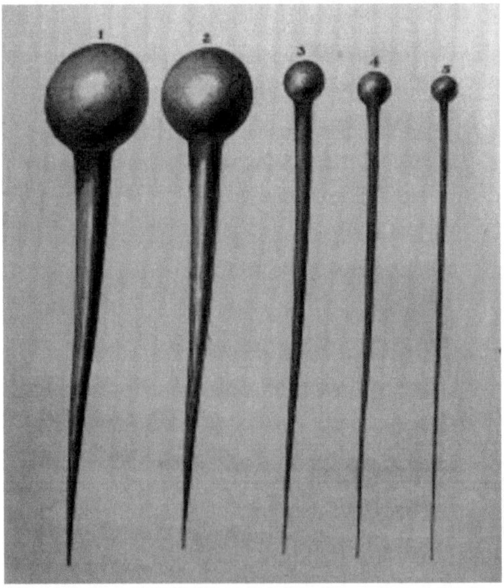

fingers, this plan is much better than using liquid colours. The wax is sold by the packet in almost every colour; but white, coloured to suit the taste, will be found much more natural looking, with the exception of yellow and green wax, which is used for seed-vessels and stems; it consists of three thicknesses, thick white, or lily wax, medium, and thin or rose wax. Cutters can be obtained for most flowers, but they are not always so true to shape as they should be, and flowers cut by them are unnatural looking, unless used for very small ones, as Forget-me-nots, Lilies of the Valley, and Verbenas, which are tedious to cut from patterns. A clean, sharp, and well pointed pair of scissors are requisite, in order to cut the wax properly, and that there may be no waste. The patterns are cut by placing the natural petal on white paper or cardboard and proceeding to cut it as near like it as possible, then write on each a number according to the number of petals, or rows of petals, contained in the natural flower; write also on one of the largest of every flower its proper name, and then in looking afterwards for any pattern there will be no confusion, each being kept separately. If the flowers are made in summer it will be found rather difficult to get them to set, especially when the weather is very hot, but this difficulty may be overcome by putting them in a cool place; it is well to hang the stalks on a chandelier, as by laying them down on a flat surface they get crushed; but care must be taken that nothing comes in contact with them to break them. If in winter, in order to make them well, it is requisite to be in a very warm room and near the fire; the wax must then be still more carefully used, as it is liable to crack with a very slight touch. Before being used, if it has been in a cold place previously, it should be warmed through by placing it a little distance off the fire. Spring and autumn are the best times of the year for modelling Wax Flowers; the wax is then in capital condition, although all we may wish to copy are not in flower. Orchids are capital flowers for modelling. Although they are not within the reach of everyone, yet those who have not a collection of their own can always obtain flowers to copy from any eminent horticulturist; but care must be taken to procure as good a variety as possible, for in plants of the same name the flowers vary in size and colour. Some of this genus are so wax-like that even the natural from the artificial flowers when placed together can scarcely be distinguished. After the more popular and simple flowers have been described, together with directions for making them in wax, I shall proceed to give a few instructions as plainly as possible as to the manner in which some of this extraordinary genus may be imitated. In a group of flowers, fern leaves should be introduced as a relief to the bright colours which otherwise would be too dazzling; they will be found rather tedious to model, but will well repay the trouble bestowed upon them. Much depends upon the artistic arrangement of the flowers with regard to size and colour, more particularly the latter; great care should be given in harmonising the colours well, so that all the showy ones may not be in one place, at the same time placing the largest flowers in the centre of the group. One fact of great importance in an economic point of view remains to be told, that is, with great care not a scrap of wax need be wasted; all cuttings from petals, stamens, stalks, in fact, from every sheet of wax used, must be saved and placed in a covered jar, to keep all rubbish or dust of any kind from getting mixed with it, or it will otherwise be useless for the purposes required.

TO MAKE LEAVES AND STEMS FOR WAX FLOWERS.

In my introductory remarks, I made a special point of the pupil saving all scraps of wax that remain after cutting the flower; this was done with a view to making leaves. Some are made from plaster of Paris models, the leaves being previously oiled. To obtain these, plaster of Paris made into a paste is poured on the leaf, and left until hard, when the impression is obtained; but done in this way, they are all the same size and shape, unless a great many models are made, and in nature no two things are exactly alike. They have this advantage, that in winter, when British fern leaves, or the leaves of all deciduous trees and shrubs are dead, the models are always ready for use. The green wax for leaves is made by melting all the pieces of wax saved, and colouring it with oil-colour. Three or four colours are required: Prussian Blue, Gamboge, Burnt Sienna, and Raw Sienna. These are mixed in while the wax is hot, being well stirred with an old brush or stick. If a bright green is required, just a small portion of Prussian Blue and Gamboge can be used; if an olive green, Burnt Sienna added will produce it. By pouring a small quantity of salad oil into the vessel containing the hot wax after the colouring process is accomplished, it will prevent the wax leaves from cracking, and they will be found more pliable for bending to the natural shapes. The hot wax is then poured on the plaster model after well soaking it in water; to remain until cold, and the form of the leaf is obtained. Another and more natural-looking method is by making them from the leaves. The backs of them must be wetted, and the hot wax, when well mixed, be passed once over the leaf with a brush; when this is cold another coating may be given. Two will generally be found sufficient, unless for leaves of a very thick texture. Before the second coating is given, the wire stem already prepared must be placed on the back of the wax, and the second coating passed over it; when cold remove the natural leaf, taking care not to crack the wax in so doing, or it must be re-melted, in order to get a perfect leaf. There is a way in which wax leaves are made without melting the wax, but it is a much more extravagant plan than either of the two previous ones. Get some leaves, take a paper pattern of each, and have ready some sheets of green wax, two or three placed together according to the thickness of the leaf which is to be imitated. Place the paper patterns on the wax, and cut from them; this having been done, place the leaf on the wax, so that all the veins are accurate, but they will not be quite so perfectly marked as those made according to the two previous methods. Stems of flowers are made of wire; never cut them too short, or they are not so convenient for grouping. When wire of the required thickness is obtained, cut narrow strips of green wax and roll round it with the fingers. In order to look neat, no portion of it must be left uncovered. The footstalks of leaves may be covered with the same wax as the leaves are composed of, one thickness of wax being sufficient, or the effect will be clumsy.

ARRANGEMENT OF FLOWERS FOR BASKETS, VASES, &c.

ALL wax flowers require to be covered with glass shades to keep them free from dust, and air should not be admitted, as it tends to make them brittle. The size of the shade employed must be according to the size of the group arranged; too great a space between the flowers and the glass has a bad effect, and too small a space has a cramped appearance. An important feature is the arrangement of colours, therefore all the flowers for one group should be made before commencing to arrange them for the vase; you then have a far better opportunity of displaying their beauties to the greatest advantage. The vase employed should be pure white; this tends to give the coloured flowers a much brighter appearance. Alabaster is the most suitable material for a vase, but it is expensive, and where this is an object, Parian marble can be used in its stead and at a much less cost. The vase must be filled with dry silver sand, or any other kind will answer the purpose, to prevent the weight of the flowers overbalancing it. A stem of two pieces of thick wire covered together with green wax, must be thrust down to the bottom of the vase, and be quite secure before any attempt at placing the flowers is made. When secure, they are fastened separately by means of wire covered with green silk to the thick stem. A few sprays of climbers such as small ivy, hops, Stephanotis floribunda, clematis of different kinds, honeysuckles, &c., should be placed in a drooping position in front of the vase; it takes off any stiff appearance the group may have. Baskets are also used as receptacles for wax flowers; those of an oval shape and of a neat description will be found the most suitable; they must have a handle, over which a pretty climber should be twined. The basket must be half filled with sand; above that place some dyed moss – this relieves the flowers, and it does not require as many to fill up with. No wire stem as a support is requisite; the flowers are simply but firmly placed in the moss by means of their own stems. When all are arranged they must be carefully placed under a glass shade to suit the basket of flowers, which should be very carefully removed before dusting, or if possible should not be removed at all unless greatly needed.

TO MAKE STAMENS, PISTILS, &c., FOR WAX FLOWERS.

THESE can all be purchased ready for use, but they come more expensive than those made by the pupil, and cannot be obtained for every flower. One method is by using white sewing cotton previously dipped in gum, then cut in the required lengths, and afterwards dipped in wax. The end will be found to be thicker than any part; this must be dipped whilst wet, in powder of the required colour. Another mode is by cutting a strip of wax and colouring it according to the natural stamens, after which it should be cut in fine lengths. Previous to this a narrow piece of wax should be neatly folded over the edge of the wax; this, when cut, forms the anthers. But the first-named method is the most natural looking. Stamens for lilies require to be made of the finest wire, and a narrow strip of wax very neatly rolled over them with the fingers. The anthers, which are made of a flat oval-shaped piece of wax, must be placed on the end of the wire and afterwards dipped in powder. Stamens for white water lilies are usually made of yellow wax; they are cut rather wider than is usual for other large flowers, as they will be found to curl slightly. This is done by placing the steel pin against the wax and turning it over, the pin having been previously wetted to prevent it sticking.

64

'THE LEAF A TYPE OF THE TREE', 'HOW AND WHAT TO COLLECT', 'MACERATION', 'SEED-VESSELS', 'BLEACHING', 'MOUNTING THE SPECIMENS'

Edward Parrish

Source: Edward Parrish, 'The Leaf a Type of the Tree', 'How and What to Collect', 'Maceration', 'Seed-vessels', 'Bleaching', 'Mounting the Specimens', *The Phantom Bouquet . . . on the Art of Skeletonising Leaves . . . and adapting them to Embellish Homes of Taste* (1862), pp. 13–24, 30–41

The Leaf a Type of the Tree.

THE veinings of the leaf exhibit striking evidence of that unity of design, combined with special adaptations, displayed in every department of nature.

What the trunk and branches are to the tree, are the veins to the exogenous leaf. By continuity and compactness of structure, the delicate spiral fibres which constitute the veins are endowed with strength and elasticity adapting them to sustain and supply the loose tissue which fills up the interstices between them, while, obeying the primal law of growth stamped upon the parent tree, each separate leaf spreads its continuous skeleton into a flattened outline of the tree itself.

A glance at the well-developed tree by twilight, stripped of its leaves, with nothing to obstruct its figure against the sky, will scarcely fail to recall, to one accustomed to observe and to compare, the general outline and arrangement of the skeleton leaf. Nor is this a mere fancied resemblance. Dr. McCosh,[1] of the Queen's University, Ireland, in his comprehensive work on "Typical Forms and Special Ends in Creation," has demonstrated the correspondence between the disposition and distribution of the branches of the tree and of the leaf-veins. The very angles at which the branches leave the trunk are shown to correspond in many individual cases with those formed by the lateral veins and midrib upon the leaf; and even the curves which give grace and contour to the tree are repeated among the veins which permeate its leaves. This is believed to be a great natural law throughout the almost infinite variety of vegetable forms.

We rarely see trees which have not been somewhat changed from their normal mode of growth by prevailing winds, or too partial sunlight, or by cattle browsing

DOI: 10.4324/9780429198809-74

on them, or man pruning or otherwise mutilating them: yet so far as the true pattern of the whole plant has been studied, it is found to correspond beautifully with that of its every single leaf.

Observe, also, how often the length of the petiole, or leaf-stalk, has a direct relation to the height reached by the stem in its unobstructed growth before branches strike off from it. In the case of the box, the privet, the laurel, the snowberry, and other favorite lawn or hedge plants, and the oak, the elm, the beech, and some other common forest-trees, which when unpruned incline to send out branches from the very base, the leaf, in botanical language, is almost sessile, sitting directly upon the stem; while those which have naturally a bare trunk, such as the cherry, the apple, the pear, the chestnut, the poplar, and, above all, our splendid American tulip poplar tree, have leaf-stalks of greater length, sometimes exceeding that of the leaf itself.

Without designing to pursue this somewhat abstruse subject farther than will suffice to vindicate its interest and importance and lead the amateurs of our art to trace out these relations for themselves, it will suffice, in conclusion, to direct attention to a peculiarity in the arrangement of certain divided and compound leaves, which accords beautifully with the idea we have been attempting to illustrate.

Our familiar sycamore or buttonwood tree has a very characteristic growth. Its bare trunk, rising eight or ten feet from the ground, commonly divides itself at once into four or five branches of great length, the whole aspect of the tree in its branching and leafage being loose and open; so its leaf, branching out from a long foot-stalk, is sustained by five midribs, the intervening structure being extremely delicate, so much so that no skeletonizer with whom I am acquainted has succeeded in separating and preserving it. The horse-chestnut presents a similar instance. The seven branches into which its leaf-stalk divides, constituting as many separate leaflets, follow the same type as the perfect tree which spreads from the top of its bare trunk a whorl of ascending and spreading branches.

Reader, we may now be charged with having trenched somewhat upon the domain of the imagination; and if the stern logic of unexplained or conflicting facts is brought in contact with these inspiring ideas, sustained as they are by many observations of undoubted import, let us not contend, but rather confess that, like all human strivings after the archetypes of nature, this may contain such errors as are inseparable from imperfect observing powers and the finite reason of human beings, whose "knowledge[2] is patchwork," after all. We may, however, derive from such observations much pleasure; they will impart new and unexpected interest to the objects to which they relate, and give to the process of skeletonizing, which unmasks the structure of the leaf, a practical and a scientific importance little anticipated by those who have pursued it with an exclusive view to its ornamental and æsthetic results.

How and What to Collect.

Skeletonizing may be out of season when this essay for the first time falls into the hands of the reader; but as soon as the summer foliage has fully expanded in luxuriant variety, the collection of material may commence. It is important that the amateur should make the proper choice of leaves. It will be observed that those which have just opened upon the fresh-grown terminations of the limbs are relatively softer, of a lighter color and more delicate fibre, than those which expanded at the first summons of the spring, in close proximity to the solid woody branches. The little suckers which are apt to grow up from the origin of the roots often bear enormous leaves, but with very little stamina.

Every leaf for our purpose should be fully formed, having attained the firmness belonging to complete maturity, and should be without a blemish. The punctures made by insects are often fatal to success when every other condition has been fulfilled. In an appendix to this essay, prepared with the aid of Dr. J. Gibbons Hunt,[3] and several ladies of experience in the art, a list is given of those trees the leaves of which we know to have been successfully skeletonized. Almost any or all of our deciduous shade-trees might probably be included; and, variety as well as beauty of form being kept in view, the pretty little leaves of the box, the rose, the pear, and the apple will be as highly prized as those of the larger magnolias, tulip poplars, maples, and lindens.

The lemon, orange, camilla, caoutchouc, and other tropical trees found in our greenhouses, besides being well fitted to the process, have the advantage of being obtainable in the winter, when our own more familiar trees are bare and leafless.

No leaves need be gathered from herbs: the organs of these short-lived tenants of the meadow and copse are formed to develop only a succulent stem and soft, flabby leaf, which will soon disappear in the processes we are about to describe, leaving no skeleton to perpetuate their outlines. The various-leaved oaks are unsuited to maceration from a different cause: they have developed underneath their glossy and tough exterior a principle called tannin, which gives to the nut-gall and oak-bark their great commercial value, and which effectually obstructs the decomposition by which we propose to get at the delicate lace-work of the leaves.

Maceration.

THE first step in the process of skeletonizing is to arrange suitable vessels for conducting the maceration. Almost any utensils of wood or earthenware which will hold water will answer the purpose; a single vessel will serve, but there is an advantage in having several, so as to separate the rough calyxes from the tender leaves, and of leaves, to place such as are porous and rapidly decomposable apart from the comparatively tough and unyielding. The specimens are to be covered with soft boiling water, and to be held in place by pieces of glass or other suitable weights, and in the summer placed on a shed or veranda, preferably exposed to the sun during part of the day. A closet near a flue, or some other warm situation, will be better in the winter. Fresh water must be added from time to time, to compensate for evaporation, but otherwise our rather unsightly and presently offensive mass is best out of sight and out of mind. In four or five weeks the leaves should be examined, and upon those sufficiently softened to admit of cleaning, this somewhat delicate operation may be commenced.

In the case of the ivy, holly, and some other leaves, after maceration the tough skin will peel off from the two faces in separate layers, so as to expose the skeleton covered only by loose parenchyma, which readily washes off. Maples, linden, elm, horse-chestnut, abele, and others, on the contrary, have so slight an epidermis that it soon softens and decays. They require a different treatment; the decayed leaf, being laid smoothly upon a plate or pane of glass, should be swept lightly over with a soft tooth-brush or shaving-brush till cleaned, turning the leaf over at intervals. Sometimes it will be found necessary to return the leaf to the macerating vessel, to complete the softening process. Leaves of very lacey and delicate texture may be best cleaned without a brush, by washing them gently between the thumb and finger under water.

The holly is one of the most difficult leaves to clean. It softens by maceration in from six to twelve weeks, but the skin is very difficult to separate from the hard, thickened edge, and especially from the prickles: a small pair of forceps, used with dexterity, will assist in this operation. When obtained in a perfect condition, this skeleton, like the lemon, japonica, and some others, displays a complete double structure, as though two perfect leaves had been laid one upon the other and grown together at the edges. The leaves of the numerous species of oak, and of hazel and buckthorn, resist the process of maceration so persistently as to induce a resort to other means to obtain their shadowy outlines.

Seed-Vessels.

MANY mature seed-vessels, and the calyxes of certain flowers, may be skeletonized by maceration or by boiling. These add greatly to the interest and beauty of the phantom case; though it is an obvious fault in most of those prepared in England that they are cumbered with too many large and comparatively opaque objects, at the expense of that airy character which is one of the chief charms of these ornaments. That unsightly weed growing with rank luxuriance on almost every heap of rubbish, especially in the suburbs of all cities, – an opprobrium to both hemispheres, each of the continents disowning its origin, though all know it but too well, – the thorn-apple, or, as we call it in America, from its "discovery" in the vicinity of the first settlement in Virginia, the Jamestown weed, matures a bur in the autumn, a skeleton of which is present in almost every phantom case I have seen. This should be gathered when it has only slightly opened at the apex, then boiled for some hours or macerated for several weeks, till sufficiently softened, and scrubbed with a tooth-brush under a running stream of water till perfectly cleaned. The seeds are to be removed by shaking, and the cellular structure, called the placenta, pricked out, though without removing the beautiful ligneous structure in which it is embedded. Water should be allowed to run freely through the bur till it is perfectly cleaned, and attention should be given to the external prickles, that every particle of the skin be removed; otherwise, after bleaching, these prominent tips will be less white than the rest. So tough is the ligneous fibre in this bur that it may be freely opened and closed without breaking it or materially weakening its union with the stem.

Poppy-heads – the pericarps of *Papaver somniferum* – require only about two weeks' maceration. As imported from Turkey and the Levant, they are generally rather large for ordinary phantom cases; but the little heads produced in our gardens are quite delicate and appropriate. The former should be prepared for maceration by removing the seeds through a gimlet-hole in the stem. The delicate lace-work interspersed among the coarser fibres which give it shape is seldom perfectly preserved; but, if a little opening is found necessary in the side, to remove the interior decayed structure, it will occur to every one, in mounting the specimens, to present the fair side to view.

There is great room for experiment in the use of seed-vessels and mature calyxes, the outer floral envelopes, for this ornamental purpose. Some campanulas, mallows, nettles, and scull-caps have already been rendered tributary to it; and in the English cases considerable prominence is given to the graceful bell-shaped calyx of henbane, and of the scarcely less elegant belladonna

Hydrangea. Ptelia.

STRAMONIUM.

plants, which, though not naturalized in the United States, are cultivated, and sometimes found growing spontaneously in the neighborhood of botanic gardens. From experience in skeletonizing the inflated calyx of the *Nicandra physaloides*, or ground cherry, and of a small cultivated species of purple tomato brought to our markets, they can be recommended as furnishing most delicate additions to the phantom case.

The large heads of the common garden hydrangea yield, by maceration, fine preparations, in which each calyx is distinctly and beautifully skeletonized. These are best mounted in little clusters or as single flowers, attached to slender stems, where their extreme lightness can be displayed to the best advantage.

Bleaching.

NEXT to the perfection of the tissues, their perfect whiteness is most important. This is best secured by soaking them in a solution of chloride of lime, which may be made with water alone in a proportion of from one to four ounces of the chloride to the pint. More delicate leaves and calyxes require a weaker solution than that used for the stramonium and poppy-heads: they are to be bleached in a glass or queensware vessel, and removed in a few hours, or as soon as the requisite brilliant whiteness is reached, then washed off with clear water and laid away to be mounted.

Some sprigs of broom-corn, and the delicate petioles or leaf-stalks which have been separated from the leaves in skeletonizing, should also be bleached, as they will be needed subsequently in mounting the specimens.

Some ferns, lycopods, and mosses – especially when browned by age or etiolated by the deep shade in which they sometimes grow – are easily bleached and adapted to occupy conspicuous places in mounted collections. They contain less fibrous structure than the more highly organized leaves used for skeletonizing, and when the green coloring-matter of their cells is completely obliterated, they are so nearly transparent as not to interfere with the airy lightness at which we aim. They are to be bleached in the same solution used for the skeleton leaves, and carefully floated off on to paper, and pressed between the leaves of a book till needed for mounting.

Labarraque's solution, (sometimes called solution of chloride of soda,) a preparation occasionally used in medicine and as a disinfectant, is preferred by some for bleaching the skeletons, and, according to my experience, is preferable for ferns, lycopods, or any other vegetable structure in which the green coloring-matter has not been previously removed. This green coloring-matter, called by vegetable physiologists chlorophyle, is an organized material, always present in plants grown in the light. Its relations to the life of the plant are not unlike those of the red corpuscles of the blood to that of red-blooded animals. It is hard to destroy this principle by chemical means; and hence, when we desire to bleach a fern or lycopod, as before stated, we shall succeed better after the partial bleaching produced by time and the slow process of drying. Old ferns, brought as relics from places visited long ago, and which have been laid away between the leaves of a book, generally bleach with facility with either of the chlorinated solutions. Some operators are in the habit of macerating ferns for a few days and boiling them occasionally in the water, by which means the chlorophyle appears to be partially decomposed, and bleaching becomes practicable.

As much disappointment is apt to occur to the inexperienced in bleaching, it is necessary to observe the quality of the bleaching-materials. Chloride of lime should be dry, or nearly so, and should have a strong chlorinous odor, and care should be taken in dissolving it that it becomes thoroughly incorporated with the water by stirring or trituration. The powder should be poured into the water, not the water on to the powder, and after the solution is made it is to be poured off clear, or filtered, before the skeletons or ferns are introduced. The solution should not be prepared until it is wanted. It is even more important that Labarraque's[4] solution should be fresh and properly prepared; it is a more active bleaching-agent than the solution of chloride of lime. When of good quality, it should be diluted with from four to eight parts of water for bleaching ferns or skeleton leaves, but it is frequently so weak as to require scarcely any dilution. It is always important to remove the preparations as soon as they are perfectly bleached to prevent their becoming too soft and tender.

Mounting the Specimens.

THE information contained in the foregoing pages, besides its general scope, has been specific enough to enable any whose tastes and instincts incline them to attempt skeletonizing to begin with a fair prospect of success. These will naturally desire some hints towards rendering the specimens they have prepared most available for parlor-ornaments. It will hardly be expected, however, that any attempt should be made to enter into detail on this part of the subject.

The English cases, as before hinted, usually contain too many opaque seed-vessels for our fancy; and many of those arranged in this country seem to have been grouped too compactly, with the aim to display a great variety of objects. Our preference is for a loose, airy-looking arrangement of the leaves, rising at the summit almost to contact with the glass under which they are displayed. This glass shade should be abundantly large and of about the same height as its diameter. The cushion that underlies the skeletons, to our taste, may best be of green velvet, recalling the image of verdure in contrast with frost.

An opened and inverted stramonium-bur, with its tough and hollow stem, may serve as a substantial pedestal for the airy superstructure of leaves, while the more solid objects of the collection may be interspersed around and upon its base. One of the most appropriate surmountings is a spray of fine fern, its top curving gracefully under the shade, its fronds slightly pendent as they fall on either side the curved stem.

In a model phantom case, arranged by a medical friend, – himself a model naturalist, "humble[5] that he knows no more," – a delicate fern, rising to the summit, trembles with electric vibrations on every touch of a silk handkerchief to the glass, while a little tuft of hydrangea-flowers, loosed from its moorings, rises to the top, like a balloon, whenever the unseen electric flash is wakened even by dusting the surface of the shade.

Although the usual form of phantom case is that of a bouquet under a glass shade, the fashion has lately obtained of arranging the specimens in an oval frame behind a convex glass. The background may then be of some rich, dark color which displays the white skeleton leaves in strong contrast. A grouping of seed-vessels at the base of the bouquet allows of the loose ends of the stems being collected and concealed from view; while the leaves and fern-fronds are spread loosely and widely over the enclosed space, allowing an opportunity for the display of taste in their arrangement. The chief objection to this method of displaying the products of the art arises from a deficiency of light, penetrating the recesses of the frame, to give the brilliant effect obtained in the bouquet covered by a transparent shade, – a defect obviated only in part by using a very shallow frame with a deeply-convex glass. When hung against a wall, this should be always placed opposite a window, or in such position as to command the best possible light. It will then form a fitting ornament to the parlor or drawing-room.

Single seed-vessels, as of the poppy, the stramonium, or physalis, or small groupings of these, constitute neat mantel-ornaments. They may be displayed to advantage in a tall wineglass, or a small vase, preferably under a suitable shade. It was the intention to illustrate the mounting of the specimens by appropriate representations of these methods; but the Phantom Bouquet is too delicate and full of exquisite detail to be well represented in an engraving; and even photography here fails to do justice to objects which to almost microscopic fineness add a lustre nearly approaching transparency.

Editorial Endnotes

1 Mcosh, James and George Dickie, *Typical Forms and Special Ends in Creation* (New York: Carter Brothers, 1856).

2 "Knowledge is patchwork" – refers to a statement by the German philosopher Johann Georg Hamann (1730–1788).

3 Dr John Gibbons Hunt (1826–1893), also a microscopist.

4 Antoine Germain Labarraque (1777–1850), a French chemist. "Labarraque's solution" was used as a disinfectant.

5 "Humble that he knows no more" – "Wisdom is humble that he knows no more." Parrish has adapted the line from William Cowper, *The Task*, VI, 'Winter Walk at Noon', vol. 1 (1785).

65

'LEAF PRINTING FROM NATURE'

James F. Robinson

Source: James F. Robinson, 'Leaf Printing from Nature', *The Garden* (29 November 1873), pp. 432–433

THERE are several different methods in use for printing or photographing leaves, some of which are exceedingly simple. First select the leaves, &c., then carefully press and dry them. If they are placed in a botanical press, care must be exercised not to put too great a pressure on the specimens at first, or they will be spoiled for printing. I have always found an old book the best for drying the examples to be used. Fibrous leaves, such as the Birch, Oak, and Maple make the best photographs. Never think of printing leaves which are either covered with hairs or are rough and uneven; it will only lead to unsatisfactory results. Mature leaves only should be gathered, not succulent or young leaves, neither attempt to print any leaf with a thick epidermis, such as Bay leaves. The best for a beginner are the fronds (leaves) of Ferns, the Maidenhair, Parsley, Hart's-tongue, and Polypody, Sea Spleenwort, the little Wall-rue Spleenwort, and Centipede Spleenwort; nay, you can scarcely choose any Fern that will not produce excellent results. The most lovely album pictures I have ever seen were printed from Fern fronds by one of the following processes. Not only are they pretty objects, either framed for mantleshelf ornamentation, or in the album, but botanists, and more especially every gardener who is interested in pteridology, should print the fronds of every species of Fern they can procure, afterwards mount them in a blank book on the left-hand page, then on the right hand page write the name, and habitat, with a description by which it may be recognised, the method of cultivation, and any other useful memoranda; by this means, in a short time, he will have compiled a volume, full of deep interest and utility. Many persons cannot afford to purchase expensive works on pteridology, yet they may soon form a permanent record for their own reference, which will be more valuable to them than any other work. I throw in these remarks as a hint worth notice by practical horticulturists.

DOI: 10.4324/9780429198809-75

Requisites.

The really useful requisities for Fern photography are two pieces of glass of the size you intend your pictures to be, common window or crown glass without air bubbles, which may be bought from any glazier, will answer the purpose admirably. For the last process described, one of the glass squares should be cut in two equal parts and a strip of muslin pasted over the division to form a kind of hinge. Then you will require paper, which should also be of one regular size. If a photographer was asked as to the best kind, he would at once recommend the albumenised paper, which he uses for his carte-portraits, but as I wish to point out how to work cheaply as well as efficiently, I would request my readers to use only the best wove letter paper. With a sharp-pointed penknife you may afterwards cut it the size you wish. I must warn the inexperienced not to use what is known in the stationery trade as "laid paper." We often see on the quarter-ream packet of note-paper the words "the best cream laid note;" never use this, it will not succeed if you do. On the laid paper the peculiar wire marks left by the manufacturer are to be seen when held up to the light; eschew this by all means. The wove papers are devoid of these marks; white paper only, of course, should be used.

The above are the needful requisites to work with, or the "stock-in-trade." It is well, however, to have at hand a few American clothes'-pegs, or spring clips, which may be purchased for a penny each from most ironmongers. Of course a few chemicals are required. In working with these great cleanliness must be observed. Hard or spring water must not be employed; use, if you can procure it, nothing but distilled water, or what will answer equally as well, clean and pure rain water. The chemicals requisite are nitrate of silver, prussiate of potash, bichromate of potash, blue vitriol or sulphate of copper, and hypo-sulphite of soda. Keep the chemicals in bottles properly labelled to prevent any accidents and confusion; they will not be kept so cleanly if preserved in papers.

Photographic Printing.

We will first describe the most simple mode of photo-printing, and then proceed to more difficult and complex formulas. Dissolve in a clean half-pint medicine bottle, four drachms (half an ounce) of prussiate of potash in about four table-spoonfuls of rain water. When it is all dissolved, so that none is visible in the bottle, pour half of the solution in a dinner plate, and float on its surface, for a few minutes, a sheet of paper of the size intended. After sufficient of the chemical is absorbed, hang it up on the edge of a shelf, by sticking a pin through one of the corners, to dry. All these operations should be performed in a dark room. A candle, however, may be used, because the rays of light emitted by a candle, being yellow, do not effect the sensitised or prepared paper. A better plan, perhaps, is to brush the solution over the paper by means of a large camel-hair pencil. Let cleanliness be observed, or the attempt will end in failure and disappointment. Lay upon one of your glass sheets three or four folds of white blotting paper; upon this the dried and sensitised prussiate of potash paper, with the prepared surface uppermost, and upon this place the Fern frond or other selected and dried leaf. Upon the leaf lay the other sheet of glass (perform all this in the dark room), then, to keep the leaf and sheets from moving, fasten one of the clothes'-clips on each end.

Place the glass in the open air, under the direct influence of clear sunlight if possible. If the day is at all cloudy, it will take much longer to print, but, with a clear sky, &c., it will probably be half an hour or thereabouts before it is sufficiently exposed. Be sure, however, to expose it a sufficient length of time; a little practice will soon teach you how long to leave it in the sun. Now watch the paper, and you will gradually observe the part not covered by the leaf changing, from a yellowish hue to a bright blue. The latter will deepen until it is almost black. When you believe it is sufficiently exposed to the sun's rays, remove the leaf, glasses, &c., and you will find printed on the paper, a yellow outline of the leaf on a deep blue ground. Now wash the paper in water, changing it frequently, until you observe the yellow change to a white – the whiter the better. By this means all the chemical is dissolved from the paper, leaving the part where the leaf lay a clear distinct white. If the glasses are removed too soon, the picture will be pale, and if they are left too long in the sunlight, the blue ground will fade.

The Chromotype Process.

This method is similar to the last, except that different chemicals are required to work it efficiently. Speaking personally, I have not succeeded so well with it as with the last process; some of my readers may, however, succeed better. The chemicals required are pure sulphate ef copper, bichromate of potash, and nitrate of silver. The latter should be handled with care, as, if it comes in contact with linen, it stains it black; the hands, too, suffer much, so that the skin is blackened and destroyed; it is, therefore, advisable to wear kid gloves when using the solution. The same kind of paper and glasses answer for this as for the former process. Dissolve about ½ ounce of sulphate of copper in half-a-pint of rain-water, and ½ ounce bichromate of potash also in half-a-pint of rain-water. The nitrate of silver is very expensive; very little, however, is required in any of the processes, and, for this, 1 drachm should be dissolved in an ounce of clean water. First, coat over, as in the first method, one side of the paper with the copper solution, then, when it is dry, coat it with the bichromate solution. After drying the paper, expose it as before to the sun for about an hour, then, if you think it sufficiently exposed, coat it over with the silver solution; or the processes may be reversed if albumenised paper is employed. The veins, &c., in this method, should be a bright scarlet. The papers may be prepared for use some time before they are required.

The best Process.

The best process[1] of Nature printing is, without doubt, the following, although it requires much more care, patience, and neatness than the foregoing. As the prepared papers are exceedingly sensitive to light, the directions in respect to a dark room must be strictly adhered to. The chemicals, which should always be dissolved in bottles ready for use, are as follows: – First a solution of nitrate of silver, containing 1 drachm or 60 grains to each ounce of pure or filtered rain-water. Any druggist will prepare this solution in a few minutes with distilled water, which can be relied upon for good results. Then a solution of hypo-sulphite of soda is needed, containing two ounces to each half-pint of water. To work well, and secure reliable prints, use only the albumenised paper, which is kept ready for sale by every photographic chemist, and, if possible, procure a glass rod, to save the hands from being stained. Always have the bottles containing the solutions correctly labelled. When you have cut the paper to the size required, pour the nitrate of silver solution on a shallow dish, and float the pieces of paper separately with the prepared surface on the liquid for about ten minutes. In doing so, take care no air bubbles are beneath the paper, or it is useless for printing; when you detect any bubbles, gently lift up the paper with the glass rod, and taking hold of the corner with the finger and thumb of the left hand, remove the bubbles by touching the part with the rod. A little practice will soon make you perfect in sensitising the paper. After floating the papers for about ten minutes take them out, and drain them for a few minutes on the dish, so that no silver may be lost; then hang them up in a dark closet by pinning the corner on a string suspended across the room. In sensitising the paper always wear old kid gloves, as well as any old garments you may happen to have at hand, the solution being so liable to spoil the clothes and injure the skin, that this advice is very needful. It is well to prepare a quantity of papers beforehand; they will keep for several days; the only difference is that newly-sensitised papers print much more rapidly, some people say they give better pictures, and probably it is so; however, if they are to be kept, take care no sunlight is admitted to them. The printing part of the process is exactly similar to the first method, only it needs watching more carefully, and in using the glasses, place the one cut into two parts and hinged at the bottom beneath the blotting papers and Fern frond; when the printing is going on, you can turn them up and examine the picture. Only do not displace the leaves ever so slightly, or the picture will be useless. A good rule is to print the picture much darker than you intend it to be when complete, because in the fixing process it becomes much lighter. After being exposed, wash the print in clean water to take away the excess of silver; you cannot wash it too well. Then having placed the hypo-sulphite of soda solution in another dish, place the picture beneath it. Do not float it as in the bath, but constantly keep it moving about with the fingers for about fifteen minutes, not longer. This is to fix the impression and make it a permanent picture. Then wash it, changing the water frequently. Do not allow a drop of the fixing solution to remain on the paper, or it will stain it yellow, which, if nothing more, will make it unsightly. All you have now to do, is to dry the picture either in a book or betwixt folds of blotting-paper beneath a gentle pressure. The last process, is after all, the best, yielding permanent and truthful results. Some Fern variety, which perhaps has never been figured in any book, by this means can be printed "true to nature" and can be referred to with pleasure in after years. I have seen a fair-sized volume containing on one page a nature-printed or photographed Fern frond, and on the opposite page, as described above, a short but correct and faithful description, &c. This was the best volume on British Ferns and their varieties I have ever seen.

Editorial Endnote

1 The best process – refers to the negative, positive paper printing photographic process initially developed by Henry Fox Talbot. It might be that the process lent itself to the activity of the individual or wealthy amateur, which included, by the 1870s, middle-class women, and contrasts with the industrial template of division of labour prefigured in the series of stages involved in the production of the daguerreotype.

66

'HOW TO SKELETONISE LEAVES AND SEED-VESSELS FOR WINTER DECORATIONS'

F. W. Burbidge

Source: F. W. Burbidge, 'How to Skeletonise Leaves and Seed-vessels for Winter Decorations', *Domestic Floriculture* (1874), pp. 195–199

Skeletonised foliage and seed-vessels, when carefully prepared, and arranged tastefully beneath a glass shade, form charming ornaments for the drawing-room table or sideboard. Being of an ivory whiteness, they show well, arranged on either crimson or black velvet, their texture rivalling that of the finest lace. The preparation of these beautiful objects forms a pleasing amusement during leisure hours, especially for our lady friends; nor are they without scientific interest, since a comprehensive series, carefully prepared and correctly named, would be specially useful to the botanist. They are very pretty arranged on coloured paper in an ornamental album, or formed into elegant designs by gluing them on a Bristol board, along with dried Fern fronds and light feathery Grasses.

"The old method, as most of my readers are aware, was simply to immerse the leaves beneath water for several weeks until the epidermis and parenchyma had decayed; then, taking them out, to rub off the decayed fleshy or cellular matter in a bowl of clean water. To say the least of this method, it was very unsatisfactory, and often yielded results far from pleasing, without taking into consideration the great amount of patience needful to complete the process. Now, thanks to chemistry, we have another and a better plan, not occupying as many hours as the old decaying method took weeks to accomplish. An excellent recommendation for processes of this kind is their simplicity; as anything complicated, or requiring expensive requisites in its performance, is sure to be scouted, or, at most, to gain but few adherents. Most of my lady friends to whom I have recommended the undermentioned process for skeletonising leaves, have fallen so much in love with it as to follow it up constantly in the autumn, merely for amusement. The result has been the production of many an elegant drawing-room ornament, either placed in the vase or mounted for framing beneath glass, as a permanent record of their industry." The following excellent mode of preparation was originally invented by Dr G. Dickson of Edinburgh, and first

 DOI: 10.4324/9780429198809-76

published in the 'Ladies' Own Journal and Miscellany' for October 20, 1866.[1] It was read at a meeting of the Edinburgh Botanical Society on November 8th of the same year, and was subsequently published in the Transactions of that Society. I am induced to mention these facts because it has so often made its appearance as original in different periodicals; and it is highly desirable that Dr Dickson should be credited with its discovery, especially as the method, simple as it appears, was only arrived at after many tedious experiments. "First dissolve four ounces of common washing-soda in a quart of boiling water, then add two ounces of slaked quicklime, and boil for about fifteen minutes. Allow this solution to cool; afterwards pour off all the clear liquor into a clean saucepan. When the solution is at the boiling-point, place the leaves carefully in the pan, and boil the whole together for an hour. Boiling water ought to be added occasionally, but sufficient only to replace that lost by evaporation. The epidermis and parenchyma of some leaves will more readily separate than in others. A good test is to try the leaves after they have been gently simmering (boiling) for about an hour, and if the cellular matter does not easily rub off betwixt the finger and thumb beneath cold water, boil them again for a short time. When the fleshy matter is found to be sufficiently softened, rub them separately, but very gently, beneath cold water until the perfect skeleton is exposed. The skeletons at first are of a dirty white colour: to make them pure white, and therefore more beautiful, all that is necessary is to bleach them in a weak solution of chloride of lime. I have found the best solution is a large teaspoonful of chloride of lime to a quart of water; if a few drops of vinegar are added to the bleaching solution it is all the better, for then the chlorine is liberated. Do not allow them to remain too long in the bleaching liquor, or they will become very brittle, and cannot afterwards be handled without injury. About fifteen minutes is sufficient to make them white and clean-looking. After the specimens are bleached, dry them in white blotting-paper beneath a gentle pressure. Of course in this, as in other things, a little practice is needful to secure perfection. Simple leaves are the best for young beginners to experiment upon: Vine, Poplar, Ficus, Beech, and Ivy leaves make excellent skeletons. Seed-vessels of the Thorn-apple (Datura Stramonium), and the Cape Gooseberry, Physalis edulis, and P. alkakengi are easily prepared, and afford variety. Care must be exercised in the selection of leaves, as well as the period of the year and the state of the atmosphere when the specimens are collected, otherwise failure will be the result. The best months to gather the specimens are July to September. Never collect specimens in damp weather, and none but perfectly-matured leaves ought to be gathered."

Very pretty rustic picture-frames may be made by cutting out the shapes (round, oval, or square, as desired) in stout cardboard. This forms a groundwork on which to glue Moss, small Fir or Larch cones, Beechmast, Nuts, or bright-coloured seeds, according to taste. After gluing on these, give the whole a thin coat of transparent varnish. I have seen some beautiful shell and pebble ornaments made in this manner, and as the process is easy, and gives good results, it may serve to employ a leisure hour.

Editorial Endnote

1 The advice of Dr. George Dickson (d.1909) concerning leaf skeletonisation is considerably shorter (and differently worded) in the sources Burbidge identifies than in those we have seen – *Ladies Own Journal* (20 October 1866) and the *Transactions of the Botanical Society of Edinburgh*, vol. 9 (November 1868), p. 48.

BIBLIOGRAPHY OF SOURCES

Broughton, Rhoda, *Not Wisely, But too Well* (London: Tinsley Brothers, 1867).

Burbidge, Frederick William, *Domestic Floriculture: Window Gardening and Floral Decorations, Being Practical Directions for the Propagation, Culture, and the Arrangement of Plants and Flowers as Domestic Ornaments*, rev. 2nd edn. (Edinburgh and London: William Blackwood and Sons, 1875).

Conder, Josiah, *Flowers of Japan and the Art of Floral Arrangement: . . . with Illustrations by Japanese Artists* (Tokio: Hakubunsha/London: Sampson, Low, 1891).

Dickens, Charles, *Great Expectations*, 3 vols., 3rd edn. (London: Chapman and Hall, 1861).

Downing, Andrew Jackson, *A Treatise on the Theory and Practise of Landscape Gardening: Adapted to North America, with a View to the Improvement of Country Residences* (New York and London: Wiley and Putnam, 1841).

Edgarton, Sarah Carter (later Mayo), *The Floral Fortune-Teller: A Game for the Season of Flowers* (Boston: A Tomkins, 1847).

Flaubert, Gustave, *Bouvard and Pécuchet* (1881), trans. by D. F. Hannigan (London: H. S. Nichols, 1896).

Gardeners' Magazine Special Commissioners, *The Wasted Orchards of England: Some Truths about British Fruit Growing* (London, Collingridge, 1896).

Hassard, Annie, *Floral Decorations for the Dwelling House: A Practical Guide to the Home Arrangement of Plants and Flowers* (London: Macmillan, 1875).

Hibberd, Shirley, *The Town Garden: A Manual for the Management of City and Suburban Gardens* (London: Groombridge, 1855).

——, *Rustic Adornments for Homes of Taste: And Recreations for Town Folk, in the Study and Imitation of Nature* (London: Groombridge, 1856).

——, *The Amateur's Greenhouse and Conservatory: A Handy Guide to the Construction and Management of Plant-Houses, and the Selection, Cultivation and Improvement of Ornamental Greenhouse and Conservatory Plants* (London: Groombridge, 1873).

Howitt, William, *The Rural Life of England*, 2 vols. (London: Longman, Orme, Brown, Green & Longmans, 1838).

Jekyll, Gertrude, *Wood and Garden: Notes and Thoughts, Practical and Critical, of a Working Amateur . . . With Illustrations from Photographs by the Author* (London: Longmans, Green, 1899).

Lachaume, Jules, *Les Fleurs Naturelles: Traité sur l'Art de Composer les Couronnes, les Parures, les Bouquets etc. de Tous Genres pour Bals et Soirées, Suivis du Langage des Fleurs* (Paris: chez l'Auteur, 1847)

Loudon, John Claudius.

——, *The Green-House Companion: Comprising a General Theory of Green-House and Conservatory Practice throughout the Year; A natural Arrangement of all Green-House Plants in Cultivation with a Descriptive Catalogue . . .* (London: Harding, Triphook, and Lepard, 1824).

——, *The Suburban Gardener and Villa Companion: Comprising the Choice of a Suburban or Villa Residence, or of a Situation on Which to Form One . . .* (London: Printed for the Author/Edinburgh: A and C Black, 1838).

——, *The Suburban Horticulturist: Or, an Attempt to Teach the Science and Practice of the Culture and Management of the Kitchen, Fruit, & Forcing Garden to Those Who Have Had No Previous Knowledge or Practice in These Departments of Gardening* (London: Printed for the Author, 1842).

Maling, Elizabeth Anne, *Flowers and Foliage for In-Door Plant Cases: Or, Hints for Arranging and Preserving Flowers in Rooms* (London: Smith, Elder, 1861).

——, *Flowers for Ornament and Decoration, and How to Arrange Them* (London: Smith, Elder 1862).

March, Thomas Charles, *Flower and Fruit Decoration: With Some Remarks on the Treatment of Town Gardens, Terraces, etc; and with Many Illustrations of Colour and Contrast* (London: Harrison, 1862).

Nicol, Walter, *The Gardener's Kalendar: Or, Monthly Directory of Operations in Every Branch of Horticulture* (1810), 2nd edn. (Edinburgh: Archibald Constable, 1812).

Parrish, Edward, *The Phantom Bouquet: A Popular Treatise on the Art of Skeletonising Leaves and Seed-Vessels and Adapting Them to Embellish Homes of Taste* (1862) (Philadelphia: J. B. Lippincott reprint, 1864).

Peachey, Emma, *The Royal Guide to Wax Flower Modelling* (London: Published and sold by Mrs. Peachey, 1851).

Perkins, John, *Floral Designs for the Table: Being Directions for Its Ornamentation with Leaves, Flowers and Fruit . . .* (London: Wyman & Sons, 1877).

Phillips, Henry, *Floral Emblems* (London: Saunders and Otley, 1825).

Pratt, Ann and Thomas Miller, *The Language of Flowers: The Association of Flowers, Popular Tales of Flowers* (London: Simpkin, Marshall [187-?]).

Repton, Humphry, *Observations on the Theory and Practice of Landscape Gardening: Including Some Remarks on Grecian and Gothic Architecture. Collected from Various Manuscripts of the Different Noblemen and Gentlemen for Whom They Were Originally Written . . .* (London: J. Taylor, 1805).

Ward, Nathaniel Bagshaw, *On the Growth of Plants in Closely-Glazed Cases* (London: John Van Voorst, 1842).

Watts, Elizabeth, *Vegetables and How to Grow Them* (London: Frederick Warne, 1866).

——, *The Orchard and Fruit Garden* (London: Frederick Warne, 1867).

Williams, Annie M., *Wax Flower Modelling Made Easy, Comprising the Practical Principles of Modelling and Colouring Flowers in Wax* (London: Bradbury, Evans, 1871).

LIST OF PRESS SOURCES

Generally, we refer to the shorter titles of magazines and newspapers as the longer titles varied over the years

Some press scources listed below are available through online libraries – the Biodiversity Heritage Library (BHL), and the HathiTrust for example. The abbreviation OL indicates that the journal, or some portion of it, is available from one or more such websites.

Amateur Gardening (1884 – to date)
Cottage Gardener (1848–1861) continued as *Journal of Horticulture* (1861–1915) – OL
Flore des Serres et des Jardins de l'Europe (1845–1880) – OL
Floricultural Magazine (1836–1842)
Gardeners' Magazine of Botany (1850–1851); continued as *Garden Companion* (1852)
The Garden (1871–1927) incorporated into *Homes and Gardens* – OL in part
Gardeners' Chronicle (1841 – to date) continued since 1983 as *Horticulture Week*) – OL in part
Gardener's Magazine (1826–44) – OL
Gardener's Magazine (1865–1916), formerly *Floricultural Cabinet* (1833–1859), then *Gardener's Weekly Magazine and Floricultural Cabinet* (1860–1865) – OL in part
Horticulturalist and Journal of Rural Art and Rural Taste (1846–1875), preceded by: *The Western Pomologist*, merged with: *Gardener's Monthly and Horticultural Advertiser*; to form: *Gardener's Monthly and Horticulturist* – OL
Illustrierte Garten-Zeitung (1856–1886) – OL in part
Le Jardin (1887–1920) – OL in part
Journal of the Horticultural Society of London (1846–1855) – OL
Journal of the Royal Horticultural Society (1866 – to date, since 1876 as *The Garden*) – OL in part
Dr Neubert's Deutsches Garten-Magazin (1882 – post-1895), preceded by: *Deutsches Magazin für Garten- und Blumenkunde* (1848–1881)
Quarterly Review (1809–1967) – OL in part

BIBLIOGRAPHY OF WORKS CITED

Alcorn, Keith, 'From specimens to commodities: The London nursery trade and the introduction of exotic plants in the early nineteenth century', *Historical Research*, vol. 93, no. 262 (November 2020), pp. 715–733. https://doi.org/10.1093/hisres/htaa025

Armstrong, Isobel, *Victorian Glassworlds, Glass Culture and the Imagination* (Oxford: Oxford University Press, 2008).

Berry, Christopher, *Social Theory of the Scottish Enlightenment* (Edinburgh: Edinburgh University Press, 1997).

Bilston, Sarah, *The Promise of the Suburbs: A Victorian History in Literature and Culture* (New Haven and London: Yale University Press, 2019).

Blacker, Mary-Rose, *Flora Domestica: A History of Flower Arranging, 1500–1930* (London: National Trust Enterprises, 2000).

Bornhorst-Winslow, C., 'The important role played by household crafts in the lives of nineteenth century women in Britain and America' (MA, Wright State University, Dayton, Ohio, 2012). https://corescholar.libraries.wright.edu/cgi/viewcontent.cgi?article=1008&context=humanities

Brake, Laurel, *Print in Transition, 1850–1910: Studies in Media and Book History* (Basingstoke: Palgrave Macmillan, 2001).

Bunyard, George and Owen Thomas, *The Fruit Garden* (London: Country Life, 1904) headnote.

Cobbett, William, *Cottage Economy* (London: C. Clement, 1822).

Cohen, Deborah, *Household Gods: A History of the British and Their Possessions* (New Haven and London: Yale University Press, 2006).

Conway, Hazel, *People's Parks: The Design and Development of Victorian Parks in Britain* (Cambridge: Cambridge University Press, 1991).

Cox, Jeffrey (Pickering, Thorne), 'Were Victorian Nonconformists the worst imperialists of all?', *Victorian Studies*, vol. 46, no. 2 (Winter 2004), pp. 243–255. www.jstor.org/stable/3830292

Crisp, Zoe, 'Housing and private outside spaces in nineteenth century England', Cambridge Working Papers in Economic and Social History (CWPESH 3), 2012. www.econsoc.hist.cam.ac.uk/docs/CWPESH%20number%203%20March%202012.pdf

Davidoff, Leonore, Megan Doolittle, Janet Fink and Katherine Holden, *Family Story: Blood, Contract and Intimacy: 1830–1960* (London and New York: Longman, 1999).

Davies, Jennifer, *Saying It with Flowers: The Story of the Flower Shop* (London: Headline Publishing, 2000).

Dewis, Sarah, *The Loudons and the Gardening Press: A Victorian Cultural Industry* (Farnham: Ashgate, 2014).
Downing, Andrew Jackson, 'Reviews', *Horticulturalist*, vol. 3 (October 1848), p. 190.
Elliott, Brent, *The Royal Horticultural Society: A History, 1804–2004* (Chichester: Phillimore, 2004).
——, 'The Royal Horticultural Society and its orchids', *Occasional Papers from the RHS Lindley Library*, vol. 2 (March 2010), pp. 3–53. www.rhs.org.uk/about-the-rhs/pdfs/publications/lindley-library-occasional-papers/volume-2-march-2010.pdf
——, 'The history of garden history', *Occasional Papers from the RHS Lindley Library*, vol. 9 (December 2012), pp. 3–84. www.rhs.org.uk/about-the-rhs/pdfs/publications/lindley-library-occasional-papers/volume-9-dec-2012.pdf
——, 'The Victorian language of flowers', *Occasional Papers from the RHS Lindley Library*, vol. 10 (April 2013), pp. 3–94. www.rhs.org.uk/about-the-rhs/pdfs/publications/lindley-library-occasional-papers/volume-ten.pdf
Floud, Roderick, *An Economic History of the English Garden* (London: Allen Lane, Penguin Books, 2019).
Fosberg, Laura, *Worlds Beyond: Miniatures and Victorian Fiction* (New Haven and London: Yale University Press, 2021).
Fowler, Corinne, *Green and Unpleasant Land: Creative Responses to Rural England's Colonial Connections* (Leeds: Peepal Tree Press, 2020).
Hager, Kelly and Talia Schaffer, 'Special Issue: "Extending Families"', *Victorian Review*, vol. 39, no. 2, (Fall 2013). www.jstor/stable/24497057
Hall, Catherine, *White, Male and Middle-Class*: *Explorations in Feminism and History* (Cambridge: Polity Press; Oxford: Blackwell, 1992).
Hartley, David, *Observations on Man* (London: 1749).
Howe, Bea, *Antiques from the Victorian Home* (London: Spring Books, 1973).
Keen, Mary, *Paradise and Plenty: A Rothschild Family* (London: Pimpernel Press, 2015).
Lambert, Tim, 'A history of Swansea'. www.localhistories.org/swansea.html#
Laws, Bill, *Spade, Skirret and Parsnip: The Curious History of Vegetables* (Stroud: Sutton Publishing, 2004).
MeasuringWorth. www.measuringworth.com/
Mitchell, W. J., *Landscape and Power*, 2nd edn. (Chicago: University of Chicago Press, 2002).
Muthesius, Stephan, *The English Terraced House* (New Haven and London: Yale University Press, 1982).
Neiswander, Judith A., *The Cosmopolitan Interior: Liberalism and the British Home, 1870–1914* (New Haven and London: Yale University Press, 2008).
Plumptre, George, *Garden Ornament: Five Hundred Years of History and Practice* (London: Thames and Hudson, 1989).
Preston, Rebecca, 'Home landscapes: Amateur gardening and popular horticulture in the making of personal, national and imperial identities, 1815–1914' (PhD, University of London, Royal Holloway, 1999). https://ethos.bl.uk
——, '"The scenery of the torrid zone": Imagined travels and the culture of exotics in nineteenth-century British gardens', in Felix Driver and David Gilbert (eds.), *Imperial Cities: Landscape, Display and Identity* (Manchester: Manchester University Press, 1999), pp. 194–209.
—— and Lesley Hoskins, 'Chickens, ducks, rabbits and me dad's geraniums: The use and meanings of yards, gardens and other outside spaces in urban working-class homes,

1890–1930', in Joseph Harley, Vicky Holmes and Laika Nevalainen (eds.), *The Working Class at Home, 1790–1940* (London: Palgrave MacMillan, 2022).

Rutherford, Sarah, 'The landscapes of public lunatic asylums in England, 1808–1914' (PhD, De Monfort University, 2003). www.dora.dmu.ac.uk › bitstream › handle

Seaton, Beverly, *The Language of Flowers, A History* (Charlottesville and London: University Press of Virginia, 1995).

Shteir, Ann B., '"Fac-similes of nature": Victorian wax flower modelling', *Victorian Literature and Culture*, vol. 35. no. 2 (2007), Cambridge University Press, pp. 649–661.

Veblen, Thorstein, *The Theory of the Leisure Class* (London: Allen and Unwin, 1970).

Veder, Robin, 'How gardening pays: Leisure, labor and luxury in nineteenth-century Transatlantic culture' (PhD, College of William & Mary, Williamsburg, 2000). https://dx.doi.org/doi:10.21220/s2-cvgv-zy92

Wall, David, 'Andrew Jackson Downing and the tyranny of taste', *American Nineteenth Century History*, vol. 8, no. 2 (June 2007), pp. 187–203.

Waters, Michael, *The Garden in Victorian Literature* (Aldershot: Scolar Press, 1988).

Way, Twigs, *Garden Gnomes, A History* (Oxford: Shire Publications, 2009).

Whiteknight, John, 'Capturing an era under glass', *19th Century, Magazine of the Victorian Society of America*, vol. 33, no. 2 (2013), pp. 28–35.

Wilkinson, Anne, *The Victorian Gardener: The Growth of Gardening and the Floral World* (Stroud: Sutton Publishing, 2006).

Willes, Margaret, *The Gardens of the British Working Class* (New Haven and London: Yale University Press, 2014).

APPENDIX: WEIGHTS AND MEASURES

MEASUREMENT

1 perch, pole or rod = 16.5 ft (around 5 metres) with some variation
40 poles (perch and rod) = 1 furlong. 4 poles = 1 chain; 10 chains =1 furlong.
1 acre measured 1 furlong (40 poles/660ft/around 200m) in length by 4 poles (66ft/around 20 metres). This rectangular shape derived from the long strips of land into which arable land was divided.

WEIGHT

Bushels were a unit of volume of dry capacity = 4 pecks or 8 dry gallons
1 imperial peck = I/4 of an imperial bushel; = 2 X imperial gallons; = 8 imperial quarts; = 16 imperial pints
'dwts. grs'. Abbrv. A pennyweight is a unit of mass equal to 24 grains which weighs 1.514 grams or roughly one and a half grams

INDEX

Note: Numbers in *italics* indicate a figure.

For Product Safety Concerns and Information please contact our
EU representative GPSR@taylorandfrancis.com Taylor & Francis
Verlag GmbH, Kaufingerstraße 24, 80331 München, Germany